AF389523

River Ecological Restoration and Groundwater Artificial Recharge

River Ecological Restoration and Groundwater Artificial Recharge

Editors

Yuanzheng Zhai
Jin Wu
Huaqing Wang

MDPI • Basel • Beijing • Wuhan • Barcelona • Belgrade • Manchester • Tokyo • Cluj • Tianjin

Editors
Yuanzheng Zhai
Beijing Normal University
China

Jin Wu
Beijing University of Technology
China

Huaqing Wang
University of Le Havre Normandy
France

Editorial Office
MDPI
St. Alban-Anlage 66
4052 Basel, Switzerland

This is a reprint of articles from the Special Issue published online in the open access journal *Water* (ISSN 2073-4441) (available at: https://www.mdpi.com/journal/water/special_issues/artificial_recharge).

For citation purposes, cite each article independently as indicated on the article page online and as indicated below:

LastName, A.A.; LastName, B.B.; LastName, C.C. Article Title. *Journal Name* **Year**, *Volume Number*, Page Range.

ISBN 978-3-0365-3469-5 (Hbk)
ISBN 978-3-0365-3470-1 (PDF)

Contents

About the Editors

Yuanzheng Zhai, male, PhD, associate professor, faculty member of Beijing Normal University from 2011 to now. Perform research in hydrogeology and teach undergraduate postgraduate class in applied hydrogeology and modern hydrogeochemistry.

Jin Wu, male, PhD, associate professor, faculty member of Beijing University of Technology from 2017 to now. Perform research in ecological risk assessment and teach undergraduate postgraduate class in water environment model.

Huaqing Wang, male, PhD, professor, research team leader, faculty member of University of Normandy Le Havre. Perform higher education and scientific research in French universities, and main fields are water resources and environment, especially mass transport in groundwater and numerical simulation.

Preface to "River Ecological Restoration and Groundwater Artificial Recharge"

The goal of this Special Issue is to highlight the research frontier of the river ecological restoration and groundwater artificial recharge, considering the river shrinkage and groundwater depletion worldwide. In the opening period of the Special Issue, we were lucky to receive 21 submissions. After peer review organized by the journal and the selection of our Special Issue, ten papers were finally selected and published in the Special Issue. The papers can generally be classified according to the following topic areas: (1) identifying and characterizing the complex hydrological processes in the river ecological restoration or groundwater artificial recharge; (2) characterizing the impacts of the river ecological restoration or groundwater artificial recharge on the physiochemical and biological evolution and the ecological risk of aquatic environment; and (3) demonstrating the redistribution of water resources and the safety of water quality in the river ecological restoration or groundwater artificial recharge. We believe that these high-quality papers have important reference value for the sustainable management of the water resources and the ecological protection of water.

Yuanzheng Zhai, Jin Wu, Huaqing Wang
Editors

Editorial

River Ecological Restoration and Groundwater Artificial Recharge

Yuanzheng Zhai [1], Jin Wu [2,*] and Huaqing Wang [3]

1   College of Water Sciences, Beijing Normal University, Beijing 100875, China; zyz@bnu.edu.cn
2   Faculty of Architecture, Civil and Transportation Engineering, Beijing University of Technology, Beijing 100124, China
3   Laboratory of Waves and Complex Media, UMR CNRS 6294, University of Le Havre Normandy, 76600 Le Havre, France; huaqing.wang@univ-lehavre.fr
*   Correspondence: wujin@bjut.edu.cn

Citation: Zhai, Y.; Wu, J.; Wang, H. River Ecological Restoration and Groundwater Artificial Recharge. *Water* 2022, *14*, 1144. https://doi.org/10.3390/w14071144

Received: 23 March 2022
Accepted: 31 March 2022
Published: 2 April 2022

Publisher's Note: MDPI stays neutral with regard to jurisdictional claims in published maps and institutional affiliations.

There is an extensive water exchange between river water and groundwater in natural conditions. With the large-scale exploitation of river water and groundwater resources, adverse ecological impacts on the river and groundwater environment arise, such as water table depression, water quality deterioration, land subsidence, dried up rivers and dry-up, and vegetation degradation. The mechanisms underlying the relationship between the river and the groundwater and its impacts on the water resource and ecosystems, its major driving factors and the responses of the ecosystems to the water pollution are still not fully understood. On the other hand, the study of numerical simulation, risk assessment model, and water resource sustainable utilization are undergoing a revolution due to the development and application of a diverse range of new technologies and methods. However, these latest technologies and methods are still supported by relatively limited scientific evidence. There is an increasing need for understanding how climate change and human activities affect river water and groundwater, considering the river shrinkage and groundwater depletion worldwide. The Special Issue "River Ecological Restoration and Groundwater Artificial Recharge" seeks to create a platform to review and present advanced methodologies, current progress and challenges, and future opportunities.

The Special Issue comprises ten papers with three interlinked research fields. Five papers focused on the impacts of river ecological replenishment and other human activities on river and watershed ecology and groundwater quality. Three papers focused on groundwater recharge and its impacts on the groundwater regime. Two papers focused on the sustainable utilization of water resources at the regional and river basin scale.

The first published paper of the Special Issue discussed the influence radius of a pumping well, which is a parameter with little scientific and practical significance that can easily be misleading [1]. This paper offers two suggestions: (1) The influence radius should not be used in the sustainable development and protection of groundwater resources, let alone in theoretical models. (2) From the perspective of regional overall planning, the calculation and evaluation of the sustainable development and the utilization of groundwater resources should be investigated systematically [1].

Most of the authors were interested in the relationship between river water and groundwater, especially in areas with strong human activities. Che et al. [2] evaluated the hydrochemical characteristics and evolution of groundwater in an alluvial plain, and the results could be useful for the effective management and utilization of groundwater resources and provide basic support for the ecological restoration of the Yangtze River Basin of China. Two teams evaluated the impacts of the ecological water supplement on groundwater restoration using numerical simulation [3,4]. Three teams focused their research on the environmental stress impacts on the groundwater, river, and basin based on a risk assessment model, respectively [5–7]. Zhang et al. [8] conducted a case study of research on the application of the typical biological chain to control algal in the lake ecological restoration. This paper's significance is related to the issue of eutrophication. The authors suggested it can be addressed by introducing Zoop (an algal predator) and Macc to a large extent, resulting in improved ecosystem maturity.

Finally, the rest of the published papers were related to sustainably utilizing groundwater resources. Guo et al. [9] assessed the groundwater suitability for irrigation and drinking purposes in an agricultural region of the North China Plain. Zhang et al. [10] presented a water resource allocation system for the rational utilization of brackish water in a water shortage area.

These published papers provide useful scientific evidence that could lead to a better understanding of the relationship between river water and groundwater impacted by human activities and climate change. We believe that these high-quality papers have important reference value for the sustainable management of water resources and the protection of water ecological security.

We thank all the authors for contributing to this Special Issue and making it a success.

Author Contributions: Y.Z., J.W. and H.W.: Conceptualization and writing—original draft preparation, review, and editing. All authors have read and agreed to the published version of the manuscript.

Funding: This research received no external funding.

Institutional Review Board Statement: Not applicable.

Informed Consent Statement: Not applicable.

Conflicts of Interest: The authors declare no conflict of interest.

References

1. Zhai, Y.; Cao, X.; Jiang, Y.; Sun, K.; Hu, L.; Teng, Y.; Wang, J.; Li, J. Further Discussion on the Influence Radius of a Pumping Well: A Parameter with Little Scientific and Practical Significance That Can Easily Be Misleading. *Water* **2021**, *13*, 2050. [CrossRef]
2. Che, Q.; Su, X.; Wang, S.; Zheng, S.; Li, Y. Hydrochemical Characteristics and Evolution of Groundwater in the Alluvial Plain (Anqing Section) of the Lower Yangtze River Basin: Multivariate Statistical and Inversion Model Analyses. *Water* **2021**, *13*, 2403. [CrossRef]
3. Ji, Z.; Cui, Y.; Zhang, S.; Chao, W.; Shao, J. Evaluation of the Impact of Ecological Water Supplement on Groundwater Restoration Based on Numerical Simulation: A Case Study in the Section of Yongding River, Beijing Plain. *Water* **2021**, *13*, 3059. [CrossRef]
4. Zhao, M.; Meng, X.; Wang, B.; Zhang, D.; Zhao, Y.; Li, R. Groundwater Recharge Modeling under Water Diversion Engineering: A Case Study in Beijing. *Water* **2022**, *14*, 985. [CrossRef]
5. Wang, J.; Zheng, N.; Liu, H.; Cao, X.; Teng, Y.; Zhai, Y. Distribution, Formation and Human Health Risk of Fluorine in Groundwater in Songnen Plain, NE China. *Water* **2021**, *13*, 3236. [CrossRef]
6. Wang, B.; Yu, F.; Teng, Y.; Cao, G.; Zhao, D.; Zhao, M. A SEEC Model Based on the DPSIR Framework Approach for Watershed Ecological Security Risk Assessment: A Case Study in Northwest China. *Water* **2022**, *14*, 106. [CrossRef]
7. Ge, Y.; Wu, J.; Zhang, D.; Jia, R.; Yang, H. Uncertain Analysis of Fuzzy Evaluation Model for Water Resources Carrying Capacity: A Case Study in Zanhuang County, North China Plain. *Water* **2021**, *13*, 2804. [CrossRef]
8. Zhang, P.; Cui, X.; Luo, H.; Peng, W.; Gao, Y. Research on the Application of Typical Biological Chain for Algal Control in Lake Ecological Restoration—A Case Study of Lianshi Lake in Yongding River. *Water* **2021**, *13*, 3079. [CrossRef]
9. Guo, H.; Li, M.; Wang, L.; Wang, Y.; Zang, X.; Zhao, X.; Wang, H.; Zhu, J. Evaluation of Groundwater Suitability for Irrigation and Drinking Purposes in an Agricultural Region of the North China Plain. *Water* **2021**, *13*, 3426. [CrossRef]
10. Zhang, D.; Xie, X.; Wang, T.; Wang, B.; Pei, S. Research on Water Resources Allocation System Based on Rational Utilization of Brackish Water. *Water* **2022**, *14*, 948. [CrossRef]

Review

Further Discussion on the Influence Radius of a Pumping Well: A Parameter with Little Scientific and Practical Significance That Can Easily Be Misleading

Yuanzheng Zhai, Xinyi Cao, Ya Jiang *, Kangning Sun, Litang Hu, Yanguo Teng, Jinsheng Wang and Jie Li *

Engineering Research Center for Groundwater Pollution Control, Remediation of Ministry of Education of China, College of Water Sciences, Beijing Normal University, Beijing 100875, China; diszyz@vip.163.com (Y.Z.); 18844120690@163.com (X.C.); 201931470034@mail.bnu.edu.cn (K.S.); litanghu@bnu.edu.cn (L.H.); teng1974@163.com (Y.T.); bnuwangjs@163.com (J.W.)
* Correspondence: 15166587857@163.com (Y.J.); lijie_lm@163.com (J.L.); Tel.: +86-151-6658-7857 (Y.J.); Fax: +86-010-58802736 (J.L.)

Citation: Zhai, Y.; Cao, X.; Jiang, Y.; Sun, K.; Hu, L.; Teng, Y.; Wang, J.; Li, J. Further Discussion on the Influence Radius of a Pumping Well: A Parameter with Little Scientific and Practical Significance That Can Easily Be Misleading. *Water* **2021**, *13*, 2050. https://doi.org/10.3390/w13152050

Academic Editors: Francesco Fiorillo and Dongmei Han

Received: 17 June 2021
Accepted: 26 July 2021
Published: 28 July 2021

Publisher's Note: MDPI stays neutral with regard to jurisdictional claims in published maps and institutional affiliations.

Abstract: To facilitate understanding and calculation, hydrogeologists have introduced the influence radius. This parameter is now widely used, not only in the theoretical calculation and reasoning of well flow mechanics, but also in guiding production practice, and it has become an essential parameter in hydrogeology. However, the reasonableness of this parameter has always been disputed. This paper discusses the nature of the influence radius and the problems of its practical application based on mathematical reasoning and analogy starting from the Dupuit formula and Thiem formula. It is found that the influence radius is essentially the distance in the time–distance problem in physics; therefore, it is a function of time and velocity and is influenced by hydrogeological conditions and pumping conditions. Additionally, the influence radius is a variable and is essentially different from the hydrogeological parameters reflecting the natural properties of aquifers such as the porosity, specific yield, and hydraulic conductivity. Furthermore, the parameterized influence radius violates the continuity principle of fluids. In reality, there are no infinite horizontal aquifers, and most aquifers are replenished from external sources, which is very different from theory. The stable or seemingly stable groundwater level observed in practice is simply a coincidence that occurs under the influence of various practical factors, which cannot be considered to explain the rationality of applying this parameter in production calculations. Therefore, the influence radius cannot be used to evaluate the sustainable water supply capacity of aquifers, nor can it be used to guide the design of groundwater pollution remediation projects, the division of water source protection areas, and the scheme of riverbank filtration wells. Various ecological and environmental problems caused by groundwater exploitation are related to misleading information from the influence radius theory. Generally, the influence radius does not have scientific or practical significance, but it can easily be misleading, particularly for non-professionals. The influence radius should not be used in the sustainable development and protection of groundwater resources, let alone in theoretical models. From the perspective of regional overall planning, the calculation and evaluation of sustainable development and the utilization of groundwater resources should be investigated in a systematic manner.

Keywords: influence radius; Dupuit; Thiem; groundwater flow system; sustainable development

1. Origin of the Issue

Darcy's Law [1], which was obtained through laboratory seepage column experiments by Darcy in 1856, is an important milestone in the history of hydrogeology and has led to the transition of hydrogeology from qualitative descriptions to quantitative calculations. Seven years later, Dupuit established the Dupuit stable well flow model (also known as the round island model) based on Darcy's Law and derived the stable well flow formula known as the Dupuit formula [2]. In the Dupuit model, the aquifer is a finite volume cylinder placed in the sea, and the pumping well is located at the axis of the cylinder, which is an ideal model for high generalization. In reality, aquifers with a finite cylinder shape, constant head boundary at the side, and zero flux boundaries at the top and bottom are extremely rare, and actual aquifers are also difficult to generalize as assumed by the Dupuit model and calculated directly using the Dupuit formula. Consequently, the practical application of the Dupuit model has been largely limited since its inception. To solve this problem, Thiem [3] extended the Dupuit model to a horizontal infinite aquifer using an approximate hypothesis and thus established the Thiem model. This model has a parameter called the range of cone of depression. Thiem assumed that this parameter represents the horizontal distance from the pumping well to the point where the water level cannot actually be observed to drop; therefore, a large error will not occur in the replacement of the round island radius "R" with the parameter of the range of cone of depression [4]. Later, Todd [5] figuratively renamed the range of the cone of depression as the influence radius and argued that it was not necessarily observable, but rather an approximate empirical value [6], which is how the influence radius originated.

The influence radius may be confusing with regard to the Dupuit model and Thiem model, and it is believed that the round island radius (R) in the former is equivalent to the influence radius (R) in the latter. Additionally, it is believed that the influence radius is objective, immutable, and measurable [7], and has nothing to do with human impact, in a manner similar to some hydrogeological parameters such as the hydraulic conductivity, specific yield, and round island radius. This erroneous understanding has misled practical work and resulted to errors in the theories and methods of groundwater resource evaluation [8]. Thus, the development of groundwater resources and the prevention and control of groundwater pollution have been subject to misleading information. This is particularly the case for the division of groundwater source protection areas [6]. In recent years, the influence radius has occasionally been discussed in academia [9–11], but consensus has not been reached. Considering this situation in combination with the needs of theoretical research and practical application, this paper tries to use non-professional language to further discuss the issue of the influence radius through mathematical reasoning and analogy to clarify this issue.

2. Birth and Application History of Influence Radius

2.1. Birth: A Misunderstanding

Dupuit made the following assumptions when he established the stable well flow model (round island model for short, Figure 1(1)): the unconfined aquifer is a homogeneous and isotropic circular island aquifer with a horizontal lower confining bed, the lateral boundary of the well has a constant head and there is a completely penetrating well in the center of the aquifer, the aquifer does not have vertical infiltration recharge and evaporation, and the seepage is a steady flow that conforms to the linear law.

The Dupuit formula for the unconfined aquifer is derived under these hypothetical conditions:

$$Q = 1.366K \frac{H_0^2 - h_w^2}{\lg \frac{R}{r_w}} \tag{1}$$

where Q is the water yield of the pumping well, [L^{3}T^{-1}]; K is the hydraulic conductivity, [LT^{-1}]; H_0 is the thickness of the aquifer, [L]; h_w is the water level of the pumping well, [L]; R is the influence radius, [L]; and r_w is the radius of the pumping well, [L].

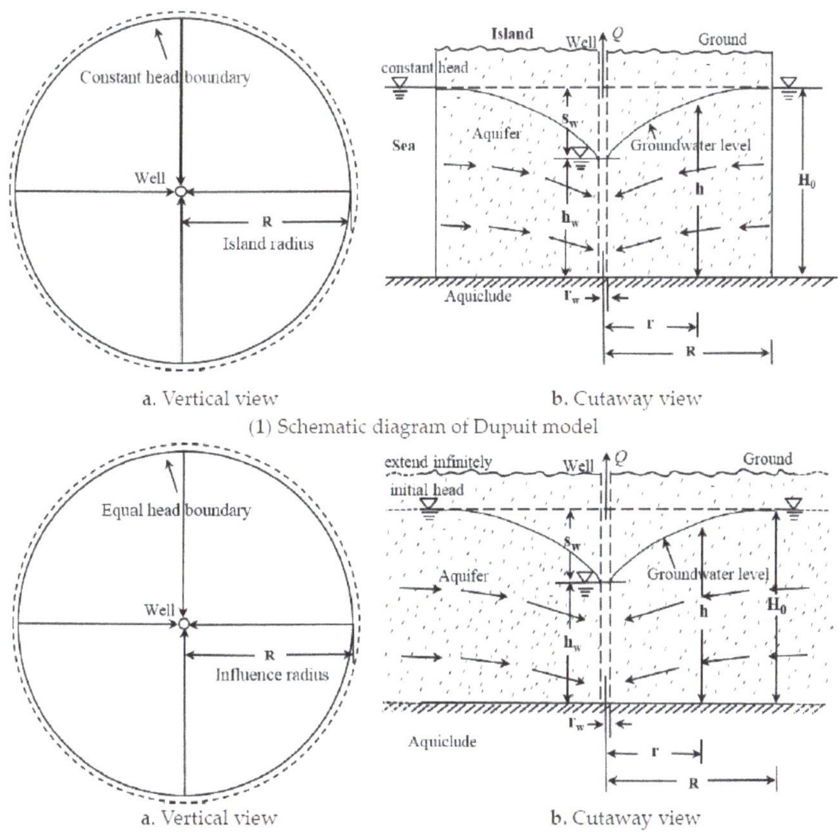

Figure 1. Schematic diagram of Dupuit model and Thiem model. Q is the water yield of the pumping well, $[L^3T^{-1}]$; s_w is the drawdown of the pumping well, $[L]$; h is the groundwater level at distance r from the pumping well, $[L]$; H_0 is the thickness of the aquifer, $[L]$; h_w is the water level of the pumping well, $[L]$; r_w is the radius of the pumping well, $[L]$; r is the distance between the pumping well and observation well, $[L]$; and R is the influence radius, $[L]$.

The relationship between Q-h_w and Q-R in Equation (1) is further analyzed in Figure 2. In the Q-h_w relationship curve (Figure 2(1)), the lower confining bed of the aquifer is considered as the base level. When h_w is zero, the drawdown in the well reaches the maximum, and the hydraulic gradient also reaches the maximum. Because the hydraulic conductivity of the aquifer remains unchanged, the flow velocity in the aquifer is also maximized according to Darcy's law. If the water sectional area is not considered, the water yield of the pumping well will be maximized. When h_w is equal to H_0, there is no drawdown in the well; therefore, the hydraulic gradient cannot be formed, the flow velocity is correspondingly zero, and the water yield of the pumping well is also zero. In the Q-R relationship curve (Figure 2(2)), when the radius of the round island (R) is the same as the radius of the pumping well (r_w), the pumping well is equivalent to pumping directly from the sea. In this case, the pumping well can extract an infinite amount of water without considering the limitation of pumping power. When the radius of the round island is infinite, the distance between the sea (source) and the pumping well is also infinite. In this case, even if the pumping time is sufficiently long, the seawater cannot reach the pumping well, and the amount of water from the sea in the pumping well will be zero.

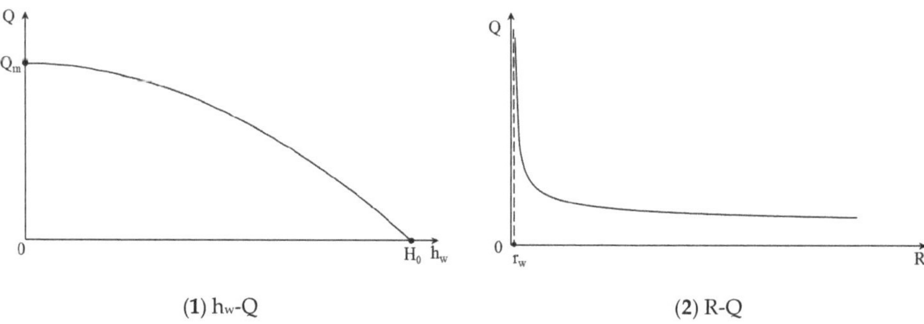

(1) h_w-Q   (2) R-Q

Figure 2. Relationship between the water yield of the pumping well and water level of the pumping well, and between the water yield of the pumping well and the influence radius in the Dupuit model. Q is the water yield of the pumping well, $[L^3T^{-1}]$; Q_m is the maximum water yield of the pumping well, $[L^3T^{-1}]$; H_0 is the thickness of the aquifer, [L]; h_w is the water level of the pumping well, [L]; r_w is the radius of the pumping well, [L]; and R is the influence radius, [L].

Thiem proposed the stable well flow model of the influence radius based on the Dupuit model (Figure 1(2)). Thiem argued that, in this model, the R value in an aquifer that extends indefinitely in the horizontal direction can be approximated as the horizontal distance from the center of the pumping well to the point where the drawdown of the groundwater level is virtually unobservable. To satisfy the Dupuit hypothesis, Thiem proposed a formula for calculating the stable well flow with one observation well and two observation wells, respectively [12]. This is known as the Thiem formula for a pumping well in unconfined aquifer and is expressed as follows:

$$\begin{cases} h^2 - h_w^2 = \frac{Q}{\pi K} \ln \frac{r}{r_w} \\ h_2^2 - h_1^2 = \frac{Q}{\pi K} \ln \frac{r_2}{r_1} \end{cases} \tag{2}$$

where h is the groundwater level at distance r from the pumping well [L]; h_w is the water level of the pumping well, [L]; Q is the water yield of the pumping well, $[L^3T^{-1}]$; K is the hydraulic conductivity, $[LT^{-1}]$; r is the distance between the pumping well and observation well, [L]; r_w is the radius of the pumping well, [L]; h_1 and h_2 are the groundwater levels of the two observation wells, respectively, [L]; and r_1 and r_2 are the distances between the two observation wells and the pumping well, respectively, [L].

Although the Thiem model was established based on the Dupuit model, there are essential differences between them. First, the aquifer in the Dupuit model is cylindrical and has a well-defined boundary, which is surrounded by water. The Dupuit formula does not work without this boundary, while the aquifer in the Thiem model extends indefinitely in the horizontal direction and does not have a boundary. Thus, there are essential differences with regard to the assumed boundary conditions. Additionally, in the Dupuit model, R refers to the radius of the cylindrical aquifer and is a geometric parameter describing the volume of an object. Moreover, R has a fixed value, which means that the cylindrical aquifer has a fixed volume. In the Thiem model, R refers to the horizontal distance from the center of the pumping well to the point at which the drawdown of the groundwater level is virtually unobservable; therefore, it is not a geometric parameter. The values of R in different directions of the aquifer may also be different because of the heterogeneity of the aquifer, and the boundary of the cone of depression in an aquifer may not be a circle [4,13]. Additionally, in the Dupuit model, the exterior of the cylindrical aquifer is full of water, and the aquifer can receive a constant supply of water. After pumping in the aquifer for a certain period of time, the water yield of the aquifer is entirely supplied by recharge from the outside of the boundary and the water is inexhaustible. Thus, a steady flow can form in the aquifer. In the Thiem model, the range beyond which R is involved is still an aquifer. When water is pumped in a horizontal infinite aquifer without leakage and external recharge, all of the water yield comes from the consumption of internal storage in the aquifer. Thus, it is impossible to form a stable flow in the aquifer [14]. Additionally, if water is pumped for a certain period of time, the water level in the pumping well will drop to the lower confining bed.

From the above analysis, Dupuit did not consider the influence radius, while Thiem only introduced the concept of the range of the cone of depression into the hypothesis. Later, Todd renamed this concept as the influence radius. Thiem also avoided the appearance of the influence radius in the formula and did not parameterize it. The present meaning of the influence radius is either the result of misunderstanding Dupuit and Thiem or laziness (saving time and effort).

2.2. Application: Crude Simplification

Based on the Dupuit model and Thiem model, studies have successively deduced various formulas for calculating the influence radius according to their own understanding (Table 1). Some of these formulas are semi-empirical [15] and do not only involve hydrogeological parameters, such as K, H_0, and μ, but also time factors. This indicates that some studies have realized that the influence radius is not a fixed hydrogeological parameter, but instead changes with time. The others are empirical formulas [15], which not only contain hydrogeological parameters but also include pumping variables such as s_w and Q in the calculation of the influence radius. These formulas only consider one of the time variables and pumping condition variables, and some even consider the influence radius as a given hydrogeological parameter [15].

Table 1. Equations for calculating the influence radius.

Equation Name	Equation	Application Condition	Author, Year	Parameter
Weber equation	$R = 74\sqrt{\frac{6KH_0t}{\mu}}$	Unconfined aquifer	Schultze, 1924	R: influence radius, [L]; K: hydraulic conductivity, $[LT^{-1}]$; H_0: thickness of aquifer, [L]; t: time from beginning of pumping to formation of stable cone of depression of groundwater level, [T]; μ: specific yield; s_w: drawdown of pumping well, [L]; Q: water yield of pumping well, $[L^3T^{-1}]$; I: hydraulic gradient of groundwater level
Kusakin equation	$R = 2s_w\sqrt{H_0K}$	Unconfined or confined aquifer	Chertousov, 1949	
	$R = 47\sqrt{\frac{6KH_0t}{\mu}}$	Unconfined aquifer	Aravin and Numerov, 1953	
Siechardt equation	$R = 10s_w\sqrt{K}$	Preliminary stage of pumping in unconfined or confined aquifer	Chertousov, 1962	
Wilbur equation	$R = 3\sqrt{\frac{KH_0t}{\mu}}$	Unconfined aquifer	Chen, 1976	
Kelgay equation	$R = \frac{Q}{2KH_0I}$	Completely penetrating well in unconfined aquifer	Chen, 1976	

To facilitate calculation, some people have introduced the influence radius into some well flow calculation formulas for the forward calculation of variables such as the water level and flow rate, or for the inversion of hydrogeological parameters such as the hydraulic conductivity and water storage coefficient (Table 2).

In some practical applications, to further simplify the calculation, the influence radius is considered empirically; that is, a quantitative relationship is established between the hydraulic conductivity and the influence radius (Table 3). This means that once the hydraulic conductivity of the aquifer is known, the influence radius of the aquifer can be obtained from the empirical value table. In other words, the influence radius is a given hydrogeological parameter that is independent of the pumping conditions. In this empirical relationship, the influence radius becomes larger as the hydraulic conductivity increases.

Table 2. Analytic solution models and equations using the influence radius.

Model/Equation Name	Equations Group	Application Condition	Author, Year	Parameter
Forward model/equation				R: influence radius, [L]; K: hydraulic conductivity, [LT^{-1}]; H_0: thickness of aquifer, [L]; t: time from beginning of pumping to formation of stable cone of depression of groundwater level, [T]; μ: specific yield; s_w: drawdown of pumping well, [L]; Q: water yield of pumping well, [L^{3}T^{-1}]; I: hydraulic gradient of groundwater level.
Plotnikov equation	$Q = e\frac{Q_0}{2R}B$	Well group pumping	Chen et al., 1976	
Dupuit–Forchheimer equation	$h^2 = H_0^2 - \frac{Q}{\pi K}\ln\frac{R}{r_w}$	Unconfined aquifer	Poehls and Smith, 2009	
s_w-calculate equation	$Q = \frac{2\pi T s_w}{\ln\frac{R}{r_w}}$	Confined aquifer	China Geological Survey, 2012	
Inversion model/equation				
Siechardt equation	$T = \frac{Q}{2\pi s_w}\ln\frac{R}{r_w}$	Confined aquifer	China Geological Survey, 2012	
Wilbur equation	$K = \frac{Q}{\pi\left[H_0^2-(H_0-s_w)^2\right]}\ln\frac{R}{r_w}$	Unconfined aquifer		

Table 3. Empirical relationship of K and R.

K (m/d)	R (m)
0.5–1	25–50
1–5	50–100
5–20	100–300
20–50	300–400
50–100	400–500
75–150	500–600
100–200	600–1500
200–500	1500–3000

This table is taken from [16]; K is the hydraulic conductivity, [LT^{-1}]; and R is the influence radius, [L].

The influence radius has been parameterized and is more widely applied in theoretical research and engineering practice [7,16], which makes relevant research and applications more convenient. However, as the research and applications become more extensive, various problems are emerging with regard to the influence radius as a hydrogeological parameter. Various studies have expressed different opinions regarding the influence radius (Table 4). Additionally, some studies have conceptually defined the influence radius as a hydrogeological parameter [15] and considered that it can be used as the basis for designing a reasonable distance between wells [12]. However, other studies have reported that this is not the case [7] and have argued that it is impossible to have a stable influence radius in an infinite aquifer [12,17] and that the time factor should be considered in the calculation of the influence radius [18]. If the R in the Dupuit and the R in the Thiem models are treated equally, this will lead to errors in the theory and methodology of groundwater resource evaluation [8].

The introduction and application of the influence radius has greatly simplified the calculations required by various engineering problems, and its empirical treatment has provided great convenience to non-professionals for carrying out relevant calculations and understanding groundwater problems. However, because coincidence in practice is not the same as correctness in theory, these crude simplifications not only harm the development of the discipline and specialty but also are misleading in practical approaches and cause irreparable losses.

Table 4. Some representative viewpoints on the influence radius (R).

Viewpoint	Reference
(1) Dupuit's R is an abstract parameter that reflects the well supply conditions and is recommended as a reference recharge radius. (2) There is still a considerable amount of drawdown beyond the range that we used to think of as R.	[4]
R should be interpreted as a parameter indicating the distance beyond which the drawdown is negligible or unobservable.	[15]
R does not exist in an infinite aquifer.	[17]
(1) In theory, R does not exist in a confined aquifer that extends indefinitely without overcurrent recharge. (2) In practice, R should be considered as the horizontal distance from the pumping well to the point where the water level cannot actually be observed to drop and can be used as the basis for designing reasonable distances between wells.	[12]
(1) Dupuit's R is different from Thiem's R. (2) Confusion between them has led to theoretical errors and incorrect methods of groundwater resource evaluation.	[8]
(1) The magnitude of R has assumed properties making it essentially the same as unsteady flow. (2) The Kusakin equation with a time factor should be applied to the calculation of R.	[18]
(1) Dupuit's R is different from Thiem's R. (2) Dupuit's R is simply the radius of the round island, while Thiem's R is a variable related to the cone of depression of the groundwater level.	[7]

3. Gap between Theory and Practice

3.1. Substance of Influence Radius: Distance

The influence radius R is essentially the distance in the time–distance problem in physics, namely R = S(v, t). Therefore, R is a function of time t and is also controlled by the velocity v and its distribution on the flow line. The influence radius is actually the influence range of the pumping well in the horizontal direction. Generally, the essence of the extension of the influence range is the velocity conduction in the process of mass transfer (water molecules) in porous media. When the distance to the pumping well is shorter, the velocity becomes higher. For a particle in an aquifer, as long as there is flow velocity to the well at that particle, the particle is within the influence range of the pumping well, unless the velocity of that particle is zero. From this viewpoint, it is reasonable to state that R is a function of velocity v. Moreover, it can also be seen from the Kusakin formulas (Table 1) that there is no stable influence radius, and the observed cone of depression will gradually expand with the extension of pumping time [19]. Thus, the influence radius is a function of time t [20,21]. Similar to water ripples (Figure 3(1)) and dominoes (Figure 3(2)), the range of influence will spread out with the advancement of time t.

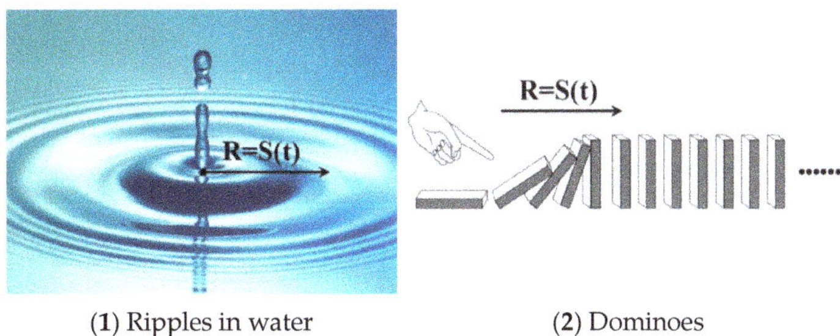

(1) Ripples in water	**(2)** Dominoes

Figure 3. Influence radius of ripples in water and dominoes (used as an analogy to describe the influence radius discussed in this paper). R is the influence radius, [L]; S is the distance in physics, [L]; and t is time, [T].

However, many studies have reported that the influence radius R is a hydrogeological parameter reflecting the natural properties of aquifers, similar to parameters such as the porosity n, specific yield μ, and hydraulic conductivity K. Therefore, the influence radius is a constant value that is not affected by the drawdown s_w and water yield Q [16]. Studies have given the empirical values of R for aquifers with different particle structures (Table 3) and have considered that greater aquifer permeability—that is, larger aquifer particles—results in a larger R value. When pumping water in an aquifer without external recharge, the cone of depression will expand with the increase of the water yield and the advancement of time. If the empirical influence radius value is used, such as in the case of the coarse gravel aquifers mentioned in some papers, the empirical value of the influence radius will be 1500–3000 m (Table 3). In other words, regardless of how large the amount of exploitation is and how long the exploitation period is, the cone of depression of the aquifer will not continue to expand outward after extending to 1500–3000 m. Accordingly, it is assumed that, in a certain pumping well group in an aquifer, the water yield of n single pumping wells is Q_{m1}, Q_{m2}, ... , Q_{mn}, respectively, their influence radius is R (Figure 4(1)), and a regional cone of depression will be formed. If the sum of the water yield of n single wells in the well group is provided by a single well, then a cone of depression with an influence radius R will be formed (Figure 4(2)). In this case, owing to the decrease in the number of wells, the area affected by pumping will be much smaller compared with when the well group is pumped. However, without external recharge, this phenomenon is completely impossible in an aquifer; otherwise, the aquifer will become an inexhaustible resource, which is contrary to common sense.

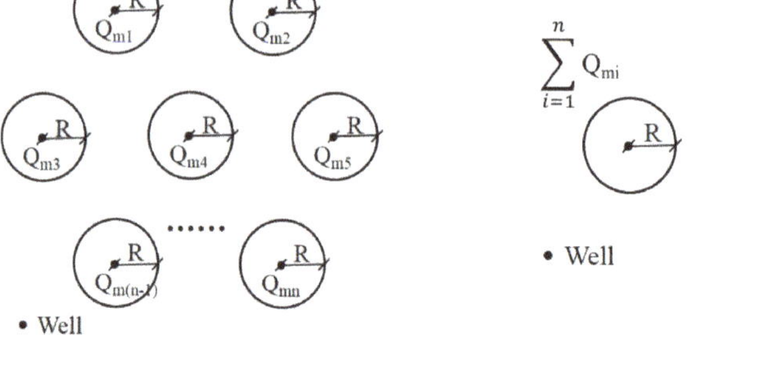

(1) Well group   (2) Equivalent diagram of single well

Figure 4. Influence radius in well group pumping and single well pumping in influence radius theory. R is the influence radius, [L]; Q_{mi} is the maximum water yield of pumping well i, $[L^3T^{-1}]$; and n is the number of pumping wells.

Further, because v = v (K, I) in R = S = S (v, t), R = S = S (v(K, I), t). For a specific aquifer, the hydraulic conductivity K is constant, and the hydraulic gradient I and t are variables. Therefore, R is affected by the two variables I and t. Additionally, the hydraulic gradient I is related to the drawdown s_w. For a particular aquifer with thickness H_0, each drawdown of a pumping well in the aquifer has a corresponding hydraulic gradient I. Notably, there are two extremes: one is that the drawdown of the well is zero. At this point, the hydraulic gradient is zero and the groundwater does not flow. In another case, the drawdown of the well reaches H_0; that is, the water level in the well drops to the lower confining bed. In this case, the hydraulic gradient reaches the maximum and the groundwater flow velocity also reaches the maximum. However, the drawdown s_w is closely related to the water yield Q; that is, it increases with the water yield. Therefore, the influence radius R is not only affected by the hydrogeological conditions reflecting the natural properties of the aquifer, such as the porosity, specific yield, and hydraulic conductivity, but is also controlled by the pumping conditions. Therefore, it is inappropriate to consider the influence radius R in the Thiem model as a fixed parameter of the aquifer, which means that this parameter is constant for a specific aquifer and does not change with the changes of the water yield and drawdown [7].

3.2. The Continuity Principle of Fluids Reverses the Rationality of the Influence Radius

In any system, fluids follow the continuity principle (conservation of mass), which means that the amount of fluid entering a region of space in a unit of time is equal to that leaving plus that stored within the region through density changes [22]. Because a horizontal infinite aquifer does not exist in practice, in the early 1960s, Tóth [23] considered the continuity principle of fluids and proposed the concept of the multi-level groundwater flow system of a basin under the assumption that the phreatic surface is the recharge boundary and the fluctuation is similar to the ground. By the 1980s, the theoretical framework had been essentially formed [24]. Tóth [23,24] pointed out that there is a difference in the elevation of the groundwater level in the basin, and a nested multi-level groundwater flow system is self-organized under the influence of gravity.

The emergence of groundwater flow system theory, which is a new paradigm of hydrogeology [25,26], has influenced the traditional groundwater movement theory, which was developed based on the well flow, whereby the groundwater converges to artificial potential sinks such as wells from potential sources [26]. Under the effect of the influence radius theory, it is commonly believed that groundwater always flows from potential sources to adjacent potential sinks within the range of the influence radius. In fact, when there are several potential sources and potential sinks with different strengths, the groundwater conforms to the theory of the minimum rate of energy dissipation and moves from different potential sources to different potential sinks [26]. Thus, it forms a multi-level nested groundwater flow system; that is, a complex flow pattern consisting of local, intermediate, and regional groundwater flow systems [23]. Of these, the scale of the regional groundwater flow system is the largest, and its flow lines can extend at great distances. Thus, the pumping well may be recharged by groundwater that is far away from it in practical situations (Figure 5). This theory suggests that we cannot use the influence radius to define the aquifer range, and water particles outside of this range do not move to the pumping well. In other words, the practical application of the influence radius is unreasonable.

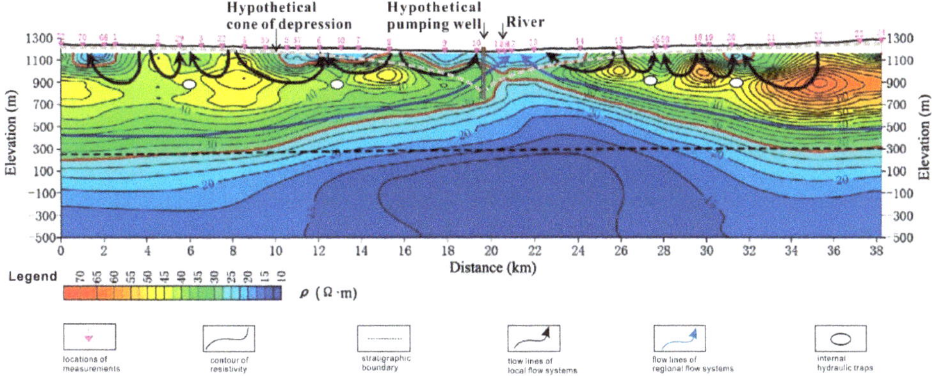

Figure 5. Schematic flow lines of a practical site according to the distribution of apparent resistivity, with a hypothetical pumping well added (modified from [27]).

3.3. Essential Difference between Theory and Practice

The seemingly stable phenomenon observed in practice known as the quasi-steady state [15] is essentially different from the influence radius considered in theory. The term "seemingly stable" means that, when pumping the water in an aquifer, the change of the groundwater piezometric head close to the well gradually slows down as the pumping time increases. Thus, within a certain range, it is close to a stable state and has the same shape as the falling curve of stable flow [12]. For a long time, the seemingly stable phenomenon observed in practice was associated with the theoretical influence radius, and the influence radius has even been used to represent the influence range when the aquifer is in a seemingly stable state. This confirms the rationality of the influence radius parameter, which is a misunderstanding of the seemingly stable phenomenon and influence radius concept.

The stable or seemingly stable groundwater level observed in practice is simply a coincidence that occurs under the influence of various practical factors and a misapprehension caused by the low accuracy of the actual measurement of the groundwater level or by the recharge of the aquifer. Therefore, this phenomenon cannot be used to explain the rationality of the influence radius parameter in theoretical models. The steady state does not exist under pumping conditions without recharge in theory. According to the continuity principle of fluids, in the cone of depression generated by pumping water, the amounts of water passing through regions A, B, and C in a unit of time are equal. However, from region A to region C, the basal area gradually increases (Figure 6(1)); therefore, the height gradually decreases (Figure 6(2)). Similarly, the drawdown at infinity will be very small but not zero, because the water yield is a concept of volume and cannot be changed from three dimensions to two dimensions. The location of the pumping well is considered as the origin of the coordinates to establish the coordinate system, and the drawdown s_w is considered as a function of x, namely $z = s_w(x)$. When pumping water, the drawdown of the aquifer is larger when the distance to the pumping well is shorter. As x approaches ∞, the drawdown tends towards zero (Figure 6). Assuming that the radius of the pumping well is $r_w = 0$, then the water yield of the pumping well is the volume of the rotating body obtained by rotating the curvilinear trapezoid bounded by a continuous curve $z = s_w(x)$, z-axis, line $x = +\infty$, and line $z = 0$ once around the z-axis. For convenience of description, the object of investigation was considered to be a unit width aquifer passing through the axis; then, the value of its volume is $V = \int s_w(x)dx, (0, \infty)$. When water is pumped continuously, x tends towards ∞; therefore, V also tends towards ∞. This means that the cone of depression will expand infinitely; therefore, a stable state cannot be formed. What is commonly referred to as the seemingly stable state does not mean that the groundwater level is stable, but that the change of the water level cannot be observed, which thus gives the illusion of stability. The phenomenon whereby the change of the water level cannot be observed is caused by the means of observation and other factors (external recharge). This is similar to the detection limit in analytical chemistry, which is a relative concept. With the improvement of the observation means, the observed drawdown range of the water level will increase. Therefore, the influence radius R cannot be considered as an intrinsic parameter of the aquifer beyond which there is no drawdown of the water level.

Therefore, in theory, the influence range will continue to expand with the extension of pumping time in an infinite aquifer without external recharge. However, there are big differences between reality and theory: (1) in reality, there are no unbounded aquifers that extend indefinitely in the horizontal direction. Therefore, for an aquifer without external recharge, the boundary of the influence range is the boundary of the aquifer when the pumping time is sufficiently long. (2) In reality, there are few aquifers without external recharge. The phenomenon whereby the observed water level reaches a stable or quasi-stable state when pumping in reality is attributed to the measurement accuracy [28]. More importantly, this phenomenon is attributed to the fact that the aquifer may have obtained unknown external recharge, such as atmospheric precipitation, surface water, and agricultural irrigation water [29], as well as to the vertical leakage recharge caused by pumping [30]. This contradicts the assumption of a lack of external recharge in the theoretical model, which is often ignored and thus results in the illusion of water level stability. Coincidences that occur in practical situations have led to the false impression that the influence radius theory has been verified, which has resulted in the wide adoption of the theory and consequently to the excessive exploitation of groundwater resources. Thus, the improved Dupuit model, which introduces external recharge, is more in line with actual needs [31].

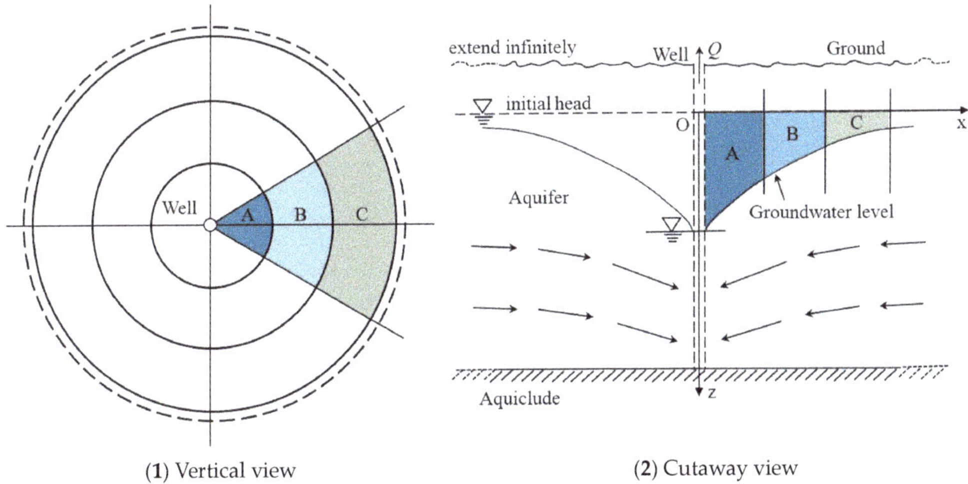

Figure 6. Formation of cone of depression of groundwater level during pumping. Q is the water yield of the pumping well, $[L^3T^{-1}]$.

4. The Dilemma of Practice

4.1. Misleading the Management of Sustainable Groundwater Development

From both a theoretical and practical viewpoint, it is proven that the influence radius cannot be used to evaluate the sustainable water supply capacity of aquifers. The influence radius is introduced for the convenience of well flow mechanics calculation to easily determine the hydraulic conductivity of an aquifer or the yield capacity of the pumping well per unit of time [4]. The yield capacity of a pumping well is actually the reflection of the working efficiency of the groundwater collecting structure, while the water supply capacity of an aquifer is an attribute of the aquifer itself. Therefore, the two concepts should not be confused. In reality, there are no aquifers that extend indefinitely in the horizontal direction; that is, actual aquifers are bounded. For an aquifer surrounded by confining boundaries, the phenomenon whereby the pumping well in the aquifer has a large yield capacity per unit of time only indicates that the aquifer can provide sufficient water during that period. In the absence of external recharge, the volume of water in an aquifer is limited. As the pumping proceeds, the water in the aquifer decreases and may not be able to continuously provide sufficient water in the next period. This is similar to a glass of water with a capacity of 600 ML: if we use a straw to draw the water from the cup and do not add water to the cup during the process, the water in the cup will gradually decrease (Figure 7(1)), and we will only obtain 600 ML of water. Thus, we must keep adding water to the cup if we desire to obtain water continuously (Figure 7(2)). This situation accurately describes the situation of pumping water in an aquifer.

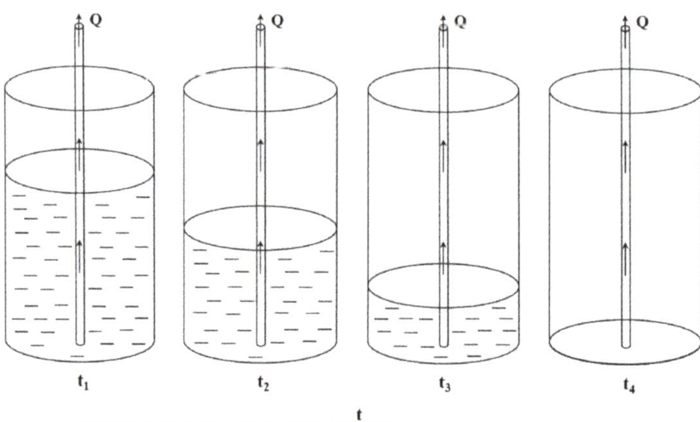

(**1**) Drawing water from cup (aquifer) without recharge (used as analogy for pumping in bounded aquifer).

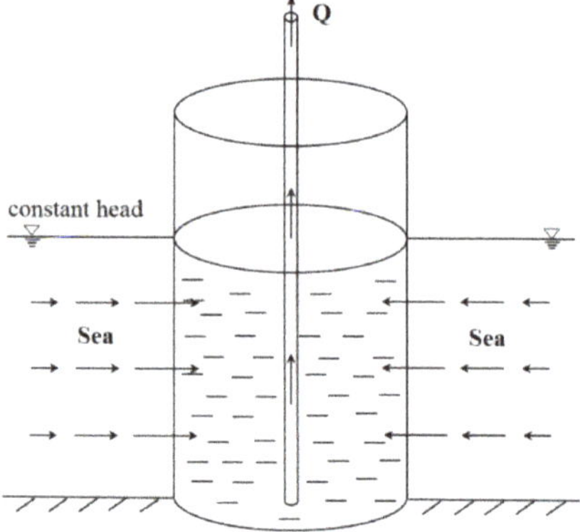

(**2**) Drawing water from cup (aquifer) with constant head on the side (used as analogy to Dupuit circular island model).

Figure 7. Drawing water from cup (used as analogy for pumping in bounded aquifer). Q is the water yield of the pumping well, [L^3T^{-1}]; and t is the time, [T].

For aquifers with a recharge boundary and drainage boundary, the increase of recharge and the reduction of discharge will occur when the influence range of pumping water extends to these boundaries [9]; that is,

$$V_w = \Delta V_r - \Delta V_d - \Delta V_s \tag{3}$$

where V_w is the total water yield in a certain period, [L^3]; ΔV_r is the increment of recharge in the same period, [L^3]; ΔV_d is the increment of discharge in the same period, [L^3]; and ΔV_s is the increment of storage in the same period, [L^3]. A riverside well is a typical example. Under natural conditions, groundwater may recharge rivers. After the addition of pumping wells close to the river, the recharge relationship between them is the fact that the river recharges groundwater and the water being pumped comes from the river and the aquifer [32]. If the recharge and discharge of the aquifer remain unchanged before and after pumping water, then $V_w = -\Delta V_s$. This indicates that all the water being pumped comes from the consumption of the aquifer storage, and it is impossible to produce an influence radius. To achieve a steady state, the sum of the increment of recharge and the reduction of discharge (sustainable yield) should always be equal to the pumping yield in the pumping state. Notably, when evaluating the amount of groundwater resources, the safe yield is usually adopted, which means that the water yield does not exceed the natural recharge amount [33]. This ignores the exploitation potential of the aquifer, which includes the increment of recharge and the reduction of discharge caused by pumping.

Because the external recharge of an aquifer directly affects the water storage of the aquifer [34], the sustainable water supply capacity of an aquifer largely depends on its ability to receive the recharge [35] and has nothing to do with the yield capacity of the pumping well, unless the extraction of the pumping well can increase the external recharge of the aquifer. From the viewpoint of sustainable development, the external recharge of the aquifer in the area and the lateral recharge inside the aquifer cannot ensure a sustainable water supply from the aquifer. First, because the lateral recharge within an aquifer is ultimately derived from the aquifer itself, its water is limited; second, from the viewpoint of resource ownership, the external recharge of the aquifer in the distance belongs to the residents in the distance, and local residents (where pumping wells are located) do not have the right to occupy it [33]. Therefore, sustainable replenishments from the upper boundary of the aquifer, such as atmospheric precipitation infiltration, river leakage, and irrigation water infiltration, play a decisive role in the sustainable water supply capacity of the aquifer. Thus, the influence radius cannot be used to evaluate the sustainable water supply capacity of the aquifer but can only be used to assess the yield capacity of the pumping well.

In practice, the influence radius theory is used to guide the development of groundwater resources in many areas. The aquifer within the range of the influence radius is considered as a "treasure basin", and it is thought that the groundwater resources are inexhaustible within this scope. However, this leads to a series of ecological and environmental problems such as the global decline of groundwater levels as a result of excessive exploitation [34,36–40], the occurrence of the cone of depression over a wide range in plain areas [41–45], the attenuation and even depletion of spring water at piedmont [46], and seawater intrusion caused by the excessive exploitation of groundwater in coastal plain areas [35,47–49]. In production practice, it is unreasonable to use the influence radius as a guideline. In engineering practice, however, various problems such as foundation pit dewatering and pumping by a well group are unavoidable and have a certain impact on the surrounding environment (such as land subsidence). Therefore, it is particularly important to develop a clear method for determining the scope of environmental impacts according to the sensitive targets and receptors around the site [6].

4.2. Misleading the Safeguarding of Groundwater Quality

The objective of water resource management is not only human water demand; rather, the quality of water is also an important factor to be considered in the process of the development of water resources [50–52]. The general deterioration of groundwater quality makes the remediation, treatment, and protection of groundwater particularly important [53–55]. Before conducting research on groundwater pollution remediation, the accurate identification of groundwater pollution sources and characteristics such as their intensity, distribution, and existence time is crucial for improving the efficiency of groundwater remediation research [56]. In most cases, pollution source analysis is carried out using the numerical simulation inversion method for groundwater. The main idea of these methods is to simulate the groundwater flow and pollutant transport process in reverse according to the monitoring data of the spatial and temporal distribution of pollutant concentration and determine the characteristics of pollution sources [55,57,58]. In the most widely used optimization simulation method [57], it is always assumed that the numbers and locations of pollution sources in the groundwater solute migration model are known, and the monitoring data of the spatial and temporal distribution of groundwater pollutant concentrations are obtained within a limited study area [57,59–61], which limits the influence range of the pollution sources within a certain space. However, in aquifer systems, pollutants migrate with the groundwater [48]. Without considering the degradation and attenuation of pollutants in the aquifer, the pumping wells are recharged from the aquifers in the distance under long-term pumping conditions. At this time, the pollutants in the distance will still be affected by pumping and migrate with the groundwater until they reach the pumping well [62,63]. In other words, not all pollutants in the pumping wells come from sources close to the pumping wells, but some also come from the aquifers outside the range of the influence radius. Therefore, the analytical method of the pollution source, which is restricted by the existing influence radius theory, cannot ensure the accuracy of analytical results, and thus its reference value is reduced. Similar problems also exist in the remediation of groundwater pollution. In the ex-situ pump and treat technique [64,65], if the layout of the pumping well group is determined based on the influence radius during the pumping design and then parameters such as the water yield, drawdown of water level, and pumping time are calculated, the groundwater outside of the polluted range may be pumped out, which will affect the treatment efficiency and increase the treatment cost. Similarly, there is also a problem regarding the injection well's influence radius when the groundwater is restored by in-situ remediation with the injection of remediation agents into the groundwater [66–68]. Without considering the chemical reactions of the remediation agents in the migration process with groundwater, these remediation agents may not be restricted to interact with pollutants within the designed influence range but may instead migrate to a further range along with the flow and thus change the primary groundwater environment and cause secondary pollution.

The influence radius is also widely used as the basis for the division of groundwater source protection areas to prevent the pollution of water sources [69,70]. However, the flow line follows the range of influence and should be covered by the influence radius. The delineation of a protection area near a pumping well artificially cuts off the flow line, which violates the flow continuity principle. Such protection measures cannot guarantee the sustainable supply from groundwater sources and lead to dangers that may compromise water quality. Additionally, the influence radius is often used as the basis for designing a reasonable distance between wells in the planning of riverbank filtration systems [12]. Although the external recharge of the river can stabilize the influence radius [32,71–73], leakage recharge may occur when pumping is carried out on the side of a incompletely penetrating river. Pollutants in the groundwater on the other side of the river also migrate to the pumping wells with the groundwater [74]. Therefore, pumping water in riverbank filtration systems, which are planned based on the influence radius, cannot ensure the water quality of pumped water and may affect the groundwater on the other side of the river.

Thus, it is reasonable to state that any methodology involving the influence radius must be reconsidered. Moreover, effective methods of groundwater remediation, treatment, and protection should be investigated to make the influence radius theory obsolete. To this end, new technology should be used, such as the MODPATH module in GMS, which is used to track the virtual particle beam of a pumping well to obtain the capture area of the well in a given time [75] and is considered to be an effective research tool.

5. Summary and Prospects

The introduction of the Dupuit model has placed a focus on the quantitative calculation of well flow. To solve the computing problem of actual aquifers, Thiem introduced the concept of the range of cone of depression, which Todd later renamed as the influence radius. In later production practice, the round island radius was confused with the influence radius, and the parameterized influence radius has been widely used to evaluate the sustainable water supply capacity of aquifers and in the planning of groundwater source areas, which has led to the emergence of a series of ecological and environmental problems.

The influence radius was originally used in the calculation of some hydrogeological parameters but, owing to various coincidences that occur in practical situations, it has been considered that the parameterized influence radius is reasonable and convenient for calculations pertaining to actual production problems, and this misconception has perpetuated. However, by considering the continuity principle of flow, it can be proven that the parameterized influence radius does not exist. The influence radius is essentially the distance in the time–distance problem in physics and is influenced by the hydrogeological conditions and pumping conditions, which is different from the hydrogeological parameters reflecting the natural properties of aquifers, such as the porosity, specific yield, and hydraulic conductivity. In reality, infinite horizontal aquifers do not exist, and most aquifers are replenished from external sources, which is very different from the theoretical considerations. The stable or seemingly stable groundwater level observed in practice is a misapprehension caused by low accuracy in the actual measurement of the groundwater level or by aquifer recharge.

In the past, the well flow model was only used to solve local water supply problems, and attention was only given to local groundwater problems due to economic and technological limitations. The influence radius theory has provided incorrect guidance in the analysis of groundwater pollution sources, the division of groundwater source protection areas, and the planning of riverbank filtration. With the development of the social economy, in addition to scientific and technological developments, attention has gradually shifted to regional and even global groundwater flow systems. Moreover, focus is increasingly being placed on sustainable development and the protection of the ecological environment. Because groundwater is a local and regional resource and is increasingly considered as a global resource, the evaluation and rational development of groundwater resources should consider hydrogeologic units as a whole, as in the case of basin management for surface water. Instead of solving the local practical problems of production, the long-term, comprehensive, and systematic topics of sustainable development should be given attention, and problems in resource ecology and environmental disasters should be addressed holistically in a systematic and methodical manner. Additionally, some methods may be useful in groundwater resource evaluation. For example, the concept of the "scope of environmental impacts" is suitable for different industries. Additionally, the original theoretical model can be improved such that it can be applied to current aquifer calculations. Finally, the combination of numerical simulation with new technologies, such as isotopic methods, remote sensing, and big data, can improve the accuracy of models.

Author Contributions: Conceptualization, Y.Z.; methodology, Y.Z. and Y.J.; formal analysis, Y.J. and X.C.; investigation, Y.Z. and Y.J.; data curation, K.S. and L.H.; writing—original draft preparation, Y.J. and X.C.; writing—review and editing, Y.Z. and J.L.; supervision, Y.T. and J.W.; funding acquisition, Y.Z. and J.W. All authors have read and agreed to the published version of the manuscript.

Funding: This work was supported by the National Natural Science Foundation of China (Nos. 42077170 and 41831283), Major Science and Technology Program for Water Pollution Control and Treatment of China (2018ZX07101005-04), Beijing Advanced Innovation Program for Land Surface Science of China, and the 111 Project of China [B16020].

Institutional Review Board Statement: Not applicable.

Informed Consent Statement: Not applicable.

Data Availability Statement: Not applicable.

Conflicts of Interest: The authors declare no conflict of interest.

Abbreviations

Symbol	Description	Dimension
R	Influence radius	L
K	Hydraulic conductivity	LT^{-1}
H_0	Thickness of aquifer	L
s_w	Drawdown of pumping well	L
Q	Water yield of pumping well	L^3T^{-1}

r_w	Radius of pumping well	L
T	Transmissibility coefficient of aquifer	L^2T^{-1}
t	Time from beginning of pumping to formation of stable cone of depression of groundwater level	T
μ	Specific yield	/
I	Hydraulic gradient of groundwater level	/
r	Distance between pumping well and observation well	L
h	Groundwater level at distance r from pumping well	L
h_w	Water level of pumping well	L
e	Empirical coefficient	/
Q_0	Water yield of single pumping well	L^3T^{-1}
B	Width of aquifer	L
Q_m	Maximum water yield of pumping well	L^3T^{-1}
S	Distance in physics	L
v	Seepage velocity	LT^{-1}
n	Porosity of porous media (aquifer)	/

References

1. Darcy, H. *Les Fontaines Publiques de la Ville de Dijon [The Public Fountains of the City of Dijon]*; Victor Dalmont: Paris, France, 1856.
2. Dupuit, J. *Etudes Théoriques et Pratiques sur le Mouvement des Eaux dans les Canaux Découverts et à Travers les Terrains Perméables [Theoretical and Practical Studies on the Movement of Water in Open Channels and Permeable Ground]*; Dunod: Paris, France, 1863.
3. Thiem, G.A. *Hydrologische Methoden*; Gebhardt: Leipzig, Germany, 1906.
4. Chen, Y.S.; Yao, D.S.; Hua, R.R. Try to discuss the influence radius. *Geotech. Invest. Survey* **1976**, *6*, 5–33. (In Chinese)
5. Todd, D.K. *Ground Water Hydrology*; Wiley: New York, NY, USA, 1959.
6. Wang, J.H.; Wang, F. Discussion on the range of groundwater depression cone, radius of influence and scope of environmental impacts during pumping. *J. Hydraul. Eng.* **2020**, *51*, 827–834. (In Chinese with English abstract)
7. Wang, X.M.; Wang, X.H.; Wen, W.; Li, G.Y. Model analysis of Dupuit's steady well flow formula. *Coal. Geol. Explor.* **2014**, *6*, 73–75, (In Chinese with English abstract).
8. Chen, C.X.; Lin, M.; Cheng, J.M. *Groundwater Dynamics*, 5th ed.; Geological Publishing House: Beijing, China, 2011. (In Chinese)
9. Chen, C.X. Stable well flow model of influence radius and sustainable yield: Differences of a basic theoretical problem in Groundwater Dynamics—Discuss with academician Xue Yuqun. *J. Hydraul. Eng.* **2010**, *41*, 1003–1008. (In Chinese)
10. Xue, Y.Q. Discussion on the stable well flow model and Dupuit formula—Reply to professor Chen Chongxi's article "Discussion". *J. Hydraul. Eng.* **2011**, *39*, 1252–1256. (In Chinese)
11. Chen, C.X. Rediscuss on stable well flow model of influence radius and sustainable yield: Differences of a basic theoretical problem in Groundwater Dynamics—Reply to academician Xue Yuqun's discussion. *J. Hydraul. Eng.* **2014**, *45*, 117–121. (In Chinese)
12. Xue, Y.Q.; Wu, J.C. *Groundwater Dynamics*, 3rd ed.; Geological Publishing House: Beijing, China, 2010. (In Chinese)
13. Wu, C.M.; Yeh, T.C.J.; Zhu, J.; Lee, T.H.; Hsu, N.S.; Chen, C.H.; Sancho, A.F. Traditional analysis of aquifer tests: Comparing apples to oranges? *Water Resour. Res.* **2005**, *41*. [CrossRef]
14. Theis, C.V. The relation between the lowering of the piezometric surface and the rate and duration of discharge of a well using ground-water storage. *Eos Trans. AGU* **1935**, *16*, 519–524. [CrossRef]
15. Bear, J. *Hydraulics of Groundwater*; McGraw-Hill, Inc.: New York, NY, USA, 1979.
16. China Geological Survey. *Handbook of Hydrogeology*, 2nd ed.; Geological Publishing House: Beijing, China, 2012. (In Chinese)
17. Zhang, H.R. The steady state of groundwater movement. *Hydrogeol. Eng. Geol.* **1986**, *6*, 18–21. (In Chinese)
18. Zhu, D.T. Discussion on Dupuit formula and Forchheimer formula. *J. Hydraul. Eng.* **2012**, *43*, 502–504. (In Chinese)
19. Zhang, H.R. A couple of interesting inference of Theis's equation. *Hydrogeol. Eng. Geol.* **1985**, *5*, 30–32. (In Chinese)
20. Chang, P.Y.; Hsu, S.Y.; Chen, Y.W.; Liang, C.; Wen, F.; Lu, H.Y. Using the resistivity imaging method to monitor the dynamic effects on the vadose zone during pumping tests at the Pengtsuo site in Pingtung, Taiwan. *Terr. Atmos. Ocean. Sci.* **2016**, *27*, 059. [CrossRef]
21. Chang, P.Y.; Chang, L.C.; Hsu, S.Y.; Tsai, J.P.; Chen, W.F. Estimating the hydrogeological parameters of an unconfined aquifer with the time-lapse resistivity-imaging method during pumping tests: Case studies at the Pengtsuo and Dajou sites. *Taiwan. J. Appl. Geophys.* **2017**, *144*, 134–143. [CrossRef]
22. Craik, A.D.D. "Continuity and change": Representing mass conservation in fluid mechanics. *Arch. Hist. Exact Sci.* **2013**, *67*, 43–80. [CrossRef]
23. Tóth, J. A theoretical analysis of groundwater flow in small drainage basin. *J. Geophys. Res.* **1963**, *68*, 4795–4812. [CrossRef]
24. Tóth, J. Cross-formation gravity flow of groundwater: A mechanism of the transport and accumulation of petroleum (The generalized hydraulic theory of petroleum migration). *AAPG Stud. Geol.* **1980**, *10*, 121–167.
25. Mádl-Szónyi. *From the Artesian Paradigm to Basin Hydraulics: The Contribution of József Tóth to Hungarian Hydrogeology*; Publishing Company of Budapest University of Technology and Economics: Budapest, Hungary, 2008.

26. Zhang, R.Q.; Liang, X.; Jin, M.G.; Wan, L.; Yu, Q.C. *Fundamentals of Hydrogeology*, 7th ed.; Geological Publishing House: Beijing, China, 2018. (In Chinese)
27. Jiang, X.W.; Wan, L.; Wang, J.Z.; Yin, B.X.; Fu, W.X.; Lin, C.H. Field identification of groundwater flow systems and hydraulic traps in drainage basins using a geophysical method. *Geophys. Res. Lett.* **2014**, *41*, 2812–2819. [CrossRef]
28. Zha, Y.Y.; Shi, L.S.; Liang, Y.; Tso, C.H.M.; Zeng, W.Z.; Zhang, Y.G. Analytical sensitivity map of head observations on heterogeneous hydraulic parameters via the sensitivity equation method. *J. Hydrol.* **2020**, *591*, 125282. [CrossRef]
29. Liu, Y.P.; Yamanaka, T.; Zhou, X.; Tian, F.Q.; Ma, W.C. Combined use of tracer approach and numerical simulation to estimate groundwater recharge in an alluvial aquifer system: A case study of Nasunogahara area, central Japan. *J. Hydrol.* **2014**, *519*, 833–847. [CrossRef]
30. Yuan, R.Q.; Song, X.F.; Han, D.M.; Zhang, L.; Wang, S.Q. Upward recharge through groundwater depression cone in piedmont plain of North China Plain. *J. Hydrol.* **2013**, *500*, 1–11. [CrossRef]
31. Chen, C.X. Improvement of Dupuit model: With infiltration recharge. *Hydrogeol. Eng. Geol.* **2020**, *47*, 1–4. (In Chinese with English abstract).
32. Zhu, Y.G.; Zhai, Y.Z.; Du, Q.Q.; Teng, Y.G.; Wang, J.S.; Yang, G. The impact of well drawdowns on the mixing process of river water and groundwater and water quality in a riverside well field, Northeast China. *Hydrol. Process.* **2019**, *33*, 945–961. [CrossRef]
33. Wood, W.W. Groundwater "durability" not "sustainability"? *Groundwater* **2020**. [CrossRef]
34. Ahmed, K.; Shahid, S.; Demirel, M.C.; Nawaz, N.; Khan, N. The changing characteristics of groundwater sustainability in Pakistan from 2002 to 2016. *Hydrogeol. J.* **2019**, *27*, 2485–2496. [CrossRef]
35. Jia, X.Y.; Hou, D.Y.; Wang, L.W.; O'Connor, D.; Luo, J. The development of groundwater research in the past 40 years: A burgeoning trend in groundwater depletion and sustainable management. *J. Hydrol.* **2020**, *587*, 125006. [CrossRef]
36. White, E.K.; Costelloe, J.; Peterson, T.J.; Western, A.W.; Carrara, E. Do groundwater management plans work? Modelling the effectiveness of groundwater management scenarios. *Hydrogeol. J.* **2019**, *27*, 1–24. [CrossRef]
37. Famiglietti, J.S. The global groundwater crisis. *Nat. Clim. Chang.* **2014**, *4*, 945–948. [CrossRef]
38. Giordano, M. Global groundwater? Issues and solutions. *Ann. Rev. Environ. Resour.* **2009**, *34*, 153–178. [CrossRef]
39. Konikow, L.; Kendy, E. Groundwater depletion: A global problem. *Hydrogeol. J.* **2005**, *13*, 317–320. [CrossRef]
40. Wada, Y.; van Beek, L.P.H.; van Kempen, C.M.; Reckman, J.W.T.M.; Vasak, S.; Bierkens, M.F.P. Global depletion of groundwater resources. *Geophys. Res. Lett.* **2010**, *37*, L20402. [CrossRef]
41. Erban, L.E.; Gorelick, S.M.; Zebker, H.A.; Fendorf, S. Release of arsenic to deep groundwater in the Mekong Delta, Vietnam, linked to pumping-induced land subsidence. *Proc. Natl. Acad. Sci. USA* **2013**, *110*, 13751–13756. [CrossRef]
42. Galloway, D.L.; Burbey, T.J. Review: Regional land subsidence accompanying groundwater extraction. *Hydrogeol. J.* **2011**, *19*, 1459–1486. [CrossRef]
43. Gleeson, T.; Alley, W.M.; Allen, D.M.; Sophocleous, M.A.; Zhou, Y.X.; Taniguchi, M.; VanderSteen, J. Towards sustainable groundwater use: Setting long-term goals, backcasting, and managing adaptively. *Groundwater* **2012**, *50*, 19–26. [CrossRef]
44. Qian, H.; Chen, J.; Howard, K.W.F. Assessing groundwater pollution and potential remediation processes in a multi-layer aquifer system. *Environ. Pollut.* **2020**, *263*, 114669. [CrossRef] [PubMed]
45. Scanlon, B.R.; Reedy, R.C.; Gates, J.B.; Gowda, P.H. Impact of agroecosystems on groundwater resources in the Central High Plains, USA. *Agric. Ecosyst. Environ.* **2010**, *139*, 700–713. [CrossRef]
46. Feng, W.; Zhong, M.; Lemoine, J.M.; Biancale, R.; Hsu, H.T.; Xia, J. Evaluation of groundwater depletion in North China using the Gravity Recovery and Climate Experiment (GRACE) data and ground-based measurements. *Water Resour. Res.* **2013**, *49*, 2110–2118. [CrossRef]
47. Hakan, A. Application of multivariate statistical techniques in the assessment of groundwater quality in seawater intrusion area in Bafra Plain. *Turkey* **2013**, *185*, 2439.
48. Jia, Y.F.; Xi, B.D.; Jiang, Y.H.; Guo, H.M.; Yang, Y.; Lian, X.Y.; Han, S.B. Distribution, formation and human-induced evolution of geogenic contaminated groundwater in China: A review. *Sci. Total Environ.* **2018**, *643*, 967–993. [CrossRef] [PubMed]
49. Lei, S.; Jiao, J.J. Seawater intrusion and coastal aquifer management in China: A review. *Environ. Earth Sci.* **2014**, *72*, 2811–2819.
50. Narasimhan, T.N. Hydrogeology in North America: Past and future. *Hydrogeol. J.* **2005**, *13*, 7–24. [CrossRef]
51. Chinese Academy of Science. *Development Strategy of Chinese Subjects: Groundwater Science*; Science Press: Beijing, China, 2019. (In Chinese)
52. Zhang, F.G.; Huang, G.X.; Hou, Q.X.; Liu, C.Y.; Zhang, Y.; Zhang, Q. Groundwater quality in the Pearl River Delta after the rapid expansion of industrialization and urbanization: Distributions, main impact indicators, and driving forces. *J. Hydrol.* **2019**, *577*, 124004. [CrossRef]
53. Kurwadkar, S. Emerging trends in groundwater pollution and quality. *Water Environ. Res.* **2014**, *86*, 1677–1691. [CrossRef]
54. Zhang, B.; Song, X.; Zhang, Y.; Han, D.; Tang, C.; Yu, Y.; Ma, Y. Hydrochemical characteristics and water quality assessment of surface water and groundwater in Songnen plain, Northeast China. *Water Res.* **2012**, *46*, 2737–2748. [CrossRef] [PubMed]
55. Zhou, Y.; Khu, S.T.; Xi, B.; Su, J.; Hao, F.; Wu, J.; Huo, S. Status and challenges of water pollution problems in China: Learning from the European experience. *Environ. Earth Sci.* **2014**, *72*, 1243–1254. [CrossRef]
56. Datta, B.; Prakash, O.; Cassou, P.; Valetaud, M. Optimal unknown pollution source characterization in a contaminated groundwater aquifer—evaluation of a developed dedicated software tool. *J. Geosci. Environ. Prot.* **2014**, *2*, 41–51. [CrossRef]

57. Ayvaz, M.T. A linked simulation—optimization model for solving the unknown groundwater pollution source identification problems. *J. Contam. Hydrol.* **2010**, *117*, 46–59. [CrossRef] [PubMed]
58. Sun, A.Y.; Painter, S.L.; Wittmeyer, G.W. A constrained robust least squares approach for contaminant release history identification. *Water Resour. Res.* **2006**, *42*, 263–269. [CrossRef]
59. Mirghani, B.Y.; Mahinthakumar, K.G.; Tryby, M.E.; Ranjithan, R.S.; Zechman, E.M. A parallel evolutionary strategy based simulation—optimization approach for solving groundwater source identification problems. *Adv. Water Resour.* **2009**, *32*, 1373–1385. [CrossRef]
60. Singh, R.M.; Datta, B.; Jain, A. Identification of unknown groundwater pollution sources using artificial neural networks. *J. Water Res. Plan. Manag.* **2004**, *130*, 506–514. [CrossRef]
61. Singh, R.M.; Datta, B. Identification of groundwater pollution sources using GA-based linked simulation optimization model. *J. Hydrol. Eng.* **2006**, *11*, 101–109. [CrossRef]
62. Hintze, S.; Gaétan, G.; Hunkeler, D. Influence of surface water-groundwater interactions on the spatial distribution of pesticide metabolites in groundwater. *Sci. Total Environ.* **2020**, *733*, 139109. [CrossRef]
63. Zhang, H.; Jiang, X.W.; Wan, L.; Ke, S.; Liu, S.A.; Han, G.L.; Guo, H.M.; Dong, A.G. Fractionation of Mg isotopes by clay formation and calcite precipitation in groundwater with long residence times in a sandstone aquifer, Ordos Basin, China. *Geochim. Cosmochim. Acta* **2018**, *237*, 261–274. [CrossRef]
64. Kim, J.W. Optimal pumping time for a pump-and-treat determined from radial convergent tracer tests. *Geosci. J.* **2014**, *18*, 69–80. [CrossRef]
65. Bortone, I.; Erto, A.; Nardo, A.D.; Santonastaso, G.F.; Chianese, s.; Musmarra, D. Pump-and-treat configurations with vertical and horizontal wells to remediate an aquifer contaminated by hexavalent chromium. *J. Contam. Hydrol.* **2020**, *235*, 103725. [CrossRef]
66. Cecconet, D.; Sabba, F.; Devecseri, M.; Callegari, A.; Capodaglio, A.G. In situ groundwater remediation with bioelectrochemical systems: A critical review and future perspectives. *Environ. Int.* **2020**, *137C*, 105550. [CrossRef]
67. Margalef-Marti, R.; Carrey, R.; Vilades, M.; Carrey, R.; Viladés, M.; Jubany, I.; Vilanova, E.; Grau, R.; Soler, A.; Otero, N. Use of nitrogen and oxygen isotopes of dissolved nitrate to trace field-scale induced denitrification efficiency throughout an in-situ groundwater remediation strategy. *Sci. Total Environ.* **2019**, *686*, 709–718. [CrossRef] [PubMed]
68. Gierczak, R.F.D.; Devlin, J.F.; Rudolph, D.L. Field test of a cross-injection scheme for stimulating in situ denitrification near a municipal water supply well. *J. Contam. Hydrol.* **2007**, *89*, 48–70. [CrossRef] [PubMed]
69. Ministry of Environmental Protection, the People's Republic of China. *Technical Guideline for Delineating Source Water Protection Areas (HJ 338-2018)*; China Environment Press: Beijing, China, 2018. (In Chinese)
70. Zhou, Y. Sources of water, travel times and protection areas for wells in semi-confined aquifers. *Hydrogeol. J.* **2011**, *19*, 1285–1291. [CrossRef]
71. Ameli, A.A.; Craig, J.R. Semi-analytical 3D solution for assessing radial collector well pumping impacts on groundwater–surface water interaction. *Hydrol. Res.* **2017**, *49*, 17–26. [CrossRef]
72. Valois, R.; Cousquer, Y.; Schmutz, M.; Pryet, A.; Delbart, C.; Dupuy, A. Characterizing stream-aquifer exchanges with self-potential measurements. *Groundwater* **2018**, *56*, 437–450. [CrossRef]
73. Zhu, Y.G.; Zhai, Y.Z.; Teng, Y.G.; Wang, G.Q.; Du, Q.Q.; Wang, J.S.; Yang, G. Water supply safety of riverbank filtration wells under the impact of surface water-groundwater interaction: Evidence from long-term field pumping tests. *Sci. Total Environ.* **2020**, *711*, 135141. [CrossRef] [PubMed]
74. Qian, H.; Zheng, X.L.; Fan, X.F. Numerical modeling of steady state 3-D groundwater flow beneath an incomplete river caused by riverside pumping. *J. Hydraul. Eng* **1999**, *30*, 32–37. (In Chinese with English abstract)
75. Liu, S.; Zhou, Y.; Tang, C.; McClain, M.; Wang, X.S. Assessment of alternative groundwater flow models for Beijing Plain, China. *J. Hydrol.* **2021**, *596*, 126065. [CrossRef]

Article

Hydrochemical Characteristics and Evolution of Groundwater in the Alluvial Plain (Anqing Section) of the Lower Yangtze River Basin: Multivariate Statistical and Inversion Model Analyses

Qiaohui Che [1,2], Xiaosi Su [1,2,*], Shixiong Wang [3], Shida Zheng [1,2] and Yunfeng Li [4]

[1] Institute of Water Resources and Environment, Jilin University, Changchun 130012, China; cheqiaohui@126.com (Q.C.); shidazheng1101@126.com (S.Z.)
[2] College of Construction Engineering, Jilin University, Changchun 130021, China
[3] Hebei Provincial Prospecting Institute of Hydrogeology and Engineering Geology, Shijiazhuang 050021, China; hbskywsx@163.com
[4] Nanjing Geological Survey, China Geological Survey, Nanjing 210016, China; liyf@mail.cgs.gov.cn
* Correspondence: suxiaosi@163.com

Citation: Che, Q.; Su, X.; Wang, S.; Zheng, S.; Li, Y. Hydrochemical Characteristics and Evolution of Groundwater in the Alluvial Plain (Anqing Section) of the Lower Yangtze River Basin: Multivariate Statistical and Inversion Model Analyses. *Water* **2021**, *13*, 2403. https://doi.org/10.3390/w13172403

Academic Editors: Yuanzheng Zhai, Jin Wu and Huaqing Wang

Received: 15 July 2021
Accepted: 28 August 2021
Published: 31 August 2021

Publisher's Note: MDPI stays neutral with regard to jurisdictional claims in published maps and institutional affiliations.

Abstract: The alluvial plain (Anqing section) of the lower reaches of the Yangtze River basin is facing increasing groundwater pollution, not only threatening the safety of drinking water for local residents and the sustainable development and utilization of groundwater resources but also the ecological security of the Yangtze River Basin. Therefore, it is necessary to conduct a preliminary analysis on the hydrochemical characteristics and evolution law of groundwater in this area. This study aimed to evaluate potential hydrogeochemical processes affecting the groundwater quality of this area by analyzing major ions in groundwater samples collected in 2019. Compositional relationships were determined to assess the origin of solutes and confirm the predominant hydrogeochemical processes controlling various ions in groundwater. Moreover, factors influencing groundwater quality were evaluated through the factor analysis method, and the control range of each influencing factor was analyzed using the distribution characteristics of factor scores. Finally, reverse hydrogeochemical simulation was carried out on typical profiles to quantitatively analyze the hydrochemical evolution process along flow paths. The Piper trilinear diagram revealed two prevalent hydrochemical facies, Ca-HCO$_3$ type (phreatic water) and Ca-Na-HCO$_3$ type (confined water) water. Based on the compositional relationships, the ions could be attributed to leaching (dissolution of rock salt, carbonate, and sulfate), evaporation and condensation, and cation exchange. Four influencing factors of phreatic water and confined water were extracted. The results of this study are expected to help understand the hydrochemical characteristics and evolution law of groundwater in the alluvial plain (Anqing section) of the lower Yangtze River basin for effective management and utilization of groundwater resources, and provide basic support for the ecological restoration of the Yangtze River Basin.

Keywords: hydrogeochemistry; ionic ratios; factor analysis; inverse modeling; Yangtze River

1. Introduction

Groundwater resources are an important constituent of water resources. The temporal and spatial distribution of groundwater quality reflects the formation and evolution characteristics, geological and hydrogeological background, and influencing factors of groundwater, which are hot topics in hydrogeological and hydrogeochemical research [1–4]. An in-depth understanding of the interaction mechanism between groundwater and the environment can be obtained by investigating the spatio–temporal variation characteristics and evolution rules of groundwater hydrochemistry. The chemical composition of groundwater is a multivariable and complex function [5], and its formation and evolution are affected by the characteristics of aquifer media, chemical composition, hydrodynamic conditions, and human factors [6–9]. Therefore, the formation and geochemical evolution of groundwater are complex. Conventionally, various methods have been used for studying the geochemical evolution of groundwater, mainly including the Piper diagram method [10,11], Gibbs graph method [12], and ion ratio method [13,14]. These methods are often simple and intuitive. Multivariate statistical methods have been used to determine the relationship and influence among multivariate. By extracting the mathematical characteristics of data and ignoring the evolution mechanism of hydrochemical components, water quality factors can be described regionally to study the spatiotemporal distribution of the hydrochemical characteristics of groundwater, evaluate water quality, and identify influencing factors [15–19]. Integrating hydrochemical interpretation with inverse modeling, models with high confidence levels can be applied to quantitatively identify hydrochemical processes along a flow path [20,21]. Inverse geochemical modeling in PHREEQC [22] is based on a geochemical mole-balance model, which calculates phase mole transfer (moles of minerals and gases that must enter or leave a solution) to account for differences in initial and final water compositions along a flow path in a groundwater system. This model requires the input of at least two chemical analyses of groundwater at different points of the flow path and a set of phases (minerals and/or gases) that potentially react along this flow path [23].

The Yangtze River is the largest river in China and the third-largest river in the world. It plays an important role in the sustainable development of the regional economy and ecology [24,25]. To strengthen the protection and restoration of the ecological environment in the Yangtze River basin, facilitate the effective and rational use of resources, safeguard ecological security, ensure harmony between humans and nature, and achieve the sustainable development of the Chinese nation, the 24th Standing Committee session of the 13th National People's Congress passed the first river basin law "Yangtze River Protection Law" on 26 December 2020, and this law came into effect on 1 March 2021. Anqing is located beside the Yangtze River on the alluvial plain in the lower reaches. It is an important city in the Yangtze River Economic Belt and the Yangtze River Delta. Since the 1980s, many large-scale chemical plants have been built in Anqing, posing a serious risk of water pollution [26–28]. Therefore, the study of the interaction between surface water and groundwater in Anqing is of great practical significance to the prevention and control of water pollution and the restoration of the ecological environment in the Yangtze River.

In this study, the main controls on groundwater hydrogeochemistry and hydrochemical characteristics in the alluvial plain (Anqing) in the lower reaches of Yangtze River Basin were analyzed using the Piper diagram, ion ratio, and statistical analysis methods. A reverse hydrogeochemical simulation was also performed to quantitatively analyze the evolution process of groundwater along the groundwater flow path in certain areas. Detailed information on hydrogeochemical mechanisms affecting the concentrations and distributions of dissolved ions in complex geological and hydrogeological systems would provide a scientific basis for better groundwater resource development and management at the local scale and the restoration of the ecological environment of the Yangtze River Basin.

2. Study Area

Anqing is located in southeastern China, on the north bank of the lower reaches of the Yangtze River. It lies between 29°47′–31°16′ N and 115°45′–117°44′ E, spanning three geomorphic units: the middle and low mountains of the Dabie Mountains, low hills along the Yangtze River, and alluvial plain along the Yangtze River. The topography has a general trend of higher in the northwest and lower in the southeast. The Dabie Mountains have an altitude of more than 400 m a.s.l. in the northwest and 100–200 m a.s.l. in the middle; the alluvial plain of the Yangtze River is flat in the south. The alluvial plain of the Yangtze River was taken as the study area (Figure 1).

Anqing is located in the northern subtropical humid climate zone, with a mild climate and moderate rainfall. The annual average temperature ranges from 14.4 °C to 16.6 °C, with obvious geomorphic zonation. The annual average temperature in the Dabie Mountain area is 14.4 °C, and that in the area along the Yangtze River is 16.1 °C to 16.6 °C. The multi-year average rainfall is 1466.2 mm, and the multi-year average evaporation is 917.4 mm.

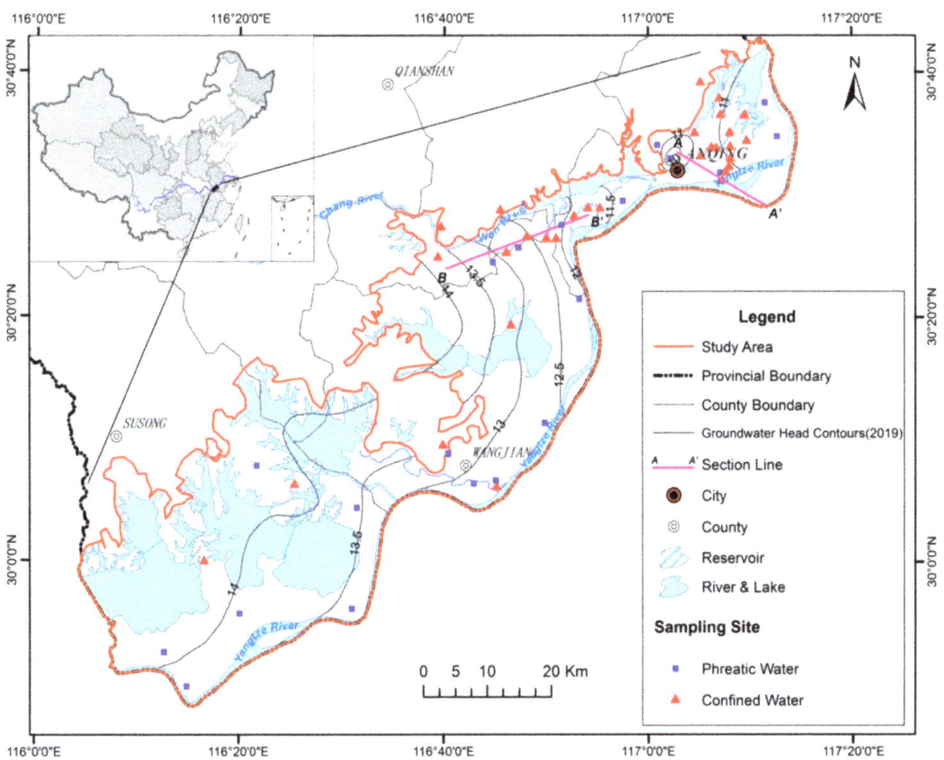

Figure 1. Alluvial plain (Anqing section) of the lower Yangtze River basin and sampling locations.

The study area has a well-developed surface water system, with many rivers and lakes. In the study area, the main stream of the Yangtze River is approximately 243 km long, and the Wan River, with a total length of 94 km, is its primary tributary. An obvious peak of river water level is observed every year. The lowest and highest water levels are observed during January–February and July–August, respectively. According to the water level of the Yangtze River monitored at the Anqing hydrological station for the period 2009–2018, the highest water level is 16.98 m (July 2016) and the lowest is 5.72 m (February 2014). The main lakes include Longgan Lake, Daguan Lake, Po Lake, and Pogang Lake.

Quaternary strata in the study area are well developed and distributed from the lower Pleistocene to Holocene. The gravel layer of the lower Pleistocene Anqing Formation is partly exposed in the third terrace and partly buried in the lower part of the second terrace. The gravel layer has a thickness of 15–30 m and unconformably overlies the Red Bed basement. The gravel is mainly composed of quartzite and quartz sandstone, with good sorting and roundness, and the particle size can reach 1–6 cm. The lower part of the Middle Pleistocene Qijiaji Formation is a 1–4 m thick mud-bearing gravel layer, and the upper part is a 3–8 m thick reticulated laterite. The lower member of the upper Pleistocene Xiashu Formation is a 3–6 m thick khaki sub-clay, containing iron and manganese, widely distributed in the second terrace; the upper member is a light yellow sub-clay, mainly distributed in the first terrace. The stratum of the Holocene Wuhu Formation is mainly distributed in the alluvial plains of the Yangtze River and the main tributary valleys of the Wan River. The stratum can be divided into 3 sections from bottom to top: the lower part comprises a gravel layer and gravel-bearing medium-coarse sand (approximately 10 m thick); the middle part comprises medium–fine sand (10–20 m thick); the upper part comprises grayish yellow–blue gray silty clay (4–10 m thick). In the area with fluvial–lacustrine sediments, the Wuhu Formation is deposited only on the shallow surface, with a thickness not more than 3 m.

Quaternary aquifers in the study area are mainly Holocene sand and gravel phreatic aquifers and lower Pleistocene gravel confined aquifers, with thicknesses of 7–50 m and 0–24 m, respectively. There is no continuous aquitard between the aquifers, but some areas have relative aquitards, as shown in Figures 2 and 3.

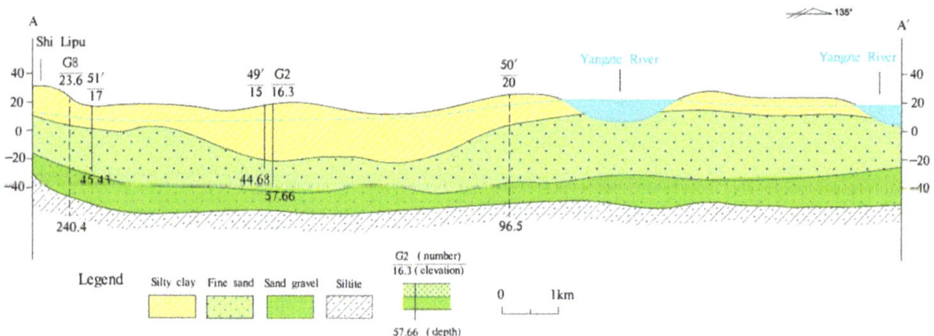

Figure 2. Hydrogeologic cross sections along the A–A′ transect in Figure 1.

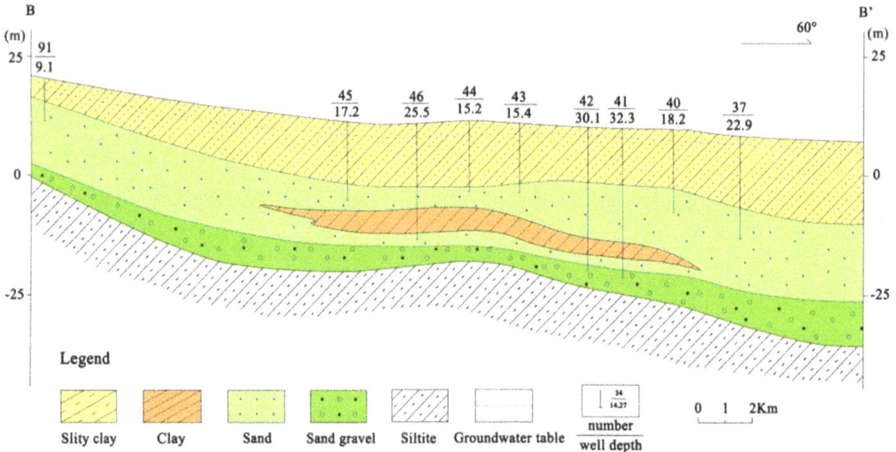

Figure 3. Hydrogeologic cross sections along the B–B′ transect in Figure 1.

In the western part of the study area, groundwater flows from the piedmont to the Yangtze River and receives lateral recharge from piedmont groundwater. Groundwater runoff occurs in the low mountain and hilly plain area with a small hydraulic gradient, discharging to surface water bodies such as lakes and rivers in the runoff path. Finally, the groundwater flows through the plain area along the Yangtze River and drains into the Yangtze River. However, under the influence of large-scale exploitation of groundwater in urban industrial areas, river water is artificially stimulated to recharge groundwater in the plain along the Anqing urban area. The contour of the groundwater level in 2019 is shown in Figure 1.

3. Materials and Methods

3.1. Sample Collection and Analysis

In this study, groundwater samples were collected in May 2019, including 20 groups of phreatic water samples and 31 groups of confined water samples. The distribution of sampling points is shown in Figure 1.

Before sampling, containers were soaked in 10% nitric acid solution for 1–2 days, then in tap water for 1–2 days, and rinsed to neutrality. Finally, they were washed with demineralized water three times and then dried at 70 °C on standby.

During sample collection, water was pumped for more than five minutes to discharge long-term residual groundwater in the well pipe. A HACH water quality rapid detector was used to measure the water temperature (T), TDS, dissolved oxygen (DO), conductivity, pH, and oxidation-reduction potential (Eh) of the water samples. The water samples were collected after the readings stabilized. Each sample was collected after rinsing the container with the water sample more than 3 times.

The concentrations of Fe and Mn were determined in the field using Hach DR1900 portable spectrophotometer. As concentration was determined by inductively coupled plasma mass spectrometry (ICP-MS, Agilent 7500C, Santa Clara, CA, USA) and hydride generation atomic fluorescence photometry (AFS 9600, Beijinghaiguang). Anions and cations were analyzed by ion chromatography (881 compact IC, Metrohm, Switzerland). The mass concentration of bicarbonate ion (HCO_3^-) was determined by acid-base titration in the laboratory. The reliability of water sample data was tested using the anion and cation balance test method, and the absolute value of the relative error of the anion and cation balance less than 5% was taken as reliable.

3.2. Qualitative Research Method

3.2.1. Ionic Ratios

In the process of groundwater circulation, the regularity of each ion component and some ion ratios will change. Therefore, the characteristics of ionic combinations and related ion ratios in groundwater can be used to assess the genesis of groundwater, and identify the source of the chemical components of groundwater and mixing process of different water bodies; this is an effective approach for analyzing the evolution of groundwater [29–31]. However, the variation of ion concentration is largely affected by mixing. When the ion composition is near the mixing line, it indicates that there is no water–rock interaction. When the ion composition deviates from the mixing line, it indicates that it is affected by water–rock interaction. The millimolar per liter (mmol/L) ratio between ($Na^+ + K^+$) and Cl^- ($\gamma(Na+K)/\gamma Cl$ can reflect the source of Na^+ and K^+. A $\gamma(Na+K)/\gamma Cl$ close to 1 indicates the dissolution of halite; a $\gamma(Na+K)/\gamma Cl$ ratio greater than 1 indicates the dissolution of silicates or cation exchange. The main sources of Ca^{2+} and Mg^{2+} in groundwater are mainly the dissolution of carbonates or silicates and evaporites. Accordingly, the mmol/L ratio between Ca^{2+} and SO_4^{2-} ($\gamma Ca^{2+}/\gamma SO_4^{2-}$, milliequivalents per liter (meq/L) ratios between Ca^{2+} and HCO_3^- ($\gamma Ca^{2+}/\gamma HCO_3^-$) and $Ca^{2+} + Mg^{2+}$ and HCO_3^- ($\gamma(Ca^{2+}+Mg^{2+})/\gamma HCO_3^-$) can be used to determine the main sources of Ca^{2+} and Mg^{2+}. A $\gamma Ca^{2+}/\gamma SO_4^{2-}$ ratio close to 1, corresponds to gypsum dissolution. A $\gamma Ca^{2+}/\gamma HCO_3^-$ ratio close to 1, corresponds to calcite dissolution. A $\gamma(Ca^{2+}+Mg^{2+})/\gamma HCO_3^-$ close to 1, corresponds to dolomite dissolution. The meq/L ratio of ($SO_4^{2-}+Cl^-$) to HCO_3^- ($\gamma(SO_4^{2-}+Cl^-)/\gamma HCO_3^-$) reflects the main source of chemical components in groundwater. A $\gamma(SO_4^{2-}+Cl^-)/\gamma HCO_3^-$ ratio greater than 1 indicates evaporite dissolution as the main contributor to the chemical composition of groundwater. A $\gamma(SO_4^{2-}+Cl^-)/\gamma HCO_3^-$ ratio less than 1 indicates carbonate dissolution as the main contributor to the chemical composition of groundwater. The ratio of $\gamma(Na^++K^+-Cl^-)/\gamma(Ca^{2+}+Mg^{2+}-HCO_3^--SO_4^{2-})$ can be used to reflect cation exchange. In the presence of cation exchange, $\gamma(Na^++K^+-Cl^-)$ will be negatively correlated to $\gamma(Ca^{2+}+Mg^{2+}-HCO_3^--SO_4^{2-})$ with a slope of -1, that is, the content of $Ca^{2+}+Mg^{2+}$ decreases with increasing Na^++K^+ content [32–34].

3.2.2. Factor Analysis

Factor analysis [35] is a multivariate statistical analysis method with dimensionality reduction. In other words, more samples or variables are replaced by fewer principal factors, which reflect as much information as possible; moreover, the principal factors are independent of each other [36,37]. According to varying research objectives, factor analysis can be divided into the Q type (correlation between samples) and R type (correlation between variables). The basic idea of R-type factor analysis is to group variables according to the correlation, such that the correlation between variables in the same group is higher, but the correlation between variables in different groups is lower. Each set of variables represents a basic structure, namely a factor, which can reflect the observed correlation. In the field of hydrogeochemistry, R-type factor analysis can eliminate independent and repetitive hydrochemical components and summarize numerous intricately interrelated variables to a few common factors. Each main factor represents a basic combination of hydrochemical components. It often indicates the origin of hydrochemical characteristics and can be used to explain complicated relationships between hydrochemical components [38–42].

3.3. Quantitative Analysis

Inverse Modeling

PHREEQC is undoubtedly the most widely used reverse hydrogeochemical simulation in the world. In this study, PHREEQC version 3 was used for reverse hydrogeochemical simulation. On the representative flow path, according to the change of sample ion concentration, possible mineral phases in the medium are ascertained, the mineral saturation index and dissolved precipitation of the mineral phase are calculated, and the formation and evolution law of regional groundwater are revealed [43–45].

4. Results and Discussion

4.1. Hydrochemical Characteristics of Groundwater

The TDS value of phreatic water in the study area ranged from 176.30 to 575.45 mg·L^{-1} with a mean value of 365.42 mg·L^{-1}, indicating that the groundwater is fresh water. The pH value ranged from 6.78 to 7.88 with a mean value of 7.39, indicating a weakly alkaline environment. The order of relative abundance of major cations in the groundwater followed $Ca^{2+} > Na^+ > Mg^{2+} > K^+$, and the corresponding average mass concentrations were 1.262 mmol·L^{-1}, 2.019 mmol·L^{-1}, 0.904 mmol·L^{-1}, and 0.079 mmol·L^{-1}, respectively. The order of relative abundance of major anions in the groundwater followed $HCO_3^- > Cl^- > SO_4^{2-} > NO_3^-$, and the corresponding average mass concentrations were 5.058 mmol·L^{-1}, 0.646 mmol·L^{-1}, 0.423 mmol·L^{-1}, and 0.214 mmol·L^{-1}, respectively. The dominant cations were Ca^{2+} and Na^+, and the dominant anions were HCO_3^- in phreatic water. Table 1 shows that the variation coefficients of mass concentrations of Fe, Mn, and As in phreatic water were all greater than 100%, indicating that they are more sensitive and unstable to external inputs, such as hydrological conditions, topography, and human activities. The mass concentrations of Fe, Mn, and As were 0.000–0.427 mmol·L^{-1}, 0.002–0.065 mmol·L^{-1}, and 0.000–0.165 μmol·L^{-1}, respectively. The contents of Fe, Mn, and As in some areas exceeded the WHO drinking water quality standard [46], which stipulates mass concentrations of Fe $\leq$ 0.3 mg·L^{-1}, Mn $\leq$ 0.1 mg·L^{-1}, and As $\leq$ 10 μg·L^{-1}. The chemical groundwater types of the study area were distinguished and grouped by their position on a Piper diagram (Figure 4). Based on the major cation and anion, the following two major hydrochemical facies were identified: Ca-HCO$_3$ and Ca-Na-HCO$_3$ types.

Previous isotopic studies confirmed that phreatic water in the study area is mainly recharged by lake water and rainfall [47], Therefore, the hydrochemistry can be safely presumed to be affected by mixing. The ion concentration of the sample after mixing was calculated based on the oxygen stable isotope ^{18}O and compared with the measured data. In this manner, the influence of water-rock interaction on each ion component was determined. In Table 1, the values in bold indicate increases in ion concentration due to water-rock interaction. The specific calculation process is shown in Table S1. The concentrations of HCO_3^-, Na^+, Ca^{2+}, and Mg^{2+} components were higher than their mixed concentrations, indicating that water–rock interaction generally leads to the dissolution of minerals containing C, Na, Ca, and Mg. The concentrations of Cl^- and NO_3^- also increased in most cases, which is related to halite dissolution and human activities in some areas. K^+ concentration changed slightly, indicating that minerals containing K are in equilibrium.

Table 1. Statistical characteristics of the chemical composition of phreatic water.

Sample ID	$\rho(Cl^-)$	$\rho(NO_3^-)$	$\rho(SO_4^{2-})$	$\rho(HCO_3^-)$	$\rho(Na^+)$	$\rho(K^+)$	$\rho(Ca^{2+})$	$\rho(Mg^{2+})$	$\rho(TDS)$	$\rho(Fe)$	$\rho(Mn)$	$\rho(As)$
11	0.287	0.041	**0.279**	**4.227**	**1.371**	0.014	**1.425**	0.277	265.640	1.071×10^{-3}	3.636×10^{-3}	4.698×10^{-2}
12	**0.676**	**0.070**	0.988	**5.566**	**2.132**	**0.077**	**1.683**	1.103	428.950	3.571×10^{-4}	1.818×10^{-3}	0.000
16	**1.156**	**0.158**	0.478	**4.686**	**3.339**	0.069	**1.403**	0.862	377.020	7.143×10^{-4}	1.818×10^{-3}	1.411×10^{-2}
18	**0.770**	**0.329**	0.539	**5.021**	**2.716**	**0.064**	**1.458**	1.130	411.650	7.143×10^{-4}	1.818×10^{-3}	0.000
48	**0.554**	0.000	0.560	**5.394**	1.827	0.028	1.571	1.155	423.590	3.036×10^{-3}	5.455×10^{-3}	0.000
52	0.369	**0.244**	0.277	2.011	1.056	0.025	1.211	0.177	176.300	1.250×10^{-3}	7.273×10^{-3}	3.975×10^{-2}
54	0.365	**0.324**	0.302	3.682	0.882	**0.079**	0.926	0.786	285.660	1.071×10^{-3}	3.636×10^{-3}	1.262×10^{-2}
55	**1.001**	**0.208**	0.623	**6.943**	**2.123**	0.049	**1.753**	1.103	491.930	5.357×10^{-4}	3.636×10^{-3}	2.080×10^{-5}
58	0.493	**0.273**	0.343	4.150	1.433	**0.190**	1.244	0.798	320.890	3.750×10^{-3}	7.273×10^{-3}	1.127×10^{-4}
60	**0.586**	**0.622**	0.476	4.361	2.983	0.078	1.227	0.848	400.650	5.357×10^{-4}	5.455×10^{-3}	6.623×10^{-3}
62	**1.120**	**0.621**	0.342	6.512	2.369	0.003	0.683	1.232	508.890	5.357×10^{-4}	1.818×10^{-3}	0.000
68	**0.466**	**0.432**	**0.488**	1.999	1.678	**0.290**	0.667	0.634	253.540	2.500×10^{-3}	7.273×10^{-3}	2.223×10^{-2}
69	**0.528**	**0.102**	**0.465**	8.138	1.776	0.254	1.764	1.157	393.670	1.571×10^{-2}	2.727×10^{-2}	5.066×10^{-2}
70	**0.699**	**0.437**	**0.499**	8.176	2.579	0.003	**1.835**	1.232	524.420	1.071×10^{-3}	1.091×10^{-2}	1.404×10^{-4}
71	**0.717**	0.054	**0.498**	2.967	2.209	0.041	0.767	0.921	245.060	7.143×10^{-4}	5.455×10^{-3}	0.000
73	0.344	**0.125**	0.250	4.724	1.055	**0.069**	1.294	0.656	282.700	1.071×10^{-3}	3.636×10^{-3}	8.819×10^{-3}
74	**1.115**	0.077	0.207	2.630	2.438	0.014	0.135	0.875	249.460	7.143×10^{-4}	1.818×10^{-3}	1.373×10^{-2}
78	**0.544**	0.037	**0.362**	7.727	1.985	0.000	**1.699**	1.114	450.080	6.071×10^{-3}	5.455×10^{-3}	6.800×10^{-3}
83	**0.602**	0.000	0.137	**9.640**	2.527	0.006	1.902	1.481	575.450	4.268×10^{-1}	6.545×10^{-2}	1.605×10^{-1}
86	**0.531**	**0.120**	**0.354**	2.611	1.901	**0.224**	0.595	0.547	242.887	1.250×10^{-3}	7.273×10^{-3}	9.173×10^{-2}
Minimum	0.287	0.000	0.137	1.999	0.882	0.000	0.135	0.177	176.300	0.000	0.002	0.000
Maximum	1.156	0.622	0.988	9.640	3.339	0.290	1.902	1.481	575.450	0.427	0.065	0.161
Mean	0.646	0.214	0.423	5.058	2.019	0.079	1.262	0.904	365.422	0.023	0.009	0.024
S.D.	0.265	0.193	0.185	2.205	0.658	0.088	0.489	0.330	112.684	0.095	0.014	0.040
C.V.	0.411	0.904	0.437	0.436	0.326	1.121	0.388	0.364	0.308	4.047	1.620	1.687

S.D. stands for standard deviation, C.V. stands for coefficient of variation. Except for As and TDS, which are in $\mu mol \cdot L^{-1}$ and mg/L, respectively, the mass concentrations of other ions and indicators are in $mmol \cdot L^{-1}$. The values in bold indicate increases in ion concentration due to water-rock interaction.

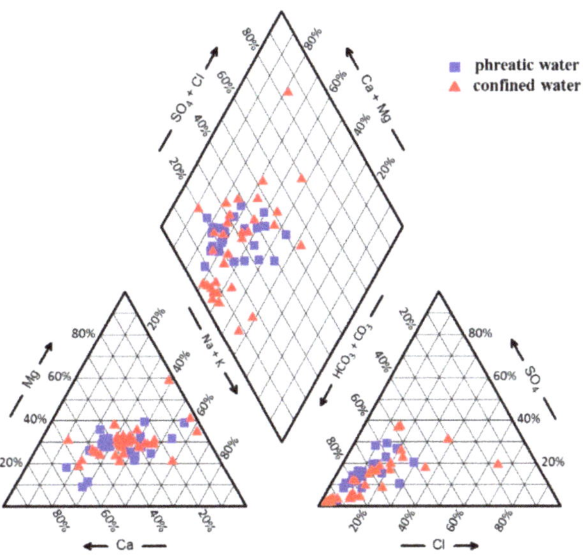

Figure 4. Piper diagram of groundwater samples in the study area.

The TDS values of confined water in the study area ranged from 104.12 to 800.00 $mg \cdot L^{-1}$ with a mean value of 363.25 $mg \cdot L^{-1}$, indicating that the groundwater is fresh water. The pH value ranged from 6.46 to 8.47 with a mean value of 7.35, indicating a weakly alkaline environment. The order of relative abundance of major cations in the groundwater followed $Na^+ > Ca^{2+} > Mg^{2+} > K^+$, and the corresponding average mass concentrations were 2.363 $mmol \cdot L^{-1}$, 1.290 $mmol \cdot L^{-1}$, 1.043 $mmol \cdot L^{-1}$, and 0.026 $mmol \cdot L^{-1}$, respectively. The order of relative abundance of major anions in the groundwater followed $HCO_3^- > Cl^- > SO_4^{2-} > NO_3^-$, and the corresponding average mass concentrations were 5.106 $mmol \cdot L^{-1}$, 0.727 $mmol \cdot L^{-1}$, 0.313 $mmol \cdot L^{-1}$, and 0.251 $mmol \cdot L^{-1}$, respectively. In confined water, the dominant cations were Na^+ and Ca^{2+} and the dominant anions were HCO_3^-. As shown in Table 2, the variation coefficients of mass concentrations of Fe, Mn, and As in confined water were high, indicating that they are more sensitive and unstable to external inputs, such as hydrological conditions, topography, and human activities. The mass concentrations of Fe, Mn, and As were 0.000–0.629 $mmol \cdot L^{-1}$, 0.000–0.060 $mmol \cdot L^{-1}$, and 0.00–1.254 $\mu mol \cdot L^{-1}$, respectively. The contents of Fe, Mn, and As in some areas exceeded the WHO drinking water quality standard [48]. According to the Piper diagram (Figure 4), the hydrochemical types of confined water are mainly Ca-Na-HCO_3, Ca-HCO_3, and Ca-Na-HCO_3-Cl types.

4.2. Qualitative Analysis of the Hydrochemical Evolution Mechanism

4.2.1. Ionic Ratios

Using the Mg equivalent ratio relationship between different ions, the water–rock interaction affecting the change of ion concentration can be assessed. However, changes in ion concentration are strongly affected by mixing. In previous studies, we found that lake and precipitation are the main recharge sources of groundwater in the study area. The groundwater mixing line can be obtained by using the ion concentrations of these two end elements (Table S1), as shown in Figure 5.

In terms of the $\gamma(Na+K)/\gamma Cl$ (Figure 5a), the ionic composition of (Na + K) and Cl deviates from the mixing line, indicating that they are affected by water-rock interaction. Moreover, most of the sample points are distributed in the upper left of the 1:1 line. In other words, the mmol/L concentration of (Na^++K^+) is basically greater than that of Cl^-. This suggests that Na^+ and K^+ in groundwater are mainly attributable to halite dissolution, and cation exchange occurs during runoff, resulting in the higher mmol/L concentration of Na^+ and K^+ ions. In addition, other silicate minerals containing Na and K may also be dissolved.

The $\gamma(SO_4+Cl)/\gamma HCO_3$ (Figure 5b) shows that the ionic composition of (Na + K) and Cl deviates from the mixing line, indicating that they are affected by water-rock interaction. Moreover, the sample points are all distributed below the 1:1 line, and the meq/L concentration of HCO_3^- is much larger than that of SO_4+Cl, indicating the dominance of carbonate dissolution.

The $\gamma Ca/\gamma HCO_3$ meq/L ratio relationship (Figure 5c) is consistent with the $\gamma(Na + K)/\gamma Cl$ meq/L ratio relationship; both deviate from the mixing line and 1:1 line and are located at the upper left of the two lines. These results indicate indicating that Ca and HCO_3 are affected by water-rock interaction, and the dissolution of gypsum and other calcium-containing minerals may occur in addition to calcite dissolution. The $\gamma Ca/\gamma SO_4$ mmol/L ratio relationship (Figure 5d) shows that the sampling points are distributed between the 1:1 line and the mixing line, indicating that Ca and SO_4 are not only affected by mixing but also affected by gypsum dissolution, although only to a small extent.

The $\gamma(Ca + Mg)/\gamma HCO_3$ is shown in Figure 5e. The sampling points are distributed near the 1:1 line and the mixing line, indicating that (Ca + Mg) and HCO_3 are jointly affected by mixing and dolomite dissolution.

As indicated by the $\gamma[(Na^++K^+)-Cl^-]/\gamma[(Ca^{2+}+Mg^{2+})(SO_4^{2-}+HCO_3^-)]$ (Figure 5f), the sampling points are generally distributed near the 1:1 line, suggesting a certain degree of cation exchange in groundwater.

Table 2. Statistical characteristics of the chemical composition of confined water.

Sample ID	$\rho(Cl^-)$	$\rho(NO_3^-)$	$\rho(SO_4^{2-})$	$\rho(HCO_3^-)$	$\rho(Na^+)$	$\rho(K^+)$	$\rho(Ca^{2+})$	$\rho(Mg^{2+})$	$\rho(TDS)$	$\rho(Fe)$	$\rho(Mn)$	$\rho(As)$
6	0.670	0.513	0.345	3.959	1.381	0.038	1.935	0.602	324.840	5.357×10^{-4}	1.818×10^{-3}	0.000
7	0.609	0.301	0.361	3.070	1.808	0.172	1.111	0.556	263.300	7.143×10^{-4}	3.636×10^{-3}	0.000
8	1.006	0.148	0.516	5.776	2.870	0.035	1.558	1.145	430.800	1.786×10^{-3}	7.273×10^{-3}	0.000
9	0.800	0.537	0.326	0.857	1.456	0.023	0.368	0.453	149.610	1.250×10^{-3}	3.636×10^{-3}	5.067×10^{-3}
19	1.764	0.000	0.836	4.906	3.065	0.003	1.735	1.308	473.580	7.143×10^{-4}	7.273×10^{-3}	5.867×10^{-3}
20	0.809	0.140	0.400	2.974	2.113	0.003	1.096	0.700	253.930	1.071×10^{-3}	5.455×10^{-3}	1.427×10^{-2}
21	0.655	0.000	0.166	5.346	2.779	0.000	1.433	1.181	348.560	3.571×10^{-4}	1.818×10^{-3}	9.200×10^{-3}
22	0.512	0.162	0.287	3.663	1.647	0.000	1.095	1.168	361.920	2.679×10^{-3}	5.455×10^{-3}	1.333×10^{-4}
23	0.441	1.762	0.322	6.070	2.723	0.000	2.897	1.528	539.930	1.071×10^{-3}	3.636×10^{-3}	6.533×10^{-3}
25	0.804	0.563	0.970	3.548	2.378	0.053	1.468	1.156	384.970	3.571×10^{-4}	1.818×10^{-3}	4.000×10^{-4}
36	0.880	0.037	0.507	7.364	2.894	0.026	1.795	1.010	490.310	2.500×10^{-3}	1.455×10^{-2}	5.333×10^{-3}
37	2.494	0.383	0.480	8.186	3.014	0.019	2.885	1.475	677.650	6.964×10^{-3}	1.091×10^{-2}	1.867×10^{-3}
39	2.058	0.000	1.286	10.806	3.859	0.049	1.819	1.591	800.000	1.232×10^{-2}	2.364×10^{-2}	6.147×10^{-2}
40	0.235	0.000	0.076	9.860	2.889	0.000	1.759	1.528	522.450	1.813×10^{-1}	6.000×10^{-2}	1.254
41	0.351	0.053	0.075	7.765	3.114	0.000	1.536	1.355	419.760	1.545×10^{-1}	1.636×10^{-2}	2.643×10^{-1}
42	0.236	0.064	0.077	7.459	2.833	0.005	1.500	1.434	384.230	8.839×10^{-2}	3.818×10^{-2}	1.293×10^{-2}
43	0.185	0.000	0.073	4.262	1.345	0.007	0.046	1.019	282.430	6.286×10^{-1}	2.364×10^{-2}	7.984×10^{-1}
44	0.212	0.000	0.074	5.738	2.348	0.009	0.865	1.134	329.130	3.643×10^{-1}	1.818×10^{-2}	9.984×10^{-1}
45	0.435	0.089	0.086	3.538	2.885	0.011	0.489	0.522	203.270	1.786×10^{-3}	5.455×10^{-3}	0.000
46	0.228	0.000	0.075	4.925	2.718	0.004	0.068	0.996	277.000	1.339×10^{-1}	1.273×10^{-2}	2.803×10^{-1}
72	0.378	0.615	0.233	1.066	1.215	0.012	0.295	0.405	104.120	2.321×10^{-3}	1.636×10^{-2}	1.333×10^{-4}
75	0.301	0.068	0.076	4.542	2.123	0.018	0.962	0.981	259.330	7.393×10^{-2}	2.909×10^{-2}	4.787×10^{-2}
76	0.226	0.070	0.081	5.833	3.102	0.058	1.156	1.044	311.060	7.286×10^{-2}	1.455×10^{-2}	3.907×10^{-2}
79	1.788	0.000	0.256	9.860	1.166	0.013	2.976	1.605	644.670	2.446×10^{-1}	3.818×10^{-2}	4.125×10^{-1}
81	0.709	0.067	0.140	5.585	1.909	0.000	1.243	1.335	245.670	1.473×10^{-1}	2.182×10^{-2}	9.867×10^{-3}
82	0.682	0.000	0.100	4.973	3.133	0.001	0.985	1.062	310.480	1.929×10^{-2}	5.455×10^{-3}	2.680×10^{-2}
84	0.212	0.000	0.053	6.876	3.067	0.010	1.255	1.240	361.220	3.482×10^{-2}	1.091×10^{-2}	1.023×10^{-1}
85	0.888	1.475	0.291	4.791	2.750	0.000	2.383	1.191	463.070	4.571×10^{-2}	1.455×10^{-2}	1.173×10^{-2}
88	1.056	0.222	0.270	1.511	2.075	0.003	0.047	0.580	207.170	7.143×10^{-4}	1.091×10^{-2}	4.000×10^{-3}
90	0.315	0.139	0.123	0.985	0.374	0.224	0.582	0.295	197.770	5.000×10^{-3}	1.818×10^{-3}	5.467×10^{-3}
91	0.606	0.384	0.746	2.193	2.220	0.012	0.654	0.724	238.530	3.571×10^{-4}	0.000	3.600×10^{-3}
Minimum	0.185	0.000	0.053	0.857	0.374	0.000	0.046	0.295	104.120	0.000	0.000	0.000
Maximum	2.494	1.762	1.286	10.806	3.859	0.224	2.976	1.605	800.000	0.629	0.060	1.254
Mean	0.727	0.251	0.313	5.106	2.363	0.026	1.290	1.043	363.250	0.072	0.014	0.141
S.D.	0.578	0.413	0.298	2.592	0.775	0.049	0.801	0.377	157.786	0.135	0.013	0.312
C.V.	0.795	1.644	0.953	0.508	0.328	1.893	0.621	0.362	0.434	1.873	0.961	2.209

S.D. stands for standard deviation, C.V. stands for coefficient of variation. Except for As and TDS, which are in $\mu mol \cdot L^{-1}$ and mg/L respectively, the mass concentrations of other ions and indicators are $mmol \cdot L^{-1}$.

In order to further analyze the occurrence and displacement direction of cation exchange in groundwater, chloro-alkaline indices (CAI) were applied [48,49]. The expressions of CAI are as follows:

$$CAI - 1 = \frac{Cl^- - \left(Na^+ + K^+\right)}{Cl^-}$$

$$CAI - 2 = \frac{Cl^- - \left(Na^+ + K^+\right)}{SO_4^{2-} + HCO_3^- + CO_3^{2-} + NO_3^-}$$

The results of CAI-1 and CAI-2 are both negative, indicating the occurrence of cation exchange during runoff and the replacement of Na^+ and K^+ adsorbed by rocks and soil by Ca^{2+} in groundwater, which is consistent with the $\gamma(Na + K)/\gamma Cl$, $\gamma(Ca + Mg)/\gamma HCO_3$.

4.2.2. Factor Analysis

Factor Analysis of Phreatic Water

From the geochemical dataset, principal components were extracted on the symmetrical correlation matrix computed for the 12 variables (Table 3). Before the analysis, the KMO (Kaiser–Meyer–Olkin) and Bartlett (Bartlett test of sphericity) tests were conducted to verify the suitability of the data. The KMO test showed a value of 0.613 and the Bartlett test showed a significance level of less than 0.01, which indicates that the data have a certain correlation and are suitable for factor analysis.

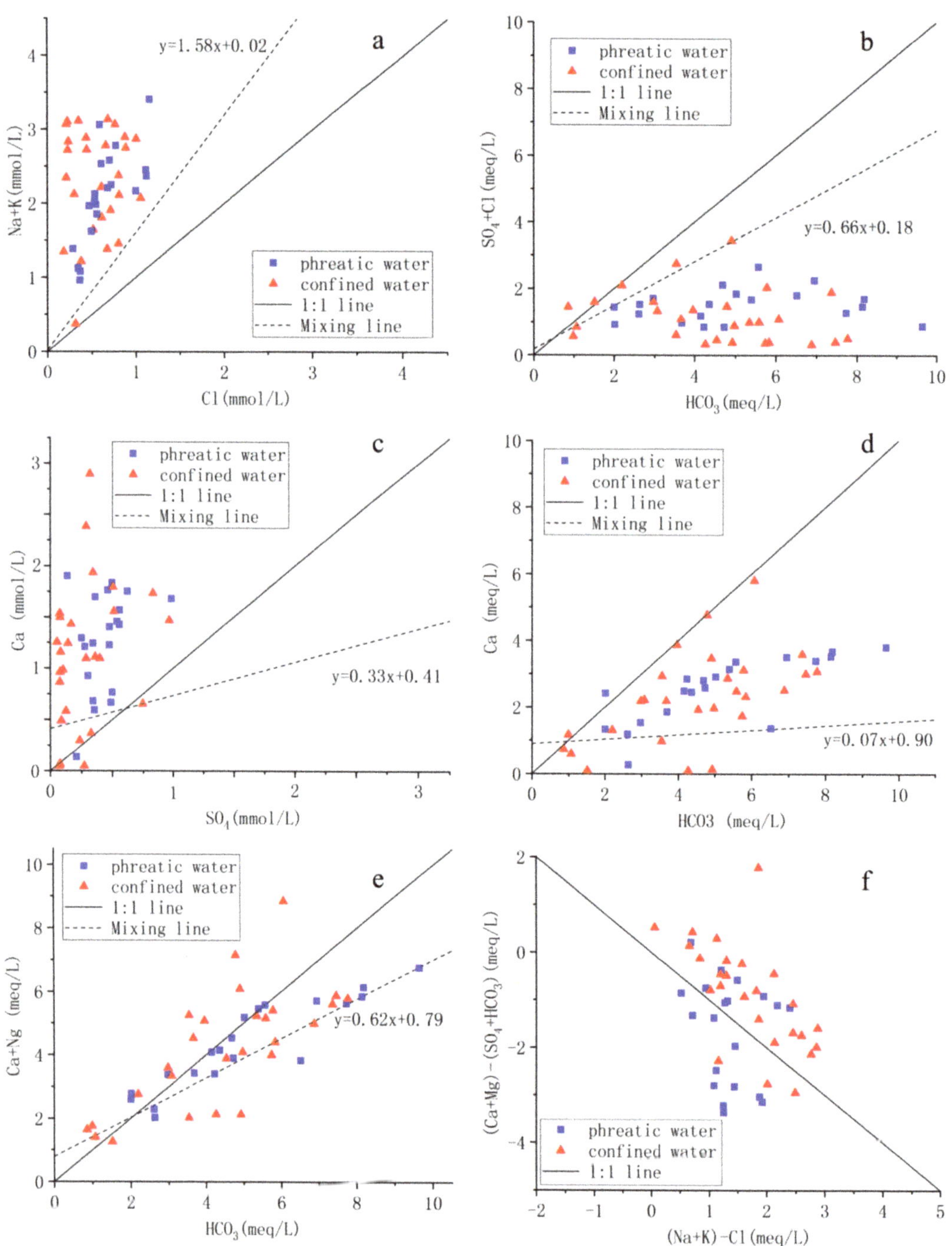

Figure 5. Relationships between the rates of the selected ions of groundwater.

Table 3. Loading for varimax rotated factor matrix of a four-factor model explaining 80.05% of the total variance.

Variable	Factor Loading			
	F_1	F_2	F_3	F_4
Cl^-	0.016	−0.159	**0.879**	−0.170
NO_3^-	−0.123	−0.236	0.446	**0.534**
SO_4^{2-}	0.502	**−0.638**	0.073	0.240
HCO_3^-	**0.857**	0.292	0.181	−0.163
Na^+	0.253	0.023	**0.831**	−0.009
K^+	−0.113	0.049	−0.232	**0.841**
Ca^{2+}	**0.884**	0.055	−0.257	−0.103
Mg^{2+}	**0.729**	0.150	0.542	−0.114
TDS	**0.825**	0.150	0.465	−0.096
Fe	0.311	**0.892**	0.084	−0.133
Mn	0.393	**0.900**	−0.018	0.025
As	0.030	**0.932**	−0.146	0.049
Eigenvalue	3.511	3.111	2.476	1.308
Explained variance%	27.011	23.928	19.048	10.063
Cumulative% of variance	27.011	50.939	69.987	80.049

Bold values: The maximum absolute value of the loadings of each index.

The main methods of factor load matrix estimation include the principal component method, principal axis factor analysis, and maximum likelihood method. In this study, the principal component method was selected to extract the eigenvalues. Four factors with eigenvalues greater than 1 were selected for analysis, and the cumulative variance contribution rate was 80.05%, indicating that the four factors reflected 80.05% of the information content of the total factors affecting water quality. To highlight typical representative variables of each common factor and explain their practical significance, the factor load matrix was rotated. After rotation, the main factor loads were converted to 1 or 0 polarization. The rotation factor load matrix is shown in Table 3.

F_1 reflects water–rock interaction, mainly carbonate dissolution. It was mainly determined by HCO_3^-, Ca^{2+}, Mg^{2+}, and TDS, and its contribution rate was 27.011%. According to the analysis of ion ratios, carbonate dissolution is widely distributed, resulting in the high contents of HCO_3^-, Ca^{2+}, and Mg^{2+} in groundwater. Figure 6 shows the interpolation of F_1 scores at each sampling point of phreatic water. Sampling points with high scores were mainly distributed in the groundwater discharge area (Anqing and Wangjiang sections) and the retention area (Wan River Valley), where groundwater runoff is slow. In these regions, the aquifers have a small grain size, the velocity of groundwater is slow, and water–rock interactions frequently occur between the groundwater and aquifer, resulting in strong carbonate dissolution. These factors contribute to the enhancement of HCO_3^-, Ca^{2+}, Mg^{2+}, and TDS in groundwater.

F_2 reflects the endogenous pollution of groundwater, which is affected by aquifer geological conditions. In F_2, the factor loads of Fe, Mn, As, and SO_4^{2-} were large, and the contribution rate of F_2 was 23.928%. The concentration of Fe and Mn in groundwater in the study area generally exceeds the WHO standard, which can be mainly attributed to the reduction and dissolution of original iron-bearing and manganese-bearing minerals in the aquifer [50]. This is consistent with the geological conditions of the aquifer medium containing iron-bearing and manganese-bearing minerals. With the reduction and dissolution of iron-bearing and manganese-bearing minerals, the content of As in groundwater exceeds the standard. Figure 7 shows the interpolation of F_2 scores at each sampling point of phreatic water. Sampling points with high scores almost covered the entire study area, indicating high contents of primary iron-bearing and manganese-bearing minerals in the aquifer medium. In the plain area, the terrain is flat and the groundwater flow rate is slow, which promotes the complete reduction and dissolution of iron-bearing and manganese-bearing minerals, releasing arsenic in the lattice and affecting groundwater quality. Thus, iron, manganese, and arsenic in groundwater are strongly correlated [51,52].

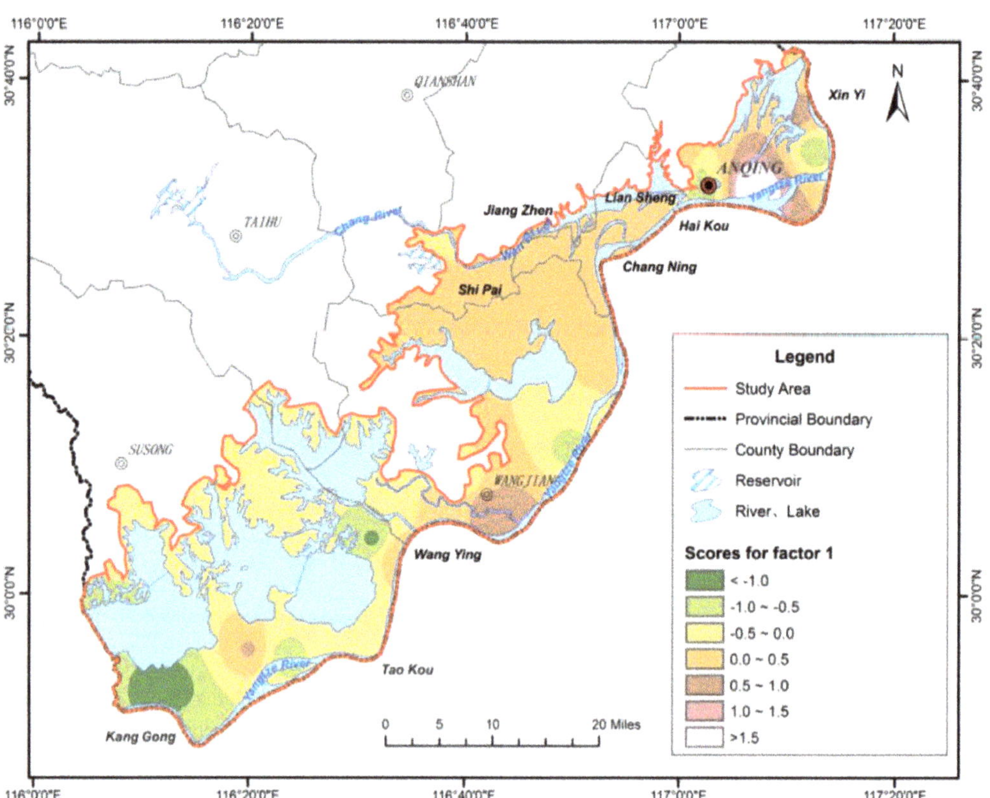

Figure 6. Distribution of scores of factor 1 for phreatic water.

F_3 reflects the effect of halite dissolution and evaporation-concentration on groundwater hydrochemistry. The loads of Na^+ and Cl^- in F_3 were large, and the contribution rate of F_3 was 19.048%. According to the ion ratio analysis, the chemical composition of phreatic water is affected by evaporation-concentration and halite dissolution in some areas. Figure 8 shows the interpolation of F_3 scores at each sampling point of phreatic water. Sampling points with high scores were mainly distributed in the plain along the Yangtze River in the Susong section. The aquifer in this area is shallow, and phreatic water is affected by evaporation concentration. In addition, this area features many lakes, and the groundwater is recharged by lake water, which is affected by evaporation-concentration, resulting in high contents of Na^+ and Cl^-.

F_4 reflects the effect of agricultural production activities on groundwater. In F_4, the factor loads of NO_3^- and K^+ were large, and the contribution rate of F_4 was 10.063%. Figure 9 shows the interpolation of F_4 scores at each sampling point of phreatic water. Sampling points with high scores were mainly distributed in the vicinity of Huang Lake and Bo Lake. With a large number of aquaculture farms in this area, fertilizers containing nitrogen and potassium were applied, resulting in the infiltration of NO_3^- and K^+ into groundwater with surface water.

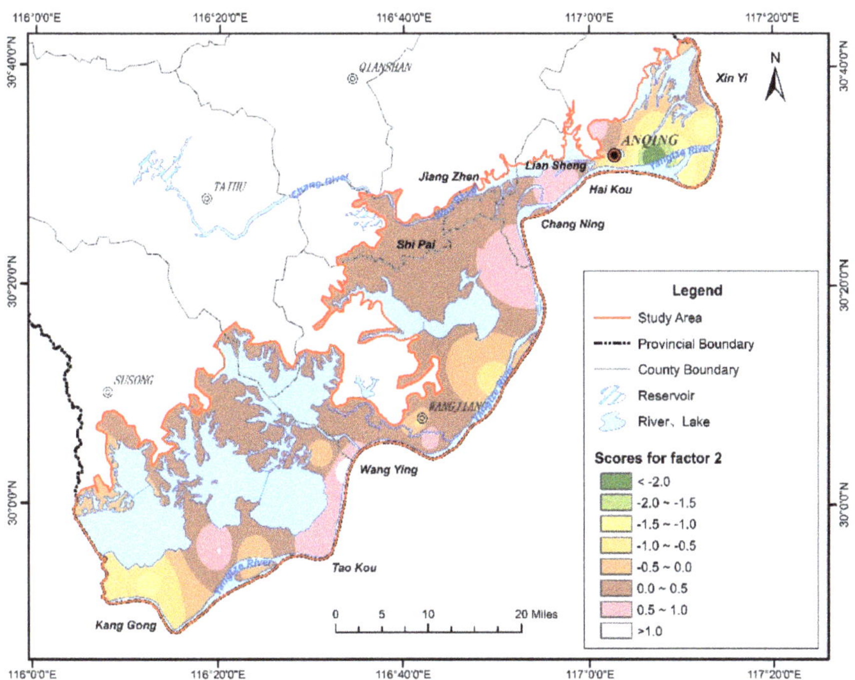

Figure 7. Distribution of scores of factor 2 for phreatic water.

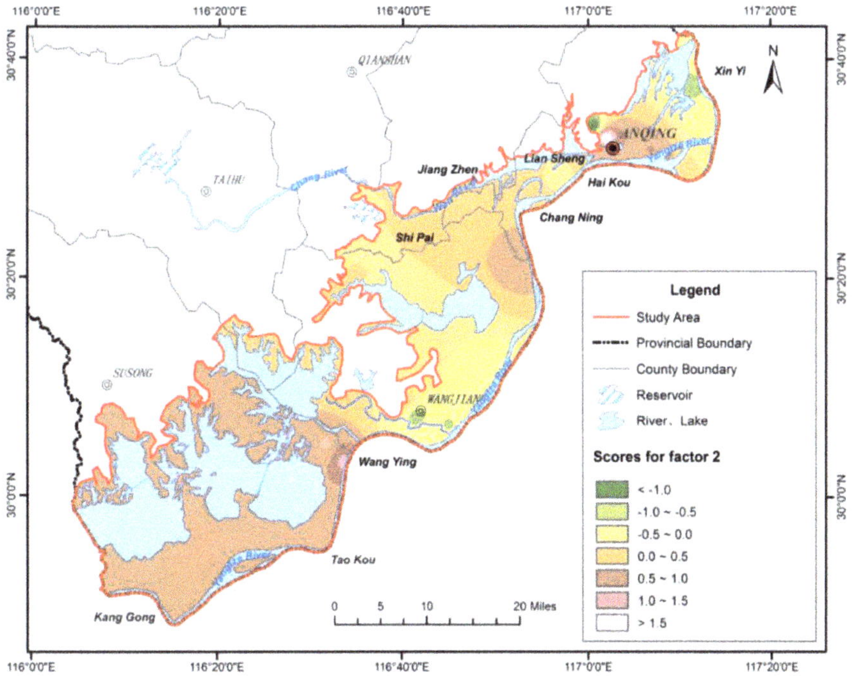

Figure 8. Distribution of scores of factor 3 for phreatic water.

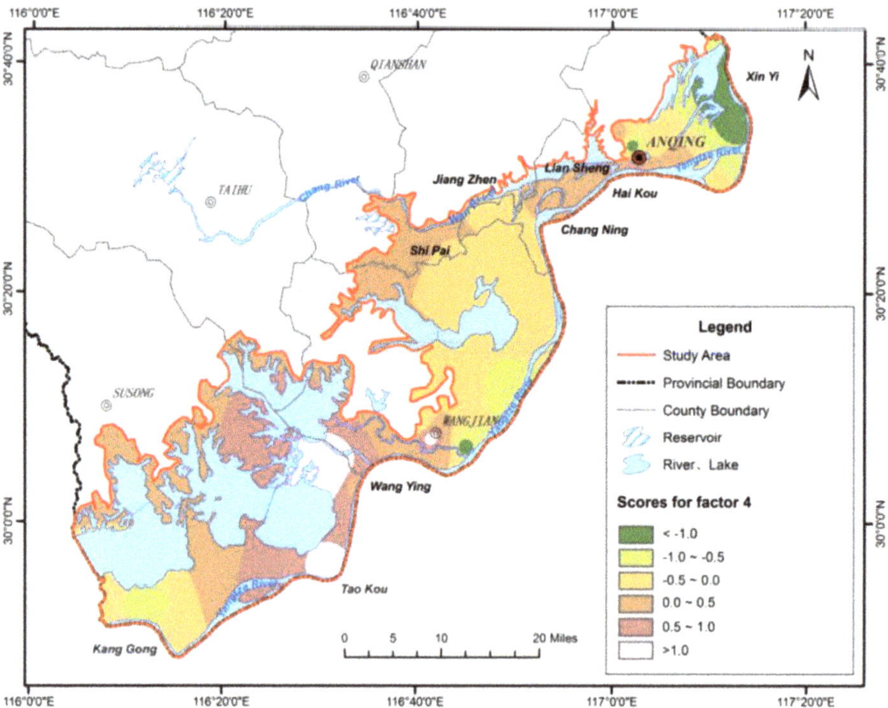

Figure 9. Distribution of scores of factor 4 for phreatic water.

Factor analysis of phreatic water revealed that the chemistry of phreatic water in the study area is mainly affected by carbonate dissolution, primary pollution of iron and manganese, halite dissolution, evaporation-concentration, and human activities.

Factor Analysis of Confined Water

Similar to the factor analysis of phreatic water, 12 chemical indexes of confined water samples were selected for analysis. The data were verified for suitability through the KMO and Bartlett tests. The KMO test produced a value of 0.721, and the Bartlett test revealed a significance level of less than 0.01, which indicates that the data have a certain correlation and are suitable for factor analysis.

Four factors with eigenvalues greater than 1 were selected for analysis by the principal component method, and the cumulative variance contribution rate was 79.63%, indicating that the four factors reflected 79.63% of the information content of the total factors affecting water quality. The rotation factor load matrix is shown in Table 4.

F_1 reflects water-rock interaction, mainly carbonate dissolution. It was mainly determined by HCO_3^-, Ca^{2+}, Mg^{2+}, and TDS, and its contribution rate was 24.367%. Figure 10 shows the interpolation of F_1 scores at each sampling point of confined water. Sampling points with high scores were mainly distributed in the plain along the Yangtze River in the Wangjiang section and the Wan River Valley. The groundwater hydraulic gradient of the riverside plain in the Wangjiang section was relatively large, and groundwater in this area is recharged by groundwater with high contents of HCO_3^-, Ca^{2+}, and Mg^{2+} from the low mountain and hilly areas. Moreover, the TDS content of phreatic water and confined water in this area is relatively high. In the Wan River Valley plain area, groundwater runoff is slow, strong water–rock interaction occurs between groundwater and aquifer, and carbonate dissolution is strong, resulting in the high contents of HCO_3^-, Ca^{2+}, Mg^{2+}, and TDS in groundwater.

Table 4. Loading for varimax rotated factor matrix of a four-factor model explaining 79.63% of the total variance.

Variable	Factor Loading			
	F_1	F_2	F_3	F_4
Cl^-	0.508	−0.139	**0.679**	0.052
NO_3^-	0.508	**−0.558**	−0.414	−0.219
SO_4^{2-}	0.306	−0.326	**0.716**	0.015
HCO_3^-	**0.619**	0.511	0.219	0.478
Na^+	0.233	−0.069	0.184	**0.855**
K^+	−0.076	−0.090	0.244	**−0.741**
Ca^{2+}	**0.942**	−0.053	0.081	0.116
Mg^{2+}	**0.690**	0.369	0.061	0.532
TDS	**0.831**	0.210	0.379	0.283
Fe	−0.035	**0.816**	−0.272	−0.080
Mn	0.240	**0.796**	−0.059	0.173
As	0.048	**0.863**	−0.218	0.006
Eigenvalue	3.168	2.949	2.258	1.977
Explained variance%	24.367	22.683	17.371	15.209
Cumulative% of variance	24.367	47.049	64.420	79.629

Bold values: The maximum absolute value of the loadings of each index.

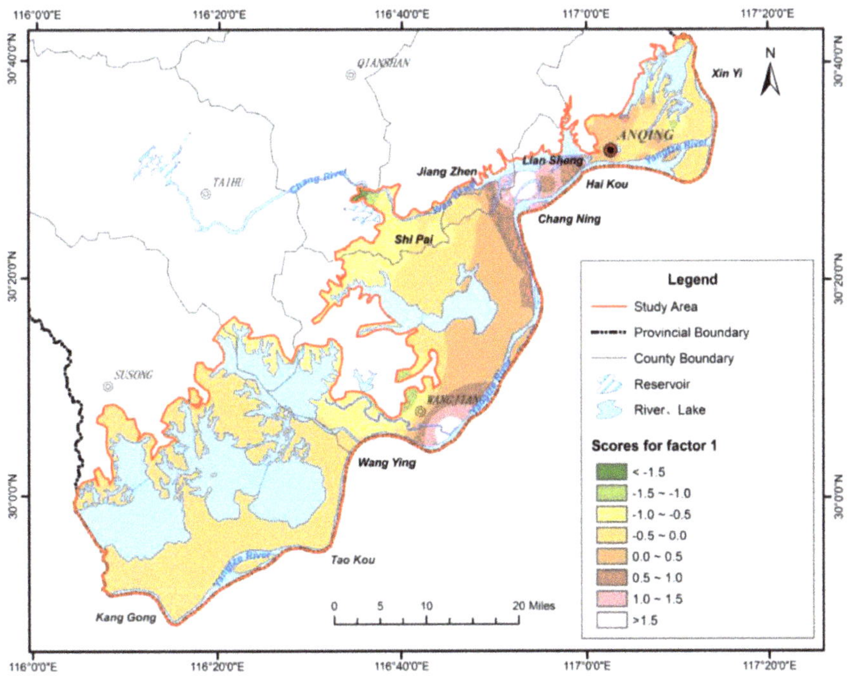

Figure 10. Distribution of scores of factor 1 for confined water.

In F_2, the factor loads of Fe, Mn, As, and NO_3^- were large, and the contribution rate of F_2 was 22.683%. The contents of Fe, Mn, and As in confined water generally exceed the standard because of the reduction and dissolution of original iron and manganese minerals in the aquifer and the release of arsenic in the lattice. Figure 11 shows the interpolation of F_2 scores at each sampling point of confined water. Sampling points with high scores are mainly distributed in the Wan River Valley plain area, where the aquifer lies at great depth and the groundwater is in a reducing environment. The dissolution of iron and manganese minerals results in the release of arsenic in the lattice. Thus, groundwater quality is controlled by the high correlation between iron, manganese, and arsenic. Moreover, this area is a crop planting area; subject to the application of agricultural nitrogen fertilizers, the content of NO_3^- is high.

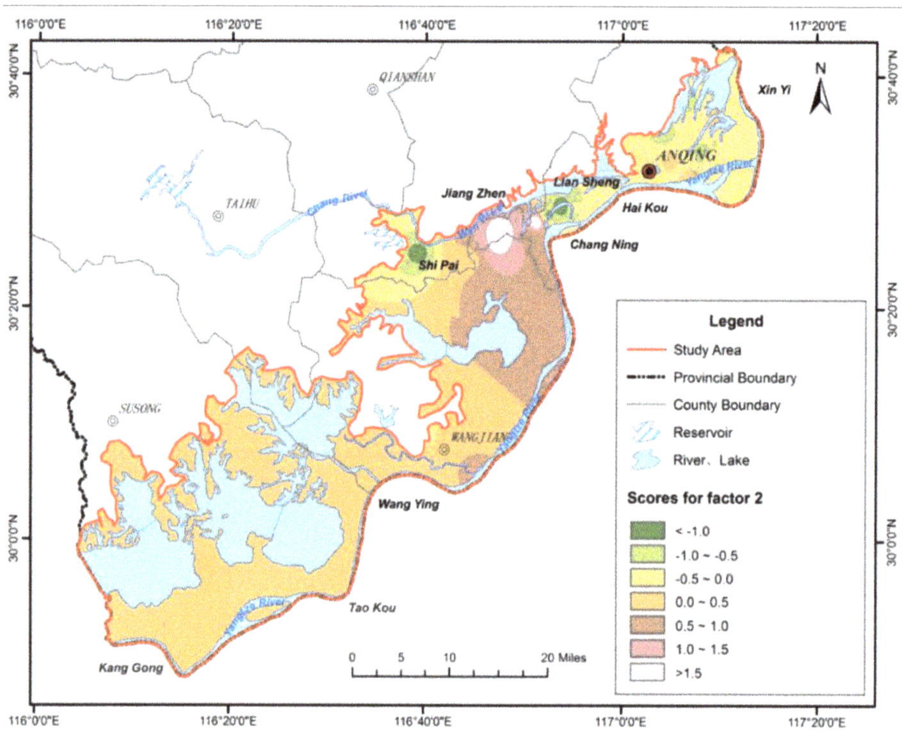

Figure 11. Distribution of scores of factor 2 for confined water.

F_3 reflects the effect of leaching of confined water. The loads of Cl^- and SO_4^{2-} in F_3 were large, and the contribution rate of F_3 was 17.371%. Figure 12 shows the interpolation of F_3 scores at each sampling point of confined water. Sampling points with high scores were mainly distributed in some areas of the riverside plain along the Anqing urban area and the Wan River Valley plain, indicating that the groundwater chemistry in this area is significantly affected by the dissolution of halite and sulfate.

F_4 reflects the effect of cation exchange on confined water. In F_4, the factor loads of Na^+ and K^+ were large, and the contribution rate of F_4 was 15.209%. Figure 13 shows the interpolation of F_4 scores at each sampling point of confined water. Sampling points with high scores were mainly distributed in the riverside plain of the Wangjiang section and the local areas of Wan River Valley plain, which indicates that halite dissolution and cation exchange are the main controlling factors, and the aquifers have a small grain size. Therefore, cation exchange is more likely to occur under such geological conditions.

Factor analysis of confined water showed that the chemistry of confined water in the study area is mainly affected by carbonate dissolution, primary pollution of iron and manganese, halite dissolution, sulfate dissolution, and cation exchange.

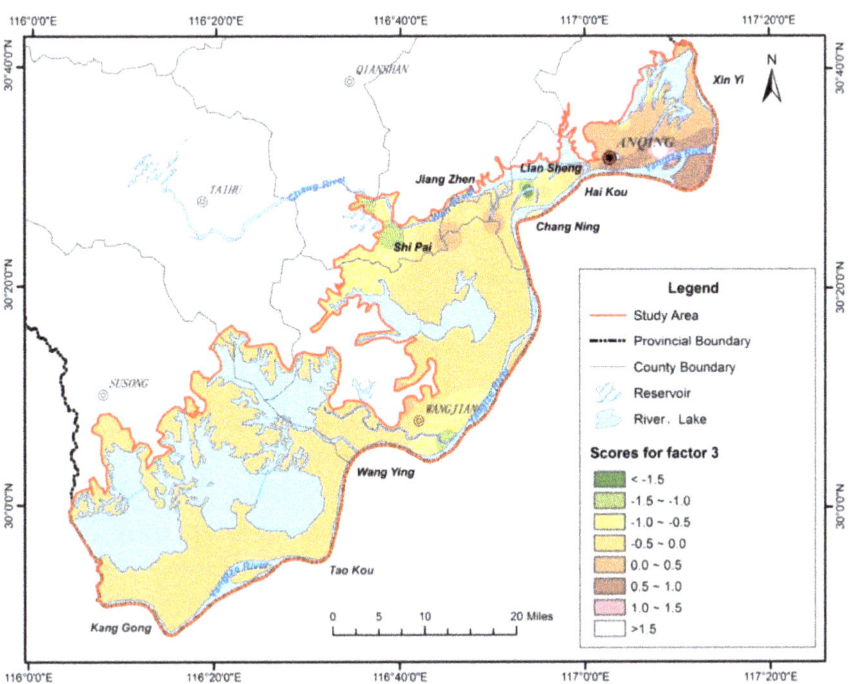

Figure 12. Distribution of scores of factor 3 for confined water.

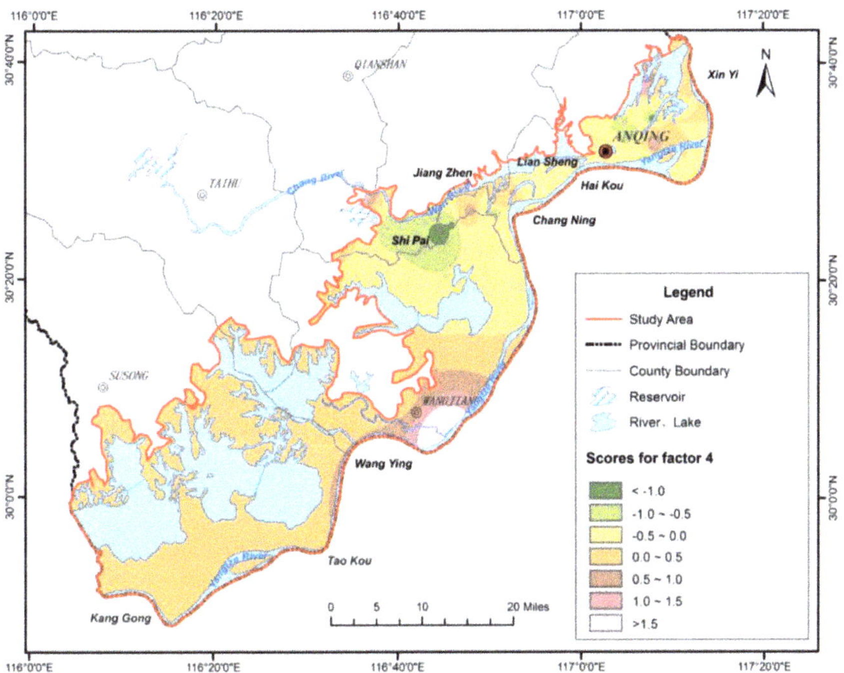

Figure 13. Distribution of scores of factor 4 for confined water.

4.3. Quantitative Analysis of Hydrochemical Evolution Mechanism: Inverse Modeling

4.3.1. Path of Simulation

The flow path of the Wan River Valley plain section (B–B′) was selected for simulation, as shown in Figure 3. According to the groundwater flow field and hydrogeological conditions in the Wan River Valley plain, groundwater flows from Wan River to the confluence of Wan River and Yangtze River. Reverse hydrogeochemical simulations were performed for the flow paths of phreatic water and confined water. The reverse hydrogeochemical simulation path of phreatic water was 45→44→40, and that of confined water was 46→42→41 (Figure 3).

4.3.2. Possible Mineral Phases

Excluding the influence of mixing, the concentration of HCO_3^-, Na^+, Ca^{2+}, Mg^{2+}, Cl^- and NO_3^- in groundwater increases due to the influence of water-rock interaction. Further analysis using the ion ratio method showed that water–rock interactions driving this phenomenon mainly occur as the dissolution of halites, sulfates and carbonates, and cation exchange. According to the scanning electron microscope results of the aqueous medium (Figure S1), typical iron-bearing minerals in the study area include hematite, siderite, and pyrite. Based on the above results, the main possible mineral phases can be determined as follows: calcite ($CaCO_3$), dolomite ($CaMg(CO_3)_2$), albite ($Na_2O \cdot Al_2O_3 \cdot 6SiO_2$), anorthite ($CaO \cdot Al_2O_3 \cdot 2SiO_2$), siderite ($FeCO_3$), fluorite ($CaF_2$), gypsum ($CaSO_4 \cdot H_2O$), halite ($NaCl$), hematite ($Fe_2O_3$), pyrite ($FeS_2$), and claudetite ($As_2O_3$). The wateq4f database was used for this simulation. Table 5 shows the variation of groundwater chemical components along the two flow paths.

Table 5. Test results of major hydrochemical components in simulated paths of phreatic water and confined water.

Sample ID	Na^+	Ca^{2+}	Mg^{2+}	Cl^-	SO_4^{2-}	HCO_3^-	F^-	Fe	As
Phreatic water simulation path									
45	66.35	19.57	12.52	15.43	8.28	215.84	0.61	0.10	0.003
44	54.00	34.59	27.21	7.54	7.14	350.01	0.84	20.40	74.88
40	66.45	70.36	36.68	8.34	7.25	601.44	0.72	10.15	94.07
Confined water simulation path									
46	62.51	2.73	23.90	8.11	7.23	300.43	0.93	7.50	21.02
42	65.17	59.98	34.41	8.39	7.40	455.01	0.82	4.95	0.97
41	71.63	61.42	32.51	12.46	7.17	473.68	0.55	8.65	19.82

Except for As, which is in $\mu g \cdot L^{-1}$, the mass concentrations of other ions and indicators are in $mg \cdot L^{-1}$.

4.3.3. Inverse Modeling Results

Through the ion component balance calculation of groundwater samples, the saturation index (*SI*) of each mineral can be obtained to further determine the occurrence of groundwater leaching. *SI* can be expressed as follows:

$$SI = \log \frac{IAP}{K}$$

In the formula, *IAP* represents the ion activity product of the mineral components of water (dimensionless); *K* is the equilibrium constant reflected by the dissolution of minerals at a certain temperature (dimensionless).

When *SI* > 0, the mineral is supersaturated relative to the aqueous solution; when *SI* = 0, the mineral is in equilibrium with the aqueous solution; when *SI* < 0, the mineral does not reach the saturation state and will dissolve. However, the mineral saturation index remains uncertain, attributable to the errors in water quality analysis and the calculation of mineral equilibrium constant and ionic activity. Therefore, in practice, the mineral is generally considered to be in equilibrium with the aqueous solution when SI = −0.5–0.5.

According to the calculation results (Table 6), water–rock interaction occurs in the study area. In the phreatic water flow path, dolomite and hematite are in the supersaturated state and may precipitate; fluorite, gypsum, rock salt, white arsenite, and CO_2 (g) are unsaturated and continue to dissolve. Siderite is close to equilibrium. On the flow path of confined water, the saturation state of each mineral is consistent with that of phreatic water as a whole, while siderite is dissolved in a more reducing environment.

Table 6. Major mineral saturation indices along the simulated path.

Sample ID	Dolomite	Siderite	Fluorite	Gypsum	Halite	Hematite	Claudetite	CO_2(g)
			Phreatic water simulation path					
45	0.55	−2.04	−1.87	−3.13	−7.54	17.76	−36.33	−2.86
44	2.06	−0.29	−1.44	−3.07	−7.96	22.32	−30.37	−3.04
40	3.16	−0.76	−1.34	−2.84	−7.83	21.65	−37.55	−3.00
			Confined water simulation path					
46	0.59	−0.45	−2.40	−4.07	−7.85	21.49	−36.22	−2.93
42	2.82	−1.09	−1.24	−2.86	−7.83	21.04	−41.19	−3.07
41	3.03	−1.12	−1.61	−2.88	−7.62	21.47	−39.52	−3.19

In the simulation path of phreatic water, the increase in Ca^{2+}, Mg^{2+}, and HCO_3^- concentrations are mainly attributable to the dissolution of calcite and dolomite, and their total dissolved amounts were 0.7582 mmol·L^{-1} and 1.1755 mmol·L^{-1}, respectively. Fluorite was dissolved first and then precipitated and its total dissolved amount was 2.905×10^{-3} mmol·L^{-1}, Ca^{2+} concentration was increased and F^- was released at the same time. The change of Na^+ concentration was mainly controlled by cation exchange. The amount of dissolved NaX was 0.8615 mmol·L^{-1}, and the amount of precipitated CaX was 0.4308 mmol·L^{-1}. The concentrations of Na^+ and Cl^- were reduced by the precipitation of halite (0.2000 mmol·L^{-1}). The variation of Fe content was mainly controlled by the dissolution of hematite (1.3346 mmol·L^{-1}) and pyrite (4.721×10^{-2} mmol·L^{-1}), and the precipitation of siderite (2.5368 mmol·L^{-1}). The content of As is mainly attributable to the release of As in the crystal lattice by the reduction and dissolution of hematite and pyrite, and the dissolution of claudetite (0.4571 mmol·L^{-1}).

In the simulation path of confined water, the increase in Ca^{2+} and Mg^{2+} concentration could be mainly attributed to the dissolution of calcite (0.2564 mmol·L^{-1}) and dolomite (0.3962 mmol·L^{-1}). The precipitation of fluorite (1.000×10^{-2} mmol·L^{-1}) reduced the concentrations of Ca^{2+} and F^-. The increase in Na^+ concentration was mainly controlled by cation exchange. The amount of dissolved NaX was 0.8615 mmol·L^{-1}, and the amount of precipitated CaX was 0.4308 mmol·L^{-1}. The precipitation of halite (0.2000 mmol·L^{-1}) reduced the concentration of Na^+ and Cl^-. The variation of Fe content is mainly controlled by the dissolution of siderite with a dissolution amount of 6.0961 mmol·L^{-1}, and the precipitation of hematite (2.8341 mmol·L^{-1}) and pyrite (0.4066 mmol·L^{-1}). The precipitation of claudetite (0.800 mmol·L^{-1}) resulted in the decrease in As content. Table 7 shows the mass exchange results of possible mineral phases on the simulated paths of phreatic and confined water.

Table 7. Mass exchange results of water samples along simulated paths (mmol·L^{-1}).

Mineral Phases	Stoichiometry	Phreatic Water Simulation Path		Confined Water Simulation Path	
		45→44	44→40	46→42	42→41
Calcite	$CaCO_3$	-	0.7582	-	0.2564
Dolomite	$CaMg(CO_3)_2$	0.7854	0.3901	0.4328	-3.663×10^{-2}
Siderite	$FeCO_3$	−3.016	0.4792	6.627	−0.5309
Fluorite	CaF_2	6.060×10^{-3}	-3.155×10^{-3}	-2.891×10^{-3}	-7.110×10^{-3}
Gypsum	$CaSO_4 \cdot 2H_2O$	−0.1703	6.516×10^{-2}	0.8854	-7.312×10^{-2}

Table 7. *Cont.*

Mineral Phases	Stoichiometry	Phreatic Water Simulation Path		Confined Water Simulation Path	
		45→44	44→40	46→42	42→41
Halite	$NaCl$	-0.2226	2.264×10^{-2}	7.959×10^{-3}	0.1149
Hematite	Fe_2O_3	1.650	-0.3154	-3.115	0.2809
Pyrite	FeS_2	7.921×10^{-2}	-3.200×10^{-2}	-0.4420	3.536×10^{-2}
Claudetite	As_2O_3	0.4570	1.283×10^{-4}	-1.339×10^{-4}	1.259×10^{-4}
CO_2	CO_2	3.337	1.869	-5.259	0.4838
Cation exchange	CaX_2	-0.1138	-0.3170	-5.783×10^{-3}	-9.034×10^{-2}
	NaX	0.2276	0.6339	1.157×10^{-2}	0.1807

5. Conclusions

Hydrogeochemical processes controlling groundwater compositions in the alluvial plain (Anqing section) of the lower Yangtze River Basin were investigated by applying conventional hydrogeochemical techniques (Piper diagram and ionic ratios), statistical methods, and inverse modeling methods to hydrochemical datasets.

The abundance of dominant cations followed the order $Ca^{2+} > Na^+ > Mg^{2+} > K^+$, and that of dominant anions followed the order $HCO_3^- > SO_4^{2-} > Cl^- > NO_3^-$. In terms of hydrochemical types of groundwater, phreatic water could be mainly classified into Ca-HCO_3 type and Ca-Na-HCO_3 type, and confined water into Ca-Na-HCO_3 type, Ca-HCO_3 type, and Ca-Na-HCO_3-Cl type.

The source of solutes was studied by determining relationships between ion ratios, and the main hydrogeochemical processes of various ions in groundwater were determined. The results show that Na^+ and K^+ in groundwater are mainly attributable to halite dissolution and cation alternating adsorption. Ca^{2+} and Mg^{2+} are mainly attributable to carbonate dissolution. Moreover, sulfate dissolution occurs during runoff, in which carbonate dissolution plays a dominant role. Cation adsorption is significant in the process of groundwater runoff, mainly manifested in the adsorption of Ca^{2+} and the release of Na^+.

Four common factors affecting the chemical composition of phreatic water in the study area were extracted through factor analysis: carbonate dissolution (F_1), primary contamination of aquifer media (F_2), halite dissolution, and evaporation-concentration (F_3), and human activities (F_4). Similarly, four common factors affecting the chemical composition of confined water were also extracted: carbonate dissolution (F_1), primary contamination of aquifer media (F_2), dissolution of halite and sulfate (F_3), and cation exchange (F_4). According to the factor scores of each sampling point, the main control range of each factor was determined.

The results of reverse hydrogeochemical simulation showed that along the flow path in a typical profile, the hydrochemical evolution of phreatic water is mainly controlled by the dissolution of calcite, dolomite, fluorite, hematite, pyrite, and claudetite, the precipitation of halite and siderite, and cation exchange (dissolution of NaX and precipitation of CaX). The hydrochemical evolution of confined water is mainly controlled by the dissolution of calcite, dolomite, gypsum, and siderite, the precipitation of fluorite, halite, hematite, pyrite, and claudetite, and cation exchange.

The regional hydrochemical evolution law of groundwater could be analyzed in a simple manner using each method, but the comprehensive application of the ion ratio, factor analysis and reverse hydrogeochemical simulation facilitated the comprehensive investigation of the regional groundwater hydrochemical characteristics and evolution law from the macro to micro scale and from the qualitative to quantitative perspectives.

These findings provide valuable information on hydrological and hydrochemical evolution processes within aquifers of the alluvial plain (Anqing section) of the lower Yangtze River Basin. This integrated approach provides deeper insight into hydrochemical and hydrological evolution processes and a reference for ground-water management where more targeted groundwater monitoring programs would be required in the future.

Furthermore, this study provides technical support for the ecological restoration of the Yangtze River Basin.

Supplementary Materials: The following are available online at https://www.mdpi.com/article/10.3390/w13172403/s1, Figure S1. SEM photos of aqueous media, Table S1. Chemical component concentrations of lake water and precipitation, Table S2. According to the ^{18}O values of lake water, precipitation and samples, the mixing concentration of each ion component is calculated. It is assumed that the ion concentration is only affected by mixing.

Author Contributions: Conceptualization, X.S. and Q.C.; methodology, Q.C.; software, Q.C.; validation, X.S., Q.C. and S.W.; formal analysis, X.S.; investigation, Q.C., S.Z., and Y.L.; resources, X.S.; data curation, Q.C., S.Z.; writing—original draft preparation, X.S., Q.C., S.W., S.Z. and Y.L.; writing—review and editing, X.S., Q.C. and S.W.; visualization, Q.C.; supervision, X.S.; project administration, X.S.; funding acquisition, X.S. and Y.L. All authors have read and agreed to the published version of the manuscript.

Funding: This research was funded by Geological Survey project of China Geological Survey, grant number DD20189250.

Institutional Review Board Statement: Not applicable.

Informed Consent Statement: Not applicable.

Data Availability Statement: The study did not provide any data.

Acknowledgments: Our deepest gratitude goes to the anonymous reviewers for their careful work and thoughtful suggestions that have helped improve this paper substantially. This work is supported by Geological Survey project of China Geological Survey with Grant DD20189250.

Conflicts of Interest: The authors declare no conflict of interest.

References

1. Fisher, R.S.; Mulican, W.F. Hydrochemical evolution of sodium-sulphate and sodium-chloride groundwater beneath the Northern Chihuahuan desert Trans-Pecos, Texas, USA. *Hydrogeol. J.* **1997**, *10*, 455–474.
2. Edmunds, W.M.; Guendouz, A.H.; Mamou, A.; Moullab, A.; Shanda, P.; Zouari, K. Groundwater evolution in the Continental Intercalaire aquifer of southernAlgeria and Tunisia: Trace element and isotope indicators. *Appl. Geochem.* **2003**, *18*, 805–822. [CrossRef]
3. Currell, M.J.; Cartwright, I. Major-ion chemistry, δ13C and 87Sr/86Sr as indicators of hydrochemical evolution and sources of salinity in groundwater in the Yuncheng Basin, China. *Hydrogeol. J.* **2011**, *19*, 835–850. [CrossRef]
4. Zhai, Y.; Zheng, F.; Zhao, X.; Xia, X.; Teng, Y. Identification of hydrochemical genesis and screening of typical groundwater pollutants impacting human health: A case study in Northeast China. *Environ. Pollut.* **2019**, *252*, 1202–1215. [CrossRef]
5. Shen, Z.; Wang, Y. Review and outlook of water-rock interaction studies. *Earth Sci. J. China Univ. Geosci.* **2002**, *27*, 127–134.
6. Kebeds, S.; Travi, Y.; Alemayehu, T.; Ayenew, T. Groud-water recharge, circulation and geochemical evolution in the source region of the blue nile river, Ethiopia. *Appl. Geochem.* **2005**, *20*, 1658–1676. [CrossRef]
7. Zhu, G.F.; Su, Y.H.; Feng, Q. The hydrochemical characteristics and evolution of groundwater and surface water in the Heihe River Basin, northwest China. *Hydrogeol. J.* **2008**, *16*, 167–182. [CrossRef]
8. Dogramaci, S.; Skrzypek, G.; Dodson, W.; Grierson, P.F. Stable isotope and hydrochemical evolution of groundwater in the semi-arid Hamersley Basin of subtropical northwest Australia. *J. Hydrol.* **2012**, *475*, 281–293. [CrossRef]
9. Zhai, Y.; Wang, J.; Zhang, Y.; Teng, Y.; Zuo, R.; Huan, H. Hydrochemical and isotopic investigation of atmospheric precipitation in Beijing, China. *Sci. Total Environ.* **2013**, *456–457*, 202–211. [CrossRef] [PubMed]
10. Piper, A.M. A graphic procedure in the geochemical interpretation of water-analyses. *Trans. Am. Geophys. Union* **1944**, *25*, 914–923. [CrossRef]
11. Zhai, Y.; Zhao, X.; Teng, Y.; Li, X.; Zhang, J.; Wu, J.; Zuo, R. Groundwater nitrate pollution and human health risk assessment by using HHRA model in an agricultural area, NE China. *Ecotoxicol. Environ. Saf.* **2017**, *137*, 130–142. [CrossRef] [PubMed]
12. Feth, J.H.; Gibbs, R.J. Mechanisms Controlling World Water Chemistry: Evaporation-Crystallization Process. *Science* **1970**, *172*, 870–872. [CrossRef]
13. Han, Y.; Wang, G.; Cravotta, C.; Hu, W.; Bian, Y.; Zhang, Z.; Liu, Y. Hydrogeochemical evolution of Ordovician limestone groundwater in Yanzhou, North China. *Hydrol. Process.* **2013**, *27*, 2247–2257. [CrossRef]
14. Farid, I.; Zouari, K.; Rigane, A.; Beji, R. Origin of the groundwater salinity and geochemical processes in detrital and carbonate aquifers: Case of Chougafiya basin (Central Tunisia). *J. Hydrol.* **2015**, *530*, 508–532. [CrossRef]
15. Olmeal, I.; Beal, J.W.; Villaume, J.F. A new approach to understanding multiple-source groundwater contamination: Factor analysis and chemical mass balances. *Water Res.* **1994**, *28*, 1095–1101.

16. Mouser, P.J.; Rizzo, D.M.; Röling, W.F.M.; van Breukelen, B. A Multivariate Statistical Approach to Spatial Representation of Groundwater Contamination using Hydrochemistry and Microbial Community Profiles. *Environ. Sci. Technol.* **2005**, *39*, 7551–7559. [CrossRef] [PubMed]
17. Ouyang, Y.; Nkedi-Kizza, P.; Wu, Q.; Shinde, D.; Huang, C. Assessment of seasonal variations in surface water quality. *Water Res.* **2006**, *40*, 3800–3810. [CrossRef]
18. Nayan, J.K.; Krishna, G.B. Hydrochemical and Multivariate Statistical Evaluation of Heavy Metals in Shallow Alluvial Aquifers of North Brahmaputra Plain, India. *Water Resour.* **2018**, *45*, 966–974.
19. Zhai, Y.; Xia, X.; Yang, G.; Lu, H.; Ma, G.; Wang, G.; Teng, Y.; Yuan, W.; Shrestha, S. Trend, seasonality and relationships of aquatic environmental quality indicators and implications: An experience from Songhua River, NE China. *Ecol. Eng.* **2020**, *145*, 105706. [CrossRef]
20. Atkinson, J.C. Geochemistry analysis and evolution of a bolson aquifer, basin and range province in the southwestern united states. *Environ. Earth Sci.* **2011**, *64*, 37–46. [CrossRef]
21. Lorite-Herrera, M.; Jiménez-Espinosa, R.; Jiménez-Millán, J.; Hiscock, K. Integrated hydrochemical assessment of the Quaternary alluvial aquifer of the Guadalquivir River, southern Spain. *Appl. Geochem.* **2008**, *23*, 2040–2054. [CrossRef]
22. Parkhurst, D.L.; Appelo, C.A.J. *User's Guide to PHREEQC (ver.2)—A Computer Program for Speciation, Batch-Reaction, One-Dimensional Transport, and Inverse Geochemical Calculations*; U.S. Geol. Survey Water Resources Investigation Report; U.S. Geological Survey: Reston, VA, USA, 1999; pp. 99–4259.
23. Charlton, S.R.; Macklin, C.L.; Parkhurst, D.D. *PHREEQCI—A Graphical User Interface for the Geochemical Computer Program PHREEQC*; U.S. Geol. Survey Water Resources Investigation Report; U.S. Geological Survey: Reston, VA, USA, 1997; pp. 97–4222.
24. Hong, Y.X. General idea and strategic framework of ecological environment protection in the Yangtze river economic belt. *Environ. Prot.* **2017**, *45*, 12–16.
25. Liu, J.; Tian, Y.; Huang, K.; Yi, T. Spatial-temporal differentiation of the coupling coordinated development of regional energy-economy-ecology system: A case study of the Yangtze River Economic Belt. *Ecol. Indic.* **2021**, *124*, 107394. [CrossRef]
26. Fang, W.W.; Zhang, L.; Ye, S.X.; Xu, R.H.; Wang, P.; Zhang, C.J. Pollution evaluation and health risk assessment of heavy metals from deposition in Anqing. *China Environ. Sci.* **2015**, *35*, 3795–3803.
27. Li, F.S.; Han, C.; Lin, D.S.; Zhou, B.H.; Xu, Z.B.; Yu, G.M.; Niu, Z.Y.; Zhao, Y.; Wang, S.Y. Pollution characteristics and ecological risk assessment of heavy metals in the sediments from lakes of Anqing City and Anqing section of Yangtze River. *J. Agro-Environ. Sci.* **2017**, *36*, 574–582.
28. Liu, Y.; Huang, H.; Sun, T.; Yuan, Y.; Pan, Y.; Xie, Y.; Fan, Z.; Wang, X. Comprehensive risk assessment and source apportionment of heavy metal contamination in the surface sediment of the Yangtze River Anqing section, China. *Environ. Earth Sci.* **2018**, *77*, 493. [CrossRef]
29. Karo, E.; Marandi, A.; Vaikmae, R. The Origin of Increased Salinity in the Cambrian-Vendian Aquifer System on the Kopli Peninsula Northern Estonia. *Hydrogeol. J.* **2004**, *12*, 424–435. [CrossRef]
30. Lee, J.-Y.; Song, S.-H. Groundwater chemistry and ionic ratios in a western coastal aquifer of Buan, Korea: Implication for seawater intrusion. *Geosci. J.* **2007**, *11*, 259–270. [CrossRef]
31. Matengu, B.; Xu, Y.; Tordiffe, E. Hydrogeological characteristics of the Omaruru Delta Aquifer System in Namibia. *Hydrogeol. J.* **2019**, *27*, 857–883. [CrossRef]
32. Zhang, B.; Zhao, D.; Zhou, P.; Qu, S.; Liao, F.; Wang, G. Hydrochemical characteristics of groundwater and dominant water-rock interactions in the Delingha Area, Qaidam Basin, Northwest China. *Water* **2020**, *12*, 836. [CrossRef]
33. Fuoco, I.; Apollaro, C.; Criscuoli, A.; De Rosa, R.; Velizarov, S.; Figoli, A. Fluoride Polluted Groundwaters in Calabria Region (Southern Italy): Natural Source and Remediation. *Water* **2021**, *13*, 1626. [CrossRef]
34. Apollaro, C.; Fuoco, I.; Bloise, L.; Calabrese, E.; Marini, L.; Vespasiano, G.; Muto, F. Geochemical Modeling of Water-Rock Interaction Processes in the Pollino National Park. *Geofluids* **2021**, *2021*, 1–17. [CrossRef]
35. Spearman, C. "General intelligence", objectively determined and measured. *Am. J. Psychol.* **1904**, *15*, 201–293. [CrossRef]
36. Farnham, I.M.; Johannesson, K.H.; Singh, A.K.; Hodge, V.F.; Stetzenbach, K.J. Factor analytical approaches for evaluating groundwater traceelement chemistry data. *Anal. Chim. Acta* **2003**, *490*, 123–138. [CrossRef]
37. Gao, S.X. *Application of Multivariate Statistical Analysis*; Beijing University Press: Beijing, China, 2005; pp. 326–335.
38. Razack, M.; Dazy, J. Hydrochemical characterization of groundwater mixing in sedimentary and metamorphic reservoirs with combined use of Piper's principle and factor analysis. *J. Hydrol.* **1990**, *114*, 371–393. [CrossRef]
39. Liu, C.-W.; Lin, K.-H.; Kuo, Y.-M. Application of factor analysis in the assessment of groundwater quality in a blackfoot disease area in Taiwan. *Sci. Total Environ.* **2003**, *313*, 77–89. [CrossRef]
40. Cortes, J.E.; Munoz, L.F.; Gonzalez, C.A.; Niño, J.E.; Polo, A.; Suspes, A.; Siachoque, S.C.; Hernández, A.; Trujillo, H. Hydrogeochemistry of the formation waters in the San Francisco field, UMV basin, Colombia-A multivariate statistical approach. *J. Hydrol.* **2016**, *539*, 113–124. [CrossRef]
41. Felaniaina, R.; Jules, R.N.N.; Zakari, M.; Eddy, H.R.; Alexis, J.N.A.; Andrew, A.A. Water quality assessment in the Bétaré-Oya gold mining area (East-Cameroon): Multivariate Statistical Analysis approach. *Sci. Total Environ.* **2018**, *610–611*, 831–844.
42. Zhai, Y.; Cao, X.; Jiang, Y.; Sun, K.; Hu, L.; Teng, Y.; Wang, J.; Li, J. Further Discussion on the Influence Radius of a Pumping Well: A Parameter with Little Scientific and Practical Significance That Can Easily Be Misleading. *Water* **2021**, *13*, 2050. [CrossRef]

43. Helstrup, T.; Jørgensen, N.O.; Banoeng-Yakubo, B.K. Investigation of hydrochemical characteristics of groundwater from the Cretaceous-Eocene limestone aquifer in southern Ghana and southern Togo using hierarchical cluster analysis. *Hydrogeol. J.* **2007**, *15*, 977–989. [CrossRef]
44. Sheila, A.C.F.; Robert, C.G.; Laurent, B.; José Pereira, d.Q.N.; Vincent, V. Mineralogy and genesis of smectites in an alkaline-saline environment of pantanal wetland, Brazil. *Clays Clay Miner.* **2008**, *56*, 579–595.
45. Liu, C.-W.; Wu, M.-Z. Geochemical, mineralogical and statistical characteristics of arsenic in groundwater of the Lanyang Plain, Taiwan. *J. Hydrol.* **2019**, *577*, 1–16. [CrossRef]
46. World Health Organization. *Guidelines for Drinking-Water Quality*, 4th ed.; Incorporating the First Addendum; WHO: Geneva, Switzerland, 2017.
47. Che, Q.; Su, X.; Zheng, S.; Li, Y. Interaction between surface water and groundwater in the Alluvial Plain (anqing section) of the lower Yangtze River Basin: Environmental isotope evidence. *J. Radioanal. Nucl. Chem.* **2021**, *329*, 1331–1343. [CrossRef]
48. Schoeller, H. Geochemistry of Groundwater. In *Groundwater Studies: An International Guide for Research and Practice*; Brown, R.H., Konoplyantsev, A.A., Ineson, J., Kovalevsky, V.S., Eds.; UNESCO: Paris, France, 1977; pp. 1–18.
49. Zhai, Y.; Han, Y.; Xia, X.; Li, X.; Lu, H.; Teng, Y.; Wang, J. Anthropogenic Organic Pollutants in Groundwater Increase Releases of Fe and Mn from Aquifer Sediments: Impacts of Pollution Degree, Mineral Content, and pH. *Water* **2021**, *13*, 1920. [CrossRef]
50. Shorieh, A.; Porel, G.; Razack, M. Assessment of Groundwater Quality in the Dogger Aquifer of Poitiers, Poitou-Charentes Region, France. *J. Water Resour. Prot.* **2015**, *7*, 171–182. [CrossRef]
51. Guo, H.; Yang, S.; Tang, X.; Li, Y.; Shen, Z. Groundwater geochemistry and its implications for arsenic mobilization in shallow aquifers of the Hetao Basin, Inner Mongolia. *Sci. Total Environ.* **2008**, *393*, 131–144. [CrossRef] [PubMed]
52. Lu, S.; Feng, X.; Su, X. Geochemical characteristics of arsenic in groundwater during riverbank filtration: A case study of Liao River, Northeast China. *Water Supply* **2020**, *20*, 3288–3300. [CrossRef]

Article

Uncertain Analysis of Fuzzy Evaluation Model for Water Resources Carrying Capacity: A Case Study in Zanhuang County, North China Plain

Yinxin Ge [1], Jin Wu [1,*], Dasheng Zhang [2], Ruitao Jia [1] and Haotian Yang [1]

[1] Faculty of Architecture, Civil and Transportation Engineering, Beijing University of Technology, Beijing 100124, China; geyinxin@emails.bjut.edu.cn (Y.G.); Jiaruitao@emails.bjut.edu.cn (R.J.); yanghaotian723@163.com (H.Y.)

[2] Hebei Institute of Water Science, Shijiazhuang 050051, China; skyzhangdasheng@126.com

* Correspondence: WuJin@bjut.edu.cn; Tel.: +86-151-1793-1639

Abstract: The scientific and accurate evaluation of water resources carrying capacity has good social, environmental and resource benefits. Reasonable selection of evaluation parameters is the key step to realize efficient and sustainable development of water resources. Taking Zanhuang County in the North China Plain as the research area, this study selected fuzzy comprehensive evaluation models with different weights in the established evaluation index framework to explore the sources of uncertainty affecting the evaluation results of water resources carrying capacity. By using the sensitivity analysis method of index weight, the index with the biggest influence factor on the evaluation result is selected to reduce the uncertainty problems such as index redundancy and small correlation degree. The results show that the correlation and reliable of comprehensive evaluation value obtained by different weight methods is different. The evaluation result obtained by using the analytic hierarchy process is more relevant than the entropy weight method, and it is more consistent with the actual load-bearing situation. The study of sensitivity index shows that water area index is the biggest factor affecting the change of evaluation results, and water resources subsystem and socio-economic subsystem play a dominant role in the whole evaluation framework. The results show that strengthening the data quality control of index assignment and weight method is helpful to reduce the error of water resources carrying capacity evaluation. It can also provide scientific basis for the improvement of fuzzy evaluation model.

Keywords: water resources carrying capacity; uncertainty analysis; fuzzy comprehensive evaluation model; weight sensitivity analysis

Citation: Ge, Y.; Wu, J.; Zhang, D.; Jia, R.; Yang, H. Uncertain Analysis of Fuzzy Evaluation Model for Water Resources Carrying Capacity: A Case Study in Zanhuang County, North China Plain. *Water* **2021**, *13*, 2804. https://doi.org/10.3390/w13202804

Academic Editor: Carmen Teodosiu

Received: 31 August 2021
Accepted: 7 October 2021
Published: 9 October 2021

Publisher's Note: MDPI stays neutral with regard to jurisdictional claims in published maps and institutional affiliations.

1. Introduction

Water resources are irreplaceable natural resources that not only restrict the sustainable development of society but also play a vital role in social development [1,2]. At present, studying the carrying capacity of water resources is a prerequisite for determining the important development relations between water resources and population, ecology and social economy in the region [3]. It is the necessity to manage the sustainable use of water resources and other related water issues [4]. Water resources and other related water issues must be managed sustainably [5]. Of great significance for promoting the development of the Chinese economy and improving the quality of life is how to effectively realize the balance and sustainable development of water resources, the water environment and the economy [6].

In recent years, the water resources carrying capacity has been discussed more than sustainable development, where the former generally refers to the development and utilization degree of natural water resources [7]. Some studies [8–10] have concluded that they often express similar meanings as indicators such as the sustainable utilization of water volume, ecological limit of water environment, limit of water resources and shortage degree [11]. Chinese research on the water resources carrying capacity was first proposed by the Xinjiang Water Resources Soft Science Research Group [12] and constituted a breakthrough in the field of water resources. In the larger theoretical context of sustainable development and water management, the most representative definition is to coordinate the reasonable scale of ecological health and sustainable development resources under certain social conditions of economic, environmental and technological development [13]. Water resource carrying capacity is defined as "the size of population and economy scale that a region's water resources can carry, which had necessary requirements for ecological environmental protection and had certain technical level and social and economic development level in a certain historical stage". Since the water resources carrying capacity involves water resources system, ecological environment system and social economic system under different regional and natural conditions, the interaction among multiple systems will further amplify the complexity and uncertainty [14]. Therefore, strengthening the study of uncertainty of water resources carrying capacity is conducive to improving the reliability of evaluation results. Among the comprehensive dynamic evaluation models, the fuzzy comprehensive evaluation model [15] is widely used. In addition, fuzzy comprehensive evaluation manages fuzzy evaluation variables through accurate mathematical methods, which can provide a more scientific and practical quantitative evaluation of hidden and fuzzy concepts. At the same time, the model can be used to verify whether the evaluation index weight and other related uncertainty issues have a great impact on the evaluation results [16]. Therefore, the fuzzy evaluation model is selected as an example to study the uncertainty of the evaluation results of the water resources carrying capacity [17,18]. From the perspective of weight data and indicator assignment, the uncertainty of water resources carrying capacity is seldom considered in the application of the fuzzy evaluation model, which makes the results of water resources carrying capacity limited in practical applications [11,15,19]. Therefore, the uncertainty study of water resources carrying capacity model will improve the accuracy of evaluation. At the same time, the research on the uncertainty of water resources carrying capacity also provides a direction for the improvement of the model [20–22].

Based on the consideration of these uncertain factors, this paper studies the fuzzy comprehensive evaluation model by comparing the weight method determined by the analytic hierarchy process and the entropy weight method [23,24] under certain technical outline standards. Correlation analysis was used to solve the uncertainty among weights, the indexes with high sensitivity coefficient and great influence on evaluation results were screened out by calculating the influence of indexes through sensitivity analysis [25]. This study provides reliable scientific basis for reducing the uncertainty of the results of the fuzzy evaluation model and improving the efficiency of water resources utilization [26,27]. Taking Zanhuang County in the North China Plain as an example, the evaluation of water resources carrying capacity is carried out with certain characteristic parameters [28,29], which can provide reference for water resources managers in the North China Plain.

2. Materials and Methods

2.1. Study Area

The study area (37°26′ to 37°46′ N, 114°20′ to 114°31′ E) is located in Shijiazhuang city in southwestern Hebei Province (Figure 1b). The district's east–west length is 44.8 km, and its north–south width is 37 km, resulting in a total area of 1210 km² (Figure 1c). It borders the surrounding counties of Shijiazhuang City and Xingtai City inside, and Shanxi Province outside. The county is located in a warm, semi-humid monsoon continental climate. The temperature difference between seasons is large, with an annual average temperature of 13.6 °C. The average annual precipitation and evaporation are 508.9 mm and 1885 mm, respectively, and precipitation mainly occurs from July to September. Regarding its topography, the region is located at the eastern foot of the Tai-hang Mountains. The landform features of the region are mainly composed of mountains and hills. On the whole, the mountain trend is higher in the west and lower in the east. The abundant basin-scale crossing of water resources from the Ziya River System and the artificial canal South-to-North Water Diversion Project in the study area can meet 68% of the county's annual water consumption. The county has 11 towns under its jurisdiction.

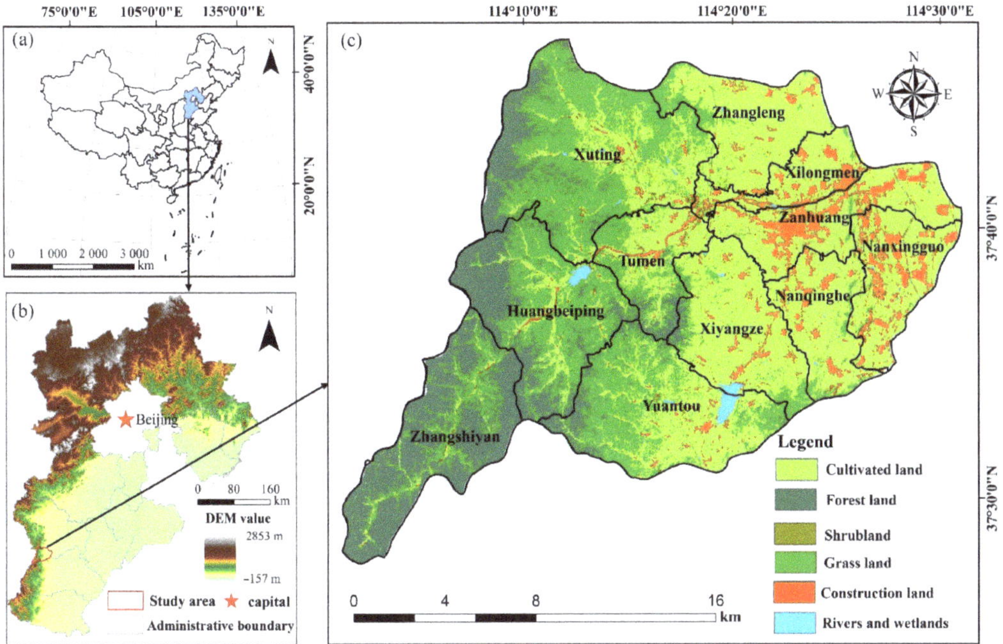

Figure 1. Maps showing (**a**) the location of Hebei in China; (**b**) the location and elevation of study area in Hebei; (**c**) the administrative boundaries in the study area combined with its land use.

2.2. Methods

2.2.1. Fuzzy Comprehensive Evaluation Model

As water resources carrying capacity evaluation index standards and evaluation system boundaries are usually uncertain and fuzzy, fuzzy comprehensive evaluation models can improve the objectivity and accuracy of evaluation results and more comprehensively reflect the situation of regional water resources more comprehensively [30]. This paper adopts this model and its basic principle is as follows: to establish evaluation index set $U = (u_1, u_2, \ldots, u_m)$ and the comment set $V = (v_1, v_2, \ldots, v_n)$, the results of fuzzy comprehensive evaluation are as follows:

$$C = (c_1, c_2, \ldots, c_m) = W \cdot R \tag{1}$$

where W is a fuzzy subset on U, $W = (w_1, w_2, \ldots, w_n)$, $0 \leq w_i \leq 1$ ($\sum_{i=1}^{n} w_i = 1$); w_i is the membership degree of U to W (the weight value of this indicator), which represents the extent to which a single element u_i plays a role in the evaluation factor; "·" is a fuzzy operator which ordinary matrix algorithm is adopted; C is a fuzzy subset of V, $C = (c_1, c_2, \ldots, c_m)$, $0 \leq c_j \leq 1$ ($\sum_{j=1}^{m} c_j = 1$); c_j is the membership degree of V to C, which represents the result of comprehensive evaluation. The membership (evaluation) matrix is as follows:

$$R = \begin{bmatrix} r_{11} & \cdots & r_{1n} \\ \ldots & \ddots & \ldots \\ r_{m1} & \cdots & r_{mn} \end{bmatrix} \tag{2}$$

In the formula, r_{ij} is the membership degree of evaluation u_i to v_j, and $R_i = (r_{i1}, r_{i2}, \ldots, r_{im})$ is the result of single factor for evaluation of u_i.

The actual value of the evaluation index u_i is compared with the classification interval, the membership degree of the corresponding level v_j, which is named the value of r_{ij}, can be calculated. In order to eliminate the level skip phenomenon in which the value of the evaluation grade changes in a small range at the end of the section, the membership function can be smoothly transitioned between each level, and the fuzzy processing is performed. When calculating the membership degree matrix R, $r_i^{(t)}$ ($t = 1, 2, 3, 4, 5,$), which means the membership of the t-th level, r_i is the actual value of the index, and $x_{max}^{(t)}$ and $x_{min}^{(t)}$ refer to the upper limit and lower limit of the t-th evaluation level, respectively. The algorithm of the membership matrix R is not elaborated in detail [31].

After obtaining fuzzy comprehensive evaluation matrix C of each town, each grade of the evaluation index reflects different situations of water resources carrying capacity [32]. A value between 0 and 1 is assigned to each grade for quantification, and the larger the value is, the stronger the water resources carrying capacity is. Generally speaking, the evaluation indexes were divided into five levels, and the comment set $V = (v_1, v_2, \ldots, v_5)$. The v_1 level indicates that the carrying capacity of water resources is in a best state, and the coordinated development of water resources with economy, society and ecology is in a state of sustainable utilization. The v_2 level indicates that the carrying capacity of water resources is in a good state, and the water resource is sufficient to support the local economic and social development level. The v_3 level indicates that the carrying capacity of water resources is in the general state and there is no obvious regional water shortage problem. The v_4 level indicates that the carrying capacity of water resources is in a poor state, but it can basically meet the water demand of various industries. The v_5 level indicates that the carrying capacity of water resources is in a very poor state and the contradiction of water resources is prominent. Take $\alpha_1 = 0.1$, $\alpha_2 = 0.3$, $\alpha_3 = 0.5$, $\alpha_4 = 0.7$ and $\alpha_5 = 0.9$ for levels v_1, v_2, v_3, v_4 and v_5, respectively. The scoring value of each grade and the final comprehensive evaluation value of water resources carrying capacity are calculated according to the formula below.

$$\theta = \frac{\sum_{t=1}^{5} b_t^k \alpha_t}{\sum_{t=1}^{5} b_t^k} \tag{3}$$

where θ is the comprehensive evaluation value of water resources carrying capacity based on the comprehensive evaluation result matrix C; b_t^k is the value of the membership degree of each evaluation index; k is the coefficient set when the dominant role needs to be highlighted, usually $k = 1$.

2.2.2. Index Weight Calculation Methods

The weight calculation methods can be divided into subjective methods and objective methods [33,34], including the binomial coefficient method and the analytic hierarchy process. The subjective method research is relatively mature, with strong subjective arbitrariness, and more dependence on the thinking of the decision analyst. However, the principal component analysis, entropy and other objective methods use decision matrices, which have a strong mathematical theoretical basis to determine weights based on relationships between the original data. Since many factors are involved in the water resources carrying capacity, and different factors have different effects on it, the actual situation of the study area should be considered when assigning weights [35]. First, the relationship between the indicators was clarified, and the corresponding index system was established. The index system was divided into three levels: target level, criterion level and index level [36] In this paper, the analytic hierarchy process and the entropy weight method were used to discuss and study this respectively, and an appropriate weight method was found.

Analytic Hierarchy Process

The analytic hierarchy process (AHP) is a systematic method of making decisions by means of qualitative indicators and fuzzy quantification [30]. According to the nature of the problem and the overall goal to be achieved, it deconstructs the problem into different constituent factors. The AHP combines the factors at different levels according to their interrelationship and the affiliation relationship, forming a multilevel analysis structure model. Thus, ultimately, the problem is attributed to the determination of the importance of the lowest level (plans, measures, etc. for decision making) relative to the highest level (the overall goal) or the arrangement of the relative order of superiority and inferiority. The main steps of the analytic hierarchy process are as follows:

(1) Establish the hierarchical structure model;
(2) Construct judgment matrix by comparing paired indexes;
(3) Calculate the maximum eigenvalue and eigenvector of the judgment matrix, and carry out the consistency test;
(4) Calculate the weight of each evaluation index.

Entropy Weight Method

The entropy weight method is used to determine the weight of each evaluation index. Generally, when the information entropy of an index is smaller, the information provided and the index weight is greater, and vice versa [37]. The main calculation steps are as follows:

The original evaluation index matrix B was obtained according to the membership:

$$B = \begin{bmatrix} b_{11} & \cdots & b_{1n} \\ \cdots & \ddots & \cdots \\ b_{m1} & \cdots & b_{mn} \end{bmatrix} \tag{4}$$

where b_{ij} is the original value of the i-th index in the j-th year. The normalized matrix A is obtained by eliminating the dimensional effect. The positive and negative indicators are treated as follows:

Positive indicators:

$$a_{ij} = \frac{b_{ij} - \min(b_{ij})}{\max(b_{ij}) - \min(b_{ij})} \tag{5}$$

Negative indicators:

$$a_{ij} = \frac{\max(b_{ij}) - b_{ij}}{\max(b_{ij}) - \min(b_{ij})} \tag{6}$$

Normalization matrix:

$$A = \begin{bmatrix} a_{11} & \cdots & a_{1n} \\ \cdots & \ddots & \cdots \\ a_{m1} & \cdots & a_{mn} \end{bmatrix} \tag{7}$$

Calculation of objective weight through the entropy weight method:

$$w_i = \frac{1 - e_i}{m - \sum_{i=1}^{m} e_i} \tag{8}$$

Information entropy:

$$e_i = -\frac{1}{\ln n} \sum_{j=1}^{n} P_{ij} \ln P_{ij}, P_{ij} = \frac{a_{ij}}{\sum_{i=1}^{n} a_{ij}} \tag{9}$$

2.2.3. Calculation Method of Weight Sensitivity

In the evaluation of water resources carrying capacity, the weight of each index in the evaluation index system can be obtained with the help of fuzzy comprehensive evaluation model, but it cannot judge which index has a high impact on the evaluation. Sensitivity analysis can explain the influence of which index weight by changing the value of relevant variables. It is an essential basic step in the process of multi-criteria decision making, because it is directly related to the accuracy and reliability of decision-making results [38].

This paper adopted the single-factor division method [39] to test the sensitivity of the index weight, which can be shown by removing a certain variable weight. The weights of the other variables were equally distributed to remove the variable weight value and to maintain the total weight value and the value of 1. The changing situation reflects the trend and regularity of the influence of single-factor weight changes on the water resources carrying capacity and then removes the weights of other indicators individually, calculates their respective sensitivities and evaluates the impact of the uncertainty of the weight of each indicator on the research results variation. If removing this index weight does not have a great impact on the score result, then the comprehensive evaluation of the water resources carrying capacity is insensitive to this index weight, and vice versa. The calculation of this method is shown in Formulas (10) and (11).

$$\text{RMSEC} = \sqrt{\frac{\sum_{i=1}^{n} \left(\frac{Y - Y_i}{Y}\right)^2}{n}} \tag{10}$$

$$\text{TF} = \sum_{i=1}^{k} F_{\text{RMSEC}} \tag{11}$$

where RMSEC is the rate of change of root mean square error (the sensitivity index); n is the number of index weight variables; Y is the comprehensive evaluation value in the original fuzzy comprehensive evaluation results; Y_i is the comprehensive evaluation value after changing the weight index variable; TF is the total sensitivity; K is the number of comprehensive evaluation results; F_{RMSEC} is the change rate of the root mean square error of the comprehensive evaluation results, that is, the sensitivity of each weight variable in each comprehensive evaluation after changing.

Firstly, the evaluation of water resources carrying capacity is calculated under the basic framework of the fuzzy comprehensive evaluation model, and the framework of the evaluation system of water resources carrying capacity is established. Then, according to the original data, it is processed and indexed according to the grading standards. Secondly, the reliability analysis of the comprehensive evaluation value obtained by different weights was compared by using correlation analysis method. The error contrast of mathematical objectivity was analyzed under a certain research report standard. The evaluation index value of water resources carrying capacity under the selected weight results was calculated. Thirdly, the sensitivity analysis method of single variable removal is used to calculate the sensitivity index of these evaluation values to screen out the index that has the greatest influence on the evaluation results. For this research, the uncertainty methods provide ideas for the systematic model study of water resource carrying capacity evaluation, Besides this, it also provides direction for the improvement of the model.

3. Results and Discussion

3.1. Construction of the Evaluation Index System and Classification Standard

The selection of indicators is directly related to the accuracy and authenticity of the evaluation results of water resources carrying capacity, so the selection of evaluation indicators should follow the principles of science, integrity, hierarchy, dynamic and operability [40,41]. To accurately reflect the status of the water resources carrying capacity in this region, indices were selected from two aspects. On the one hand, when selecting indicators, we mainly considered our available data (The Shijiazhuang Water Resources Bulletin, Statistical Yearbook of Zanhuang) [42,43] and referred to indicators in other literature [39–41,44]; on the other hand, according to the actual situation and characteristics of water resources in Hebei Province, combined with experts' opinions, the indexes of the study area were selected comprehensively [45]. Finally, the total system was divided into four subsystems: water resources, water environment, water ecology and social economy. A total of 13 evaluation indexes were selected to construct the evaluation index system of water resources carrying capacity in Zanhuang County (Table 1). If the index value increases indefinitely and approaches the V_1 standard with better carrying capacity, it is a positive index. On the contrary, as the index value increases infinitely and approaches the V_5 standard with poor carrying capacity, it is a negative index. The definition and criterion of each indicator in the index layer are shown in Table 2.

3.2. Comparison of Weight Results between the Analytic Hierarchy Process and the Entropy Weight Method

Different weights should be assigned to various evaluation indices because of their different effects on water resources carrying capacity. Weight is a very important factor in fuzzy comprehensive evaluation, and its accuracy directly affects the rationality of the evaluation results. The data sources used in this paper are mainly from the Shijiazhuang Water Resources Bulletin (2017) and Statistical Yearbook of Zanhuang in 2017. The above two calculation methods can be used to obtain the weights of the four subsystems of the analytic hierarchy process (detailed calculation process can be found in the supplementary materials section in Tables S1 and S2), which are respectively $W_A = 0.46$, $W_B = 0.14$, $W_C = 0.13$, $W_D = 0.27$; the weight of the four subsystems of the entropy weight method is $W_A' = 0.20$, $W_B' = 0.38$, $W_C' = 0.16$, $W_D' = 0.26$. The final weights calculated by each index layer are shown in Figure 2.

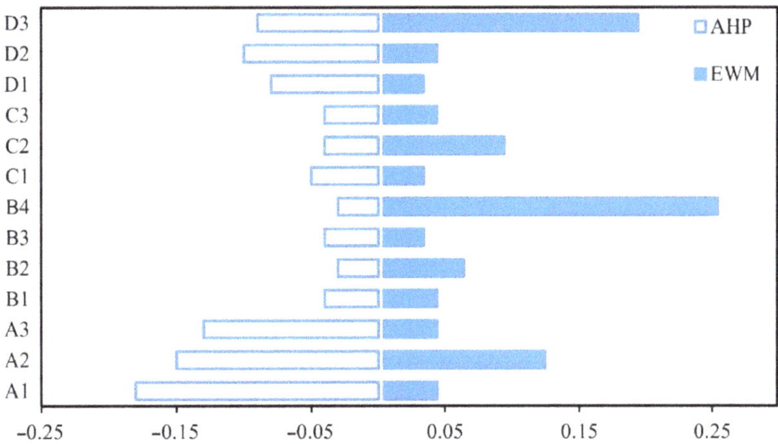

Figure 2. Comparison of weight results calculated by AHP and EWM.

From Figure 2, the five indexes ranked from high to low of the analytic hierarchy process are water resources development and utilization rate, water consumption per unit of GDP, water area index, per capita GDP and domestic water quota, with weights of 0.18, 0.15, 0.13, 0.10 and 0.09, respectively, accounting for 65% of the total contribution rate, including the water resources subsystem and the socio-economic subsystem, which should be the focus of improving the water resources carrying capacity of study area. The other indexes have relatively little influence on the evaluation results. This result verifies that the analytic hierarchy process comprehensively considers the coupling effect between multiple criteria and multiple indicators, and focuses on the importance of identifying effective indicators.

Table 1. Comprehensive evaluation index and classification standard of water resources carrying capacity in Zanhuang County.

Target Layer	Criterion Layer	Index Layer	Index Type	Code	Unit	Classification Standard				
						V_1	V_2	V_3	V_4	V_5
Comprehensive water resources carrying capacity evaluation index system	Water Resources Subsystem (A)	Water resources development and utilization rate	Negative	A_1	%	<15	15–20	20–35	35–60	>60
		Water consumption per unit of GDP	Negative	A_2	$m^3/10^4$ yuan	<50	50–75	75–80	80–100	>100
		Water area index	Positive	A_3	%	>5	4–5	3–4	2–3	<2
	Water Environment Subsystem (B)	Water environmental quality index	Positive	B_1	%	>90	80–90	70–80	60–70	<60
		Industrial wastewater discharge index	Negative	B_2	%	<10	10–20	20–40	40–50	>50
		Fertilizer intensity index	Negative	B_3	kg/hm^2	<100	100–150	150–200	200–250	>250
		Urban sewage discharge index	Negative	B_4	%	<10	10–20	20–40	40–50	>50
	Water Ecological Subsystem (C)	The vegetation coverage rate of coastal zone	Negative	C_1	%	<20	20–30	30–40	40–60	>60
		Ecological base flow guarantee rate	Positive	C_2	%	>60	40–60	30–40	20–30	<20
		River network density index	Positive	C_3	1/km	>0.8	0.6–0.8	0.4–0.6	0.2–0.4	<0.2
	Socioeconomic Sub-system (D)	Population density	Negative	D_1	$1/km^2$	<300	300–500	500–700	700–900	>900
		Per capital GDP	Positive	D_2	10^4 yuan	>7.5	6–7.5	4.5–6	3–4.5	<3
		Domestic water quota	Positive	D_3	liter/day	>130	110–130	90–110	70–90	<70

Table 2. Definition and criterion of each indicator in the index layer.

Indicator	Definition	Criterion
	Water Resources Subsystem	
Water resources development and utilization rate	Regional water consumption/regional water resources	The Shijiazhuang Water Resources Bulletin
Water consumption per unit of GDP	Regional water consumption/total regional GDP	The Shijiazhuang Water Resources Bulletin and Statistical Yearbook of Zanhuang
Water area index	Area of water area/ the total area	Statistical Yearbook of Zanhuang and Google Satellite Map
	Water Environment Subsystem	
Water environmental quality index	The rate of water quality discharge up to standard	Environmental monitoring Reports
Industrial wastewater discharge index	Regional industrial water discharge/total wastewater discharge	The Shijiazhuang Water Resources Bulletin and Environmental monitoring Reports
Fertilizer intensity index	Total amount of fertilizer applied (discounted)/cultivated area of evaluation area	Statistical Yearbook of Zanhuang
Urban sewage discharge index	Regional urban sewage discharge/total wastewater discharge	The Shijiazhuang Water Resources Bulletin and Environmental monitoring Reports
	Water Ecological Subsystem	
The vegetation coverage rate of coastal zone	Length of plant cover/length of shoreline	Statistical Yearbook of Zanhuang and Google Satellite Map
Ecological base flow guarantee rate	Average monthly actual flow/minimum ecological flow	Rain station monitoring reports
River network density index	River length/watershed area	Statistical Yearbook of Zanhuang and Google Satellite Map
	Socioeconomic Subsystem	
Population density	Regional population/regional administrative area	Statistical Yearbook of Zanhuang
Per capital GDP	Regional GDP/regional population	Statistical Yearbook of Zanhuang
Domestic water quota	Domestic water consumption/ (regional population· days)	The Shijiazhuang Water Resources Bulletin and Statistical Yearbook of Zanhuang

The weights of the first five indicators of the entropy method are urban sewage discharge index (0.25), domestic water quota (0.19), water consumption per unit of GDP (0.12), ecological base flow guarantee rate (0.09) and industrial wastewater discharge index (0.06). The first five indicators calculated by this method cover the four sub-systems of the entire criterion, namely, water resources, water environment, water ecology and social economy. First, according to the formula there is no horizontal comparison between the indicators in the calculation process, in that there is no distinction between primary and common indicators. Second, the weight value is too mathematically objective, which covers all the subsystems in the five indicators ranked from high to low. Therefore, the coupling and correlation of the main influencing indicators are often ignored, and it is limited in practical applications. The mutual influence between indicators cannot be ignored for accurate evaluation.

3.3. Comparative Study of Fuzzy Comprehensive Evaluation

It can be seen from Figure 2 that there is a big difference between the weight results of the analytic hierarchy process and the entropy weight method. Tables 3 and 4 are the comprehensive evaluation results of the analytic hierarchy method and the entropy weight method, respectively. Table 5 presents the correlation analysis, in which IBM SPSS Statistics 25 software was used to illustrate this more clearly.

Table 3. Comprehensive evaluation results of water resources carrying capacity by the analytic hierarchy process in Zanhuang County.

Sites	V_1	V_2	V_3	V_4	V_5	Comprehensive Evaluation Value θ	Theta Ranked from High to Low
Zanhuang	0.2500	0.0026	0.0879	0.1395	0.5000	0.6174	3
Xilongmen	0.1700	0.1813	0.2056	0.2107	0.3124	0.6029	6
Nanxingguo	0.3900	0.1811	0.1833	0.0956	0.2300	0.4589	10
Nanqinghe	0.3277	0.1838	0.1015	0.0971	0.2900	0.4676	9
Yuantou	0.1300	0.2708	0.2182	0.1519	0.3300	0.6067	5
Xiyangze	0.3550	0.1186	0.0364	0.0000	0.4900	0.5303	8
Tumen	0.1800	0.0950	0.1528	0.1622	0.4900	0.6774	1
Huangbeiping	0.1000	0.1932	0.2202	0.0166	0.4700	0.6127	4
Zhangshiyan	0.2433	0.0636	0.0766	0.0470	0.5700	0.6276	2
Xuting	0.2600	0.1641	0.1559	0.0400	0.4600	0.5952	7
Zhangleng	0.3750	0.1232	0.1654	0.0865	0.2500	0.4427	11

Table 4. Comprehensive evaluation results of water resources carrying capacity by the entropy weight method in Zanhuang County.

Scheme 1.	V_1	V_2	V_3	V_4	V_5	Comprehensive Evaluation Value θ	Theta Ranked from High to Low
Zanhuang	0.5434	0.0021	0.0733	0.1049	0.2582	0.3974	9
Xilongmen	0.5594	0.1436	0.1492	0.1476	0.0729	0.3426	10
Nanxingguo	0.3701	0.1275	0.0973	0.1821	0.2229	0.4520	7
Nanqinghe	0.2906	0.3073	0.2385	0.0682	0.1653	0.4370	8
Yuantou	0.4672	0.0549	0.0271	0.0000	0.4507	0.4824	6
Xiyangze	0.3245	0.0717	0.1416	0.0839	0.4507	0.5891	2
Tumen	0.0370	0.2726	0.2963	0.0055	0.3885	0.5872	3
Huangbeiping	0.1093	0.1729	0.2006	0.0409	0.4771	0.6211	1
Zhangshiyan	0.3339	0.1360	0.1247	0.0362	0.4415	0.5593	4
Xuting	0.5464	0.1255	0.1535	0.0358	0.1389	0.3190	11
Zhangleng	0.3775	0.1435	0.1193	0.2333	0.1989	0.4827	5

Table 5. Correlation analysis of the membership values of two weighting methods and evaluation values.

Weighting Methods	Evaluation Level	Mean Value	Standard Deviation	V_1	V_2	V_3	V_4	V_5	θ
AHP	V_1	0.2528	0.1003	1					
	V_2	0.1434	0.0729	−0.251	1				
	V_3	0.1458	0.0620	−0.512	−0.679 *	1			
	V_4	0.0952	0.0661	−0.303	0.073	0.391	1		
	V_5	0.3993	0.1182	−0.368	−0.575	−0.488	−0.361	1	
	θ	0.5672	0.0791	−0.815 **	0.679 *	0.871 **	0.235	0.737 **	1
EWM	V_1	0.3599	0.1715	1					
	V_2	0.1416	0.0883	−0.638 *	1				
	V_3	0.1474	0.0756	−0.717 *	0.871 **	1			
	V_4	0.0853	0.0750	0.316	−0.148	−0.266	1		
	V_5	0.2969	0.1477	−0.561	−0.159	−0.031	−0.560	1	
	θ	0.4791	0.1019	−0.833 **	0.192	0.299	−0.299	0.869 **	1

Note: *: $p < 0.05$, significant correlation; **: $p < 0.01$, extremely significant correlation.

The reason for this result is that the entropy weight method has some shortcomings. First, the number of indicators selected in this evaluation is larger than the number of objects to be evaluated, resulting in deviations. Second, the entropy weight method ignores the importance of the index and excessively relies on an objective weight assignment, which causes the dimension of the evaluation index to not be reduced and the subjective intention of the decision-maker to be ignored [32,46]. At the same time, because each township is independent and different, the weight of each index layer should be different. This factor was not well represented in this method, and many similar weight results were obtained during calculation. The analytic hierarchy process comprehensively considers the coupling effect between multiple criteria and multiple indicators according to the intention of the decision-makers and the actual local situation. Moreover, the method more effectively identifies the importance of the main influencing indicators [47].

In Table 5, we have found that among the five membership values obtained by the analytic hierarchy process, four membership values (v_1, v_2, v_3, v_5) are significantly correlated with the evaluation value theta, the contribution rate of correlation accounted for 80%. Among the five membership values obtained by the entropy weight method, only two membership values (v_1, v_5) are significantly correlated with theta and the contribution rate of correlation accounted for 40%. It indicates that the correlation contribution rate of evaluation results obtained by using the analytic hierarchy process is higher than the entropy weight method. The reliability of the analytic hierarchy process is higher. In addition, the average evaluation value theta of the analytic hierarchy process (0.5672) obtained is higher than the entropy weight method (0.4791), and the standard deviation of the former (0.0791) is lower than the latter (0.1019). It also indicates that the error result is smaller. Besides this, the evaluation value obtained by the analytic hierarchy process is also more satisfied with the degree of non-overloading of the evaluation results in the evaluation report of Carrying capacity in Hebei Province. In summary, it shows that the analytic hierarchy process in this study is more suitable for the actual situation.

According to the existing research results, combined with the water resources conditions, ecological environment characteristics and social development of Hebei Province, the comprehensive grading standards are obtained (Table 6).

Table 6. Classification criteria of comprehensive score values of water resources carrying capacity.

Evaluation Results	0–0.25	0.25–0.50	0.50–0.75	0.75–1.00
Bearing level	Unbearable	General bearing	Good bearing	Ideal bearing

3.4. Analysis of Evaluation Results

In general, the scores of the 11 towns and villages in the comprehensive evaluation value determined by the analytic hierarchy process were between 0.4 and 0.7. This result indicates that the water resources carrying capacity of the region is in a general and a good bearing capacity. Water resources remain to be exploited, so the local economy can maintain its development. Domestic and ecological water use is in a relatively balanced state, and there will not be a serious water shortage in the near term.

According to the comprehensive score, Tumen, Zhangshiyan, Zanhuang, Huangbeiping, Yuantou, Xilongmen, Xuting, Xiyangze, Nanqinghe, Nanxingguo and Zhangleng townships are ranked from high to low. In calculating the comprehensive evaluation value, the weight value of each subsystem is different. When the scores are the same, the score value of the subsystem with a larger weight value has a greater impact than does the evaluation result of the subsystem with a smaller weight value [48]. In order to better show the contribution of each subsystem's score to the comprehensive evaluation value, this paper calculates the accumulation of each subsystem in the comprehensive evaluation value of each township to explore the impact of each subsystem. The results are shown in Figure 3. The depth of the color orange on the map is used to reflect the final evaluation value of each region. The darker the color, the higher the evaluation value, and the lighter the color, the smaller the evaluation value.

Because natural water resources and the distribution of local industry are limited, various regions have different water resource conditions, ecological environments, and social and economic development patterns [49]. Therefore, the water resources carrying capacity of 11 towns shows a certain degree of spatial differentiation. To reflect the spatial divergence of various regions at the subsystem level more clearly, this paper conducted independent fuzzy comprehensive calculations on the water resources, water environment, water ecological and socioeconomic subsystems. After obtaining the comprehensive score value of each system, the results obtained were based on the Kriging interpolation method to carry out optimal unbiased interpolation research on the data results to reduce the uncertainty of the evaluation results. This process was mainly due to the consideration of spatial correlation and independence, which makes the results more reliable. The results are shown in Figure 4.

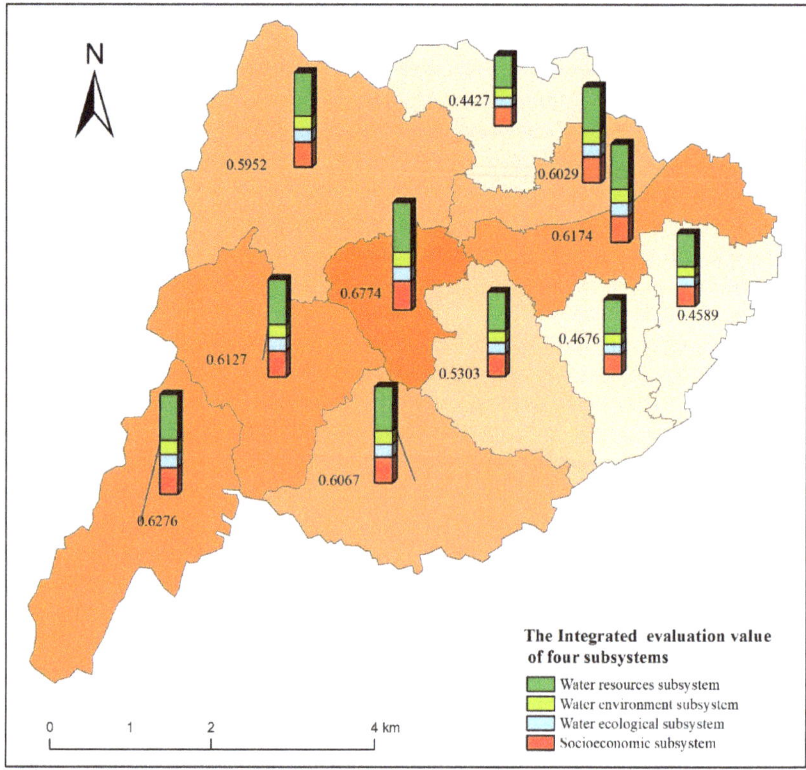

Figure 3. Comprehensive scoring values and accumulation of water resources carrying capacity in Zanhuang County.

Figure 4 shows that the spatial distribution of the four subsystems to the water resources carrying capacity was different because of the different index weights. For (a), the high scoring value was concentrated in the eastern part of the study area, which is far from the mountainous area. However, the other parts of the study area, especially Tumen, mainly consumed a large amount of water for irrigation, leading to a high utilization rate of water resources. Meanwhile, a small number of river systems pass through this area, so the water area index was low. In contrast, the score was higher in the central and western regions due to the distant distribution of industrial and urban centers for (b). The eastern part of the county had a lower score due to the discharge of domestic sewage and industrial wastewater. The difference was significant between agricultural areas and wetland rivers in the (c) regional distribution. The high and low values were more obvious when reflecting the high vegetation coverage and the construction of new river channels. The low value of the (d) region was mostly distributed in the middle and northeastern parts of the study area, which have a high per capita GDP, but the domestic water quota and population density were much higher than they were in other districts. The contribution of high scores in other regions may be influenced by the urban–rural integration.

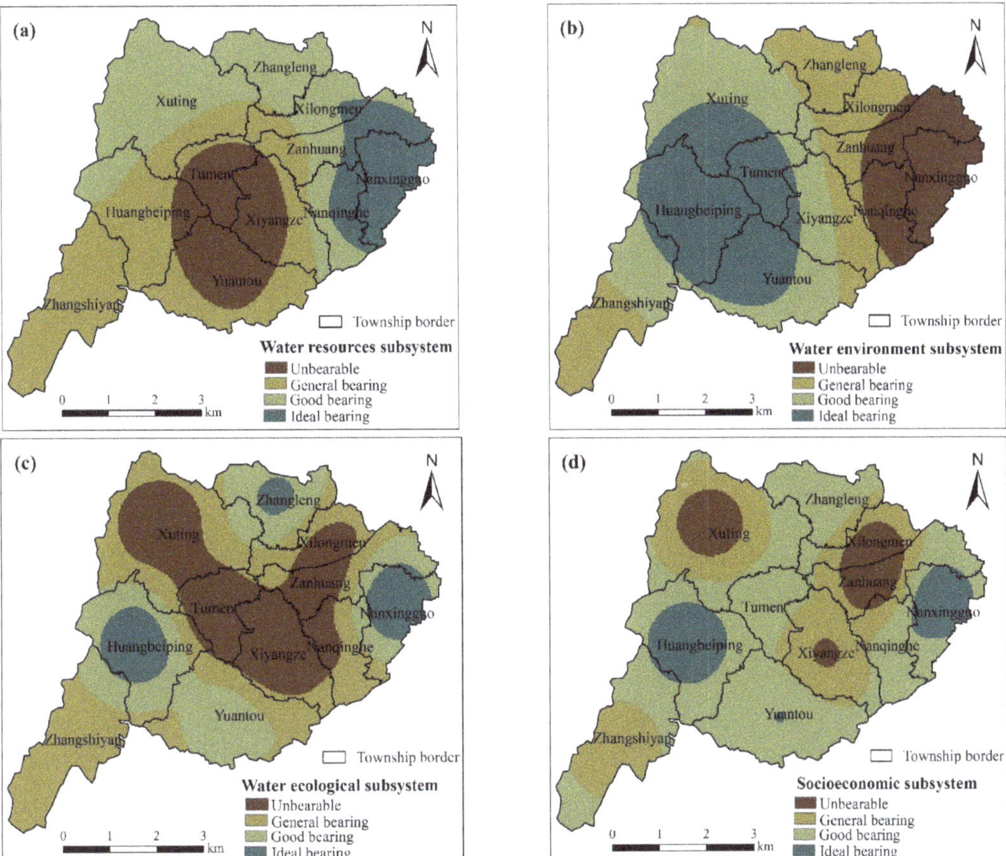

Figure 4. Maps with the spatial distribution of four subsystems' classification: (**a**) Water resources subsystem quality classification from scoring values; (**b**) Water environment subsystem quality classification from scoring values; (**c**) Water ecological subsystem quality classification from scoring values; (**d**) Socioeconomic subsystem quality classification from scoring values.

3.5. Sensitivity Analysis Results of Weight

This paper used a fuzzy comprehensive evaluation model of weight determined by the analytic hierarchy process. It determined the sensitivity distribution of the weight change of the index variable. Figure 5 compares the comprehensive evaluation value of each township with and without one indicator weight removed. "BASE" represents the result of the comprehensive evaluation value calculated under the condition of all indices, while "-xx" represents it after removing the weight of this evaluation index. Figure 5 indicates that when certain indicators were removed separately, there were many differences in the results of some comprehensive evaluation values. Therefore, the weights of these indicators are highly sensitive. To further study the sensitivity of the specific index weight to the comprehensive evaluation value of the water resources carrying capacity, the RMSEC of each township after the weight change of each index was quantitatively calculated and added to obtain the total sensitivity, TF. The calculation results are shown in Figure 6.

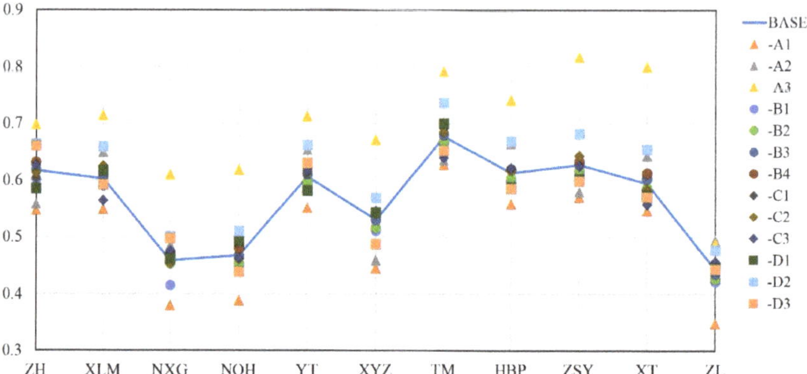

Figure 5. Comparison of the comprehensive score values of each township between original and single index weight removed. (The meaning of the abscissa in the figure is to use the first letters of the names of the locations to represent the towns. The indexes like "A1, A2...D3" express the meanings of index layer in Table 1 in turn.)

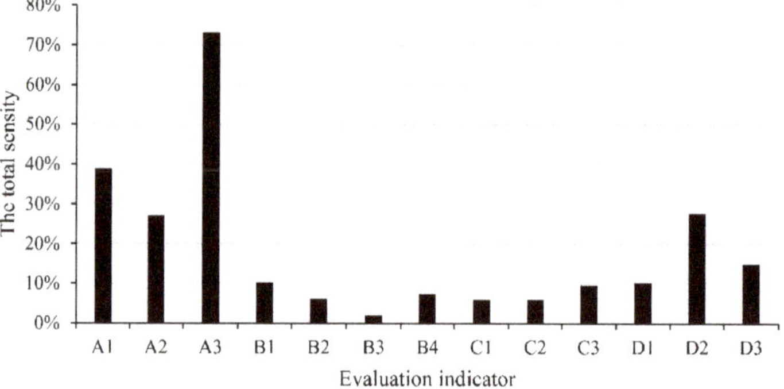

Figure 6. Distribution of total sensitivity of evaluation indexes of water resource carrying capacity. (The indexes like "A1, A2...D3" express the meanings of index layer in Table 1 in turn.)

The overall sensitivity results indicated that the index variables with high sensitivities were the water area index, water resource development and utilization rate, and per capita GDP, which reached 73.13%, 38.88% and 27.10%, respectively. All of them belong to the water resources subsystem and were significantly larger than the index variables of other systems. Therefore, special attention should be given to the water resources subsystem with a high sensitivity index in the evaluation of the water resources carrying capacity.

4. Conclusions

This paper used the fuzzy comprehensive evaluation model to evaluate the carrying capacity of water resources under the weight comparison method. The evaluation results indicated that the 11 townships in Zanhuang County had good bearings. At the same time, the sensitivity analysis method based on the index weight was used to identify the index results with the greatest influence. The evaluation results showed that the sensitivities of the water area index, water resource development and utilization ratio, per capita GDP, water consumption per unit regional GDP and domestic water quota were 73.13%, 38.88%, 27.72%, 27.10% and 15.03%, respectively. The analysis of the attribution results of the water resources carrying capacity subsystem indicated that the water resources carrying capacity of the study area was greatly affected by the regional water resources endowment and socioeconomic statuses.

The case study proved that there is a certain degree of uncertainty in the evaluation of water resources carrying capacity. The main sources of uncertainty are the uncertainty of index assignment and the uncertainty of weight. Among them, the uncertainty caused by water area index is the largest, and the uncertainty caused by fertilizer intensity index is the smallest. The uncertainty of the weight can be studied by the degree of correlation contribution. The degree of correlation contribution of the evaluation value obtained by the analytic hierarchy process is 80%, which is much higher than the 40% obtained by the entropy method. The final evaluation result obtained from the analytic hierarchy process is more in line with the actual state of good carrying capacity in Zanhuang County. Therefore, strengthening the control of the data quality of high-uncertainty indicators can help reduce errors in the evaluation of water resources carrying capacity.

Compared with previous research, the present study provides a good example of the uncertainty of water resources carrying capacity evaluation from the perspective of index sensitivity analysis and weight screening. The result is of great significance to the selection of water resources carrying capacity evaluation indicators and the practical application of evaluation results, which has great value in expanding the application of evaluation models and the development and utilization of local water resources in the future.

Supplementary Materials: The following are available online at https://www.mdpi.com/article/10.3390/w13202804/s1, Table S1: Evaluation criteria of 1–9 scale method, Table S2: The judgment matrix of four subsystems.

Author Contributions: Y.G., methodology, data analysis, software; J.W., conceptualization, formal analysis, supervision; D.Z., methodology, data provision; R.J., software, data curation; H.Y., data curation. All authors have read and agreed to the published version of the manuscript.

Funding: This study was funded by the National Natural Science Foundation of China (Grant No.: 41807344).

Institutional Review Board Statement: Not applicable.

Informed Consent Statement: Not applicable.

Data Availability Statement: Data available upon request due to privacy and ethical restrictions.

Acknowledgments: The authors would like to express their gratitude to the editors and anonymous experts for their constructive comments.

Conflicts of Interest: The authors declare no conflict of interest.

References

1. Yang, Z.; Song, J.; Cheng, D.; Xia, J.; Li, Q.; Ahamad, M.I. Comprehensive evaluation and scenario simulation for the water resources carrying capacity in Xi'an city, China. *J. Environ. Manag.* **2019**, *230*, 221–233. [CrossRef]
2. Cai, Y.P.; Huang, G.H.; Tan, Q.; Liu, L. An integrated approach for climate-change impact analysis and adaptation planning under multi-level uncertainties. Part II. Case study. *Renew. Sustain. Energy Rev.* **2011**, *15*, 3051–3073. [CrossRef]
3. Cai, Y.P.; Huang, G.H.; Wang, X.; Li, G.C.; Tan, Q. An inexact programming approach for supporting ecologically sustainable water supply with the consideration of uncertain water demand by ecosystems. *Stoch. Environ. Res. Risk Assess.* **2011**, *25*, 721–735. [CrossRef]
4. Ren, C.; Guo, P.; Li, M.; Li, R. An innovative method for water resources carrying capacity research—Metabolic theory of regional water resources. *J. Environ. Manag.* **2016**, *167*, 139–146. [CrossRef]
5. Peng, T.; Deng, H. Comprehensive evaluation on water resource carrying capacity based on DPESBR framework: A case study in Guiyang, southwest China. *J. Clean. Prod.* **2020**, *268*, 122235. [CrossRef]
6. Chi, M.; Zhang, D.; Zhao, Q.; Yu, W.; Liang, S. Determining the scale of coal mining in an ecologically fragile mining area under the constraint of water resources carrying capacity. *J. Environ. Manag.* **2021**, *279*, 111621. [CrossRef]
7. Yang, G.; Dong, Z.; Feng, S.; Li, B.; Sun, Y.; Chen, M. Early warning of water resource carrying status in Nanjing City based on coordinated development index. *J. Clean. Prod.* **2021**, *284*, 124696. [CrossRef]
8. Wheida, E.; Verhoeven, R. An alternative solution of the water shortage problem in Libya. *Water Resour. Manag.* **2007**, *21*, 961–982. [CrossRef]
9. Martin, R.; Savage, R.; Pyvis, R. Groundwater as a sustainable source of income and wealth creation. *Environ. Earth Sci.* **2013**, *70*, 1965–1969. [CrossRef]

10. Ellis, T.; Hatton, T.; Nuberg, I. An ecological optimality approach for predicting deep drainage from tree belts of alley farms in water-limited environments. *Agric. Water Manag.* **2005**, *75*, 92–116. [CrossRef]

11. Song, X.; Kong, F.; Zhan, C. Assessment of Water Resources Carrying Capacity in Tianjin City of China. *Water Resour. Manag.* **2011**, *25*, 857–873. [CrossRef]

12. Peng, T.; Deng, H.; Lin, Y.; Jin, Z. Assessment on water resources carrying capacity in karst areas by using an innovative DPESBRM concept model and cloud model. *Sci. Total Environ.* **2021**, *767*, 144353. [CrossRef] [PubMed]

13. Feng, L.; Zhang, X.; Luo, G. Application of system dynamics in analyzing the carrying capacity of water resources in Yiwu City, China. *Math. Comput. Simul.* **2008**, *79*, 269–278. [CrossRef]

14. Wang, G.; Xiao, C.; Qi, Z.; Meng, F.; Liang, X. Development tendency analysis for the water resource carrying capacity based on system dynamics model and the improved fuzzy comprehensive evaluation method in the Changchun city, China. *Ecol. Indic.* **2021**, *122*, 107232. [CrossRef]

15. Gong, L.; Jin, C. Fuzzy Comprehensive Evaluation for Carrying Capacity of Regional Water Resources. *Water Resour. Manag.* **2009**, *23*, 2505–2513. [CrossRef]

16. Li, N.; Kinzelbach, W.; Li, H.; Li, W.; Chen, F. Decomposition technique for contributions to groundwater heads from inside and outside of an arbitrary boundary: Application to Guantao County, North China Plain. *Hydrol. Earth Syst. Sci.* **2019**, *23*, 2823–2840. [CrossRef]

17. Wu, C.; Zhou, L.; Jin, J.; Ning, S.; Zhang, Z.; Bai, L. Regional water resource carrying capacity evaluation based on multi-dimensional precondition cloud and risk matrix coupling model. *Sci. Total Environ.* **2020**, *710*, 136324. [CrossRef]

18. Zhang, Q.; Xu, P.; Chen, J.; Qian, H.; Qu, W.; Liu, R. Evaluation of groundwater quality using an integrated approach of set pair analysis and variable fuzzy improved model with binary semantic analysis: A case study in Jiaokou Irrigation District, east of Guanzhong Basin, China. *Sci. Total Environ.* **2021**, *767*, 145247. [CrossRef]

19. Wu, L.; Su, X.; Ma, X.; Kang, Y.; Jiang, Y. Integrated modeling framework for evaluating and predicting the water resources carrying capacity in a continental river basin of Northwest China. *J. Clean. Prod.* **2018**, *204*, 366–379. [CrossRef]

20. Yang, J.; Lei, K.; Khu, S.; Meng, W. Assessment of Water Resources Carrying Capacity for Sustainable Development Based on a System Dynamics Model: A Case Study of Tieling City, China. *Water Resour. Manag.* **2015**, *29*, 885–899. [CrossRef]

21. Wang, C.; Hou, Y.; Xue, Y. Water resources carrying capacity of wetlands in Beijing: Analysis of policy optimization for urban wetland water resources management. *J. Clean. Prod.* **2017**, *161*, 1180–1191. [CrossRef]

22. Wang, H.; Huang, J.; Zhou, H.; Deng, C.; Fang, C. Analysis of sustainable utilization of water resources based on the improved water resources ecological footprint model: A case study of Hubei Province, China. *J. Environ. Manag.* **2020**, *262*, 110331. [CrossRef]

23. Mou, S.; Yan, J.; Sha, J.; Deng, S.; Gao, Z.; Ke, W.; Li, S. A Comprehensive Evaluation Model of Regional Water Resource Carrying Capacity: Model Development and a Case Study in Baoding, China. *Water* **2020**, *12*, 2637. [CrossRef]

24. Yu, F.; Fang, G.; Shen, R. Study on comprehensive early warning of drinking water sources for the Gucheng Lake in China. *Environ. Earth Sci.* **2014**, *72*, 3401–3408. [CrossRef]

25. Deng, L.; Yin, J.; Tian, J.; Li, Q.; Guo, S. Comprehensive Evaluation of Water Resources Carrying Capacity in the Han River Basin. *Water* **2021**, *13*, 249. [CrossRef]

26. Koop, S.H.A.; van Leeuwen, C.J. Assessment of the Sustainability of Water Resources Management: A Critical Review of the City Blueprint Approach. *Water Resour. Manag.* **2015**, *29*, 5649–5670. [CrossRef]

27. Bjørn, A.; Hauschild, M.Z. Introducing carrying capacity-based normalisation in LCA: Framework and development of references at midpoint level. *Int. J. Life Cycle Assess.* **2015**, *20*, 1005–1018. [CrossRef]

28. Matos, C.; Teixeira, C.A.; Duarte, A.A.L.S.; Bentes, I. Domestic water uses: Characterization of daily cycles in the north re-gion of Portugal. *Sci. Total Environ.* **2013**, *458–460*, 444–450. [CrossRef]

29. Ahmed, M.; Sauck, W.; Sultan, M.; Yan, E.; Soliman, F.; Rashed, M. Geophysical Constraints on the Hydrogeologic and Structural Settings of the Gulf of Suez Rift-Related Basins: Case Study from the El Qaa Plain, Sinai, Egypt. *Surv. Geophys.* **2014**, *35*, 415–430. [CrossRef]

30. Wang, Y.; Zhou, X.; Engel, B. Water environment carrying capacity in Bosten Lake basin. *J. Clean. Prod.* **2018**, *199*, 574–583. [CrossRef]

31. Ren, B.; Zhang, Q.; Ren, J.; Ye, S.; Yan, F. A Novel Hybrid Approach for Water Resources Carrying Capacity Assessment by Integrating Fuzzy Comprehensive Evaluation and Analytical Hierarchy Process Methods with the Cloud Model. *Water* **2020**, *12*, 3241. [CrossRef]

32. Han, L.; Song, Y.; Duan, L.; Yuan, P. Risk assessment methodology for Shenyang Chemical Industrial Park based on fuzzy comprehensive evaluation. *Environ. Earth Sci.* **2015**, *73*, 5185–5192. [CrossRef]

33. El Baba, M.; Kayastha, P.; Huysmans, M.; De Smedt, F. Evaluation of the Groundwater Quality Using the Water Quality Index and Geostatistical Analysis in the Dier al-Balah Governorate, Gaza Strip, Palestine. *Water* **2020**, *12*, 262. [CrossRef]

34. Zhang, J.; Zhang, C.; Shi, W.; Fu, Y. Quantitative evaluation and optimized utilization of water resources-water environment carrying capacity based on nature-based solutions. *J. Hydrol.* **2019**, *568*, 96–107. [CrossRef]

35. Liao, H.; Yang, X.; Xu, F.; Xu, H.; Zhou, J. A fuzzy comprehensive method for the risk assessment of a landslide-dammed lake. *Environ. Earth Sci.* **2018**, *77*, 750. [CrossRef]

36. Liao, X.; Ren, Y.; Shen, L.; Shu, T.; He, H.; Wang, J. A "carrier-load" perspective method for investigating regional water resource carrying capacity. *J. Clean. Prod.* **2020**, *269*, 122043. [CrossRef]

37. Wu, J.; Li, J.; Teng, Y.; Chen, H.; Wang, Y. A partition computing-based positive matrix factorization (PC-PMF) approach for the source apportionment of agricultural soil heavy metal contents and associated health risks. *J. Hazard. Mater.* **2020**, *388*, 121766. [CrossRef]

38. Zhao, Y.; Wang, Y.; Wang, Y. Comprehensive evaluation and influencing factors of urban agglomeration water resources carrying capacity. *J. Clean. Prod.* **2021**, *288*, 125097. [CrossRef]

39. Gao, F.; Wang, H.; Liu, C. Long-term assessment of groundwater resources carrying capacity using GRACE data and Budyko model. *J. Hydrol.* **2020**, *588*, 125042. [CrossRef]

40. Wang, Y.; Wang, Y.; Su, X.; Qi, L.; Liu, M. Evaluation of the comprehensive carrying capacity of interprovincial water resources in China and the spatial effect. *J. Hydrol.* **2019**, *575*, 794–809. [CrossRef]

41. Wang, X.; Li, X.; Wang, J. Urban Water Conservation Evaluation Based on Multi-grade Uncertain Comprehensive Evaluation Method. *Water Resour. Manag.* **2018**, *32*, 417–431. [CrossRef]

42. Shijiazhuang Water Bureau, 2017. *Shijiazhuang Water Bureau, the Shijiazhuang Water Resources Bulletin*; China Statistics Press: Beijing, China, 2017.

43. Statistics Bureau of Zanhuang County, Hebei Province. *Statistics Bureau of Zanhuang County, Hebei Province, Statistical Yearbook of Zanhuang*; China Statistics Press: Beijing, China, 2017.

44. Varis, O.; Vakkilainen, P. China's 8 challenges to water resources management in the first quarter of the 21st Century. *Geomorphology* **2001**, *41*, 93–104. [CrossRef]

45. Ren, L.; Gao, J.; Song, S.; Li, Z.; Ni, J. Evaluation of Water Resources Carrying Capacity in Guiyang City. *Water* **2021**, *13*, 2155. [CrossRef]

46. Kiziltan, M. Water-energy nexus of Turkey's municipalities: Evidence from spatial panel data analysis. *Energy* **2021**, *226*, 120347. [CrossRef]

47. Chi, M.; Zhang, D.; Fan, G.; Zhang, W.; Liu, H. Prediction of water resource carrying capacity by the analytic hierarchy process-fuzzy discrimination method in a mining area. *Ecol. Indic.* **2019**, *96*, 647–655. [CrossRef]

48. Teng, Y.; Zuo, R.; Xiong, Y.; Wu, J.; Zhai, Y.; Su, J. Risk assessment framework for nitrate contamination in groundwater for regional management. *Sci. Total Environ.* **2019**, *697*, 134102. [CrossRef] [PubMed]

49. Dadmand, F.; Naji-Azimi, Z.; Motahari Farimani, N.; Davary, K. Sustainable allocation of water resources in water-scarcity conditions using robust fuzzy stochastic programming. *J. Clean. Prod.* **2020**, *276*, 123812. [CrossRef]

Article

Evaluation of the Impact of Ecological Water Supplement on Groundwater Restoration Based on Numerical Simulation: A Case Study in the Section of Yongding River, Beijing Plain

Zijian Ji [1], Yali Cui [1,*], Shouquan Zhang [2], Wan Chao [3] and Jingli Shao [1]

[1] School of Water Resources & Environment, China University of Geosciences (Beijing), Beijing 100083, China; 18332606901@163.com (Z.J.); jshao@cugb.edu.cn (J.S.)

[2] Higher Institute of Finance and Economics, Capital University of Economics and Business, Beijing 100070, China; shouquanzhang@sina.com

[3] Yongding River Investment Co., Ltd., Beijing 100193, China; chaow@126.com

* Correspondence: cuiyl@cugb.edu.cn

Abstract: Ecological water supplement relies on river channels to introduce surface water, to make a reasonable supplement of groundwater, to repair the regional groundwater environment and urban river ecosystem. Evaluating the degree of groundwater restoration after ecological water supplement (by taking appropriate measures) is a critical problem that needs to be solved. Thus, based on the Yongding River ecological water supplement in 2019 and 2020, we analyzed the groundwater monitoring situates in the ecological water supplement region. We established an unstructured groundwater flow numerical model in the study area through the quadtree grids. The model was calibrated with the measured water level. The simulated results could accurately reflect the real groundwater dynamic characteristics, and it showed that the water level rise was concentrated in the 3–6 km range of the Yongding River after the ecological water supplement. In 2019, the calculated ecological water infiltration amount was 101.28×10^6 m^3, the affected area was 265.19 km^2, and the average groundwater level rise in the affected area was 2.10 m. In 2020, the calculated ecological water infiltration amount was 102.64×10^6 m^3, the affected area was 506.88 km^2, and the average groundwater level rise in the affected area was 1.25 m. While the ecological water supplement had a positive impact on groundwater level restoration, the groundwater level around the typical buildings within the study area, including Beijing West Railway Station and Beijing Daxing International Airport, would not be significantly affected.

Keywords: Yongding River; groundwater numerical simulation; artificial ecological water supplement; groundwater recharge resources amount evaluation

Citation: Ji, Z.; Cui, Y.; Zhang, S.; Chao, W.; Shao, J. Evaluation of the Impact of Ecological Water Supplement on Groundwater Restoration Based on Numerical Simulation: A Case Study in the Section of Yongding River, Beijing Plain. *Water* **2021**, *13*, 3059. https://doi.org/10.3390/w13213059

Academic Editors: Yuanzheng Zhai, Jin Wu and Huaqing Wang

Received: 29 September 2021
Accepted: 30 October 2021
Published: 2 November 2021

Publisher's Note: MDPI stays neutral with regard to jurisdictional claims in published maps and institutional affiliations.

1. Introduction

Yongding River is located in the western suburbs of Beijing and is known as the "Mother River" of Beijing. Since 1950, due to the long-term impact of numerous water conservancy projects built in the river, the ecological environment of the river basin has been continuously deteriorating [1,2]. In addition, due to the influence of social development, the over-exploitation of groundwater and the gradual seriousness of water pollution led to the deterioration of water volume and water quality in the Yongding River Basin [3,4]. In recent years, more attention has been placed on the study of river ecological restoration. Some investigators focus on the study of restoration methods [5,6], and others focus on the study of water pollution treatment [7,8]. For the ecological environment in the Yongding River Basin, some investigators also summarize the progress of the restoration and put forward specific restoration methods [9] and restoration goals [10]. Since the beginning of this century, the Beijing Municipal Government has carried out many countermeasures to restore and control the ecological environment of the Yongding River Basin. In 2016, it

63

launched the ecological water supplement project of the Yongding River, with the three provinces of Tianjin, Hebei, and Shanxi; this project has achieved ideal governance effects so far [11].

After implementation of the Yongding River ecological water supplement project, there has been concern over the amount of ecological water needed for the river basin. In-depth studies have provided a certain theoretical basis for solving the problem of water allocation in the process of the ecological water supplement. For instance, Wei Jian et al. used the environmental water demand method and the ecological water demand method to study the ecological water demand at different stages of the Yongding River mountain section [12]. Du Yong et al. also calculated the ecological water demand, which could meet the continuous flow of the entire Yongding River based on the monitoring of the groundwater level, flow, and infiltration data of the ecological water supplement in 2019 [13].

Researchers have also evaluated the impact of the ecological water supplement on the groundwater quality and quantity of the basin. Through numerical simulation, Hu et al. found the lining constructed with the geomembrane in the Yongding River can effectively control the seepage of water and the diffusion of solutes in the aquifer. They also confirmed the diffusion of potential pollutants increases with the infiltration of the Yongding River ecological water supplement [14]. Using the principal component analysis method, Luo et al. analyzed the temporal and spatial changes of 11 water quality parameters at 10 monitoring points of Yongding River in April 2011 and September 2016. Based on this, they evaluated the impact of river restoration projects on groundwater quality [15]. Based on the ecological water supplement work of the Yongding River in 2020, Hu Litang et al. used many technical methods, such as groundwater balance analysis, correlation analysis, and cluster analysis to discuss the leakage loss, groundwater dynamic changes, and control factors of each river section [16]. Ma Yao et al. used the groundwater balance analysis method to calculate the amount of groundwater supplements in different sections of the Yongding River during the ecological supplement period in 2019. They also carried out an analysis of the impact on the regional groundwater after the ecological water supplement of the Yongding River (Beijing section) [17]. Kangning Sun et al., by developing a coupled model, integrating a Muskingum method-based open channel flow model and machine learning-based groundwater model, described the dynamic changes in streamflow and groundwater levels in response to the ecological water supplement of the Yongding River [18].

In summary, many researchers have conducted relevant studies on the ecological restoration of the Yongding River Basin and have obtained many instructive study results. Studies have shown that the implementation of the ecological water supplement work in the Yongding River can significantly alleviate the problem of regional groundwater overexploitation. As a result, the quantity and quality of regional groundwater have significantly improved. Moreover, the development of ecological water supplement work can also improve the ecological environment of the Yongding River Basin and promote the re-flow of the river. However, most studies on the impact of groundwater level restoration are based on groundwater balance methods and mathematical statistics methods, and the results are insufficient to characterize the spatial variability of groundwater level restoration or forecast the repair effect. The most complete and reliable method to solve the problem is by the numerical simulation of groundwater; moreover, the simulation of groundwater by the widely used MODFLOW. Gert Ghysels et al. and A. A. El-Zehairy et al. have successfully applied MODFLOW to the simulation study of the relationship among rivers, lakes, and groundwater [19,20]. They proposed a feasible method for simulating the interaction between surface water and groundwater. In addition, MODFLOW has also been used in numerical simulation studies of saline water encroachments in coastal areas [21,22] and groundwater in complex karst areas [23]. MODFLOW is widely used in different fields and different depths of research, which also further demonstrates its full reliability.

In this paper, we analyze the dynamic characteristics of groundwater and establish a numerical model of groundwater under ecological water supplement conditions by MODFLOW, to reveal the degree of regional groundwater restoration. Meanwhile, the impact of ecological water supplement on typical buildings within the study area is also taken into account. The study conclusions can provide suggestions to identify the extent of groundwater restoration in the region, and provide a reference for future ecological water supplement work.

2. Methods

2.1. Ecological Water Supplement Information in the Study Area

The Yongding River ecological water supplement plains region (Beijing section) is located in the southwest side of Beijing (China) (Figure 1a), with a total area of about 888.72 km^2. Under the influence of climate drought, irrational exploitation, and utilization of groundwater, the Yongding River channel was in "cut-off" states for a long time [24,25]. Since the Yongding River Comprehensive Management and Ecological Restoration Project launched, two large-scale ecological water supplement tasks were carried out in 2019 and 2020. The ecological water supplement was from the Guanting Reservoir (Beijing, China) all the way to the south, by the San Jiadian barrage, into the plain region. In 2019, it only reached the 12 km downstream of the Lugouqiao barrage. However, in 2020, it flowed through the Men Tougou, Shi Jingshan, Fengtai, Fang Shan, and Da Xing, five districts, in Cui Zhihuiying town (Da Xing, China), out of Beijing's administrative boundary. Therefore, it went through the entire Beijing Plain region in 2020. In 2019 and 2020, a total volume of 313 × 10^6 m^3 and 166.42 × 10^6 m^3 was discharged from the Guanting Reservoir. The ecological water supplement volume in the plains and the duration are shown in Table 1.

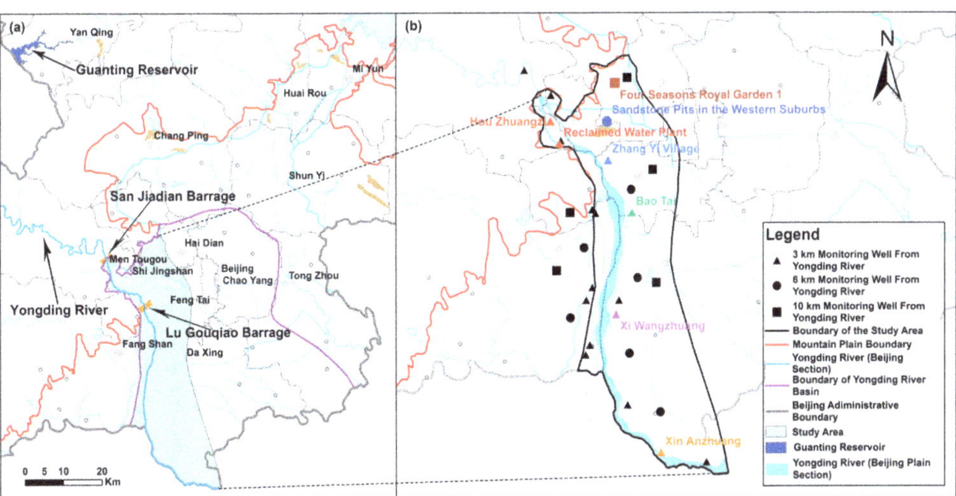

Figure 1. (a) Location map of the Yongding River ecological water supplement plains region (Beijing section). (b) Distribution map of the groundwater monitoring wells.

Table 1. Ecological water supplement in the plain region of the Yongding River in 2019 and 2020.

Ecological Water Supplement Year (a)	Start Time of Water Supplement	Lasted Times (d)	Amount of Water Supplement in San Jiadian Barrage (m^3)
2019	22 March	77	128.18 × 10^6
2020	22 April	32	142.38 × 10^6

To observe the influence of the ecological water supplement on groundwater in Yongding River, 32 automatic observation wells of groundwater within 10 km of Yongding River are shown in this study. The locations of the observation wells are shown in Figure 1b. From upstream to downstream, within 3 km of the Yongding River, the observation wells data were selected to draw the groundwater level process lines during 2018–2019 (Figure 2a). Since the Yongding River water supplement started on 22 March 2019, the groundwater table in the upper and middle reaches of Yongding River rose in early April 2019. The delay effect on the groundwater level rising in the region could be used to explain this phenomenon. After the ecological water supplement, the water table gradually decreased. Xi Wangzhuang and Xin Anzhuang were located in the middle and lower reaches of Yongding River. Because the ecological water supplement head of Yongding River did not reach here, not all were affected by the ecological water supplement. Therefore, the water table could remain unchanged. Zhang Yi village, located near the Lugouqiao barrage, affected by the San Jiadian barrage—to release water, the water level changed greatly. Hou Zhuangzi and the reclaimed water plant are located close to the San Jiadian barrage, so the ecological water supplement affected them much earlier. Compared to the Bao Tai observation well, which was located in the midstream, the water table also rose earlier. Finally, compared to the non-ecological water supplement time, the water level of the four groundwater monitoring wells located upstream recovered significantly during the ecological water supplement period. The water level gradually returned to the pre-ecological water level state after the ecological water supplement period. Only a small increase in the water level had been maintained.

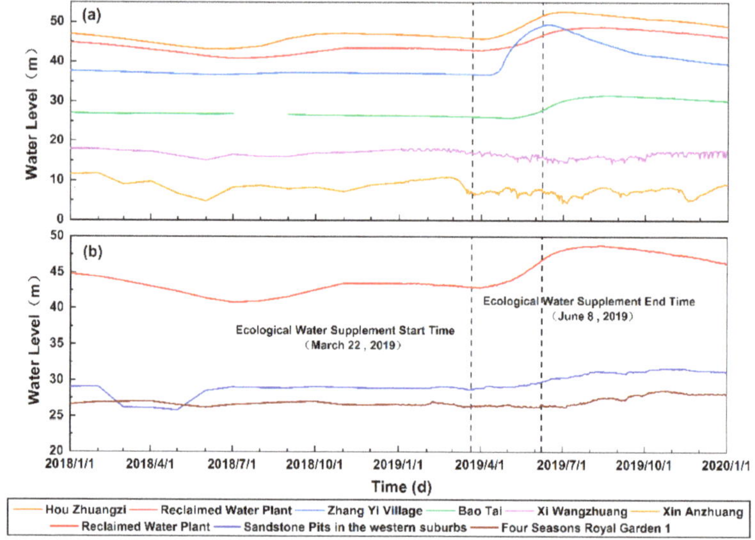

Figure 2. (a) Dynamic curve of the groundwater level of the monitoring wells along Yongding River. (b) The dynamic curve of the groundwater level of monitoring wells at different distances, with a perpendicular arrangement to the Yongding River.

According to the "principle of perpendicular" to the Yongding River, from near to far, observation well data were used to draw the groundwater level process line diagram during 2018–2019 (Figure 2b). As the distance increased, this influence was gradually weakened. For example, the Sandstone Pits in the western suburbs observation well, which was 6 km away from the Yongding River, only had a little impact on the ecological water supplement. The groundwater level between the selected monitoring wells had more obviously risen compared to the non-ecological water supplement period.

2.2. Hydrogeological Conditions in the Study Area

The study area is located in the southwestern part of the alluvial fan of the Yongding River, which is high in the northwest and low in the southeast (Figure 3a,b). The annual average atmospheric rainfall is 571.2 mm. The river in the region is mainly the Yongding River. The regional aquifer is mainly a single phreatic aquifer on top of the alluvial fan. The average thickness of the aquifer varies from 50 to 240 m, in which Quaternary sand and gravel dominate its lithology, and the hydraulic conductivity varies from 10 to 250 m/d [26,27]. The hydrogeological profile in the Yongding River alluvial fan is shown in Figure 4, which is adapted from Figure 2 in the study by Huan et al. [28]. The aquifer is mainly recharged by means of lateral runoff in the piedmont zone, precipitation infiltration, and water infiltration of the Yongding River; it is discharged by means of artificial mining and lateral outflow of the aquifer. For the water diversion channels in cities, most of them have a lining effect, and the surface water flow has no obvious replenishment effect on the groundwater. Therefore, it is not included in the source and sink items in our study. Recently, due to the continuous increase of exploiting groundwater, the depth of groundwater has been greater than 10 m, so the evaporative water loss of the diving surface can be ignored. The regional groundwater migrates from northwest to southeast and from the mountain front to the alluvial plain. The runoff intensity gradually weakened from the top to the bottom of the alluvial fan and from the upstream to the downstream of the river channel.

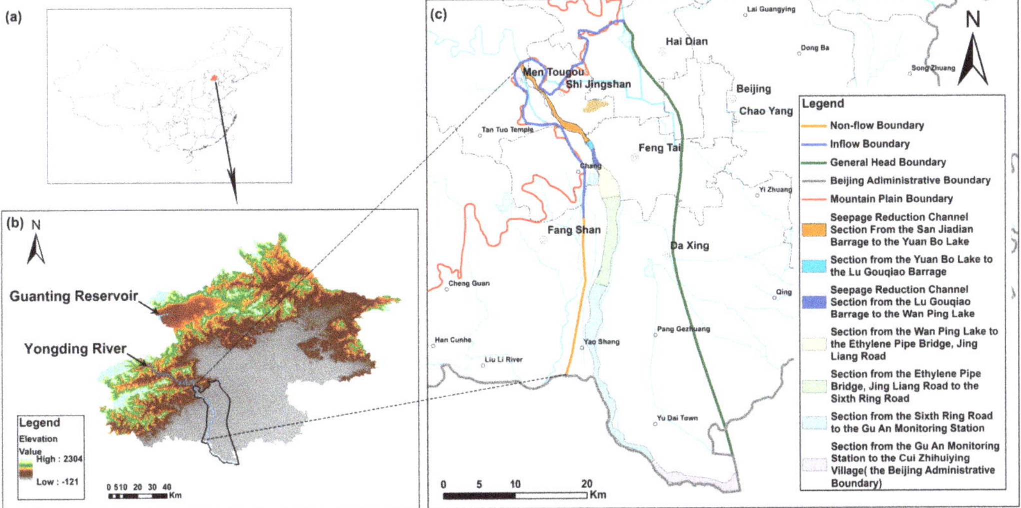

Figure 3. (**a,b**) Location and terrain distribution of the alluvial fan of the Yongding River. (**c**) Schematic diagram of generalization of the model boundary and each Yongding River ecological water supplement monitoring section.

2.3. Groundwater Flow Numerical Model

The northern model boundary was extended to the mountain-plain boundary and treated as an inflow boundary. The southern side of the western boundary was consistent with the boundary lines of the Dashi-Juma River groundwater system and the Yongding River groundwater system, divided, which was treated as a non-flow boundary. The eastern boundary considered of the impact range of groundwater restoration to artificially delimit the model boundary, and treated it as a general head boundary (Figure 3c). Based on the study objectives and the distribution characteristics of the aquifer, we generalized the model as a single-layer model. The upper boundary was the phreatic surface, which accepted external replenishment. The lower boundary considered the vertical influence of the Yongding River's ecological water supplement. Therefore, we selected the bottom boundary of the phreatic aquifer as the model boundary and regarded it as a non-flow boundary.

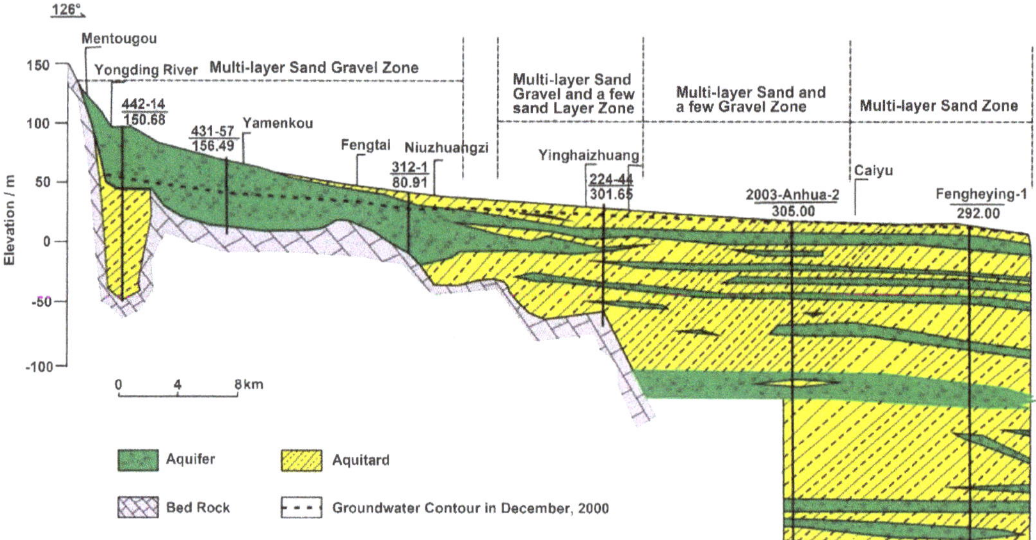

Figure 4. Hydrogeological profile in the Yongding River alluvial fan (adapted from Figure 2 in the study by Huan et al. [28]).

According to the achievements of previous studies, using quadtree grids to mesh the model could improve the calculation accuracy based on ensuring the calculation efficiency of the model [29–31]. Therefore, we also selected the quadtree grids provided by the MODFLOW-USG module in the GMS software. We set the basic grid size of the model to 200 × 200 m and carried out two different levels of local encryption. The meshing of the model was shown in Figure 5. Referring to the analysis of the dynamic characteristics of groundwater, we preliminarily estimated the impact range of Yongding River's ecological water supplement was about 3–6 km from both sides of the river. Moreover, the groundwater table within a 3 km area had obvious uplift. Thus, we refined the grids within 3 km along the Yongding River with a 50 × 50 m size. Then, we segmented the ecological water supplement infiltration volume of the Yongding River with the distribution of each monitoring section (Figure 3c). Moreover, we took the water volume, and multiplied the identified and corrected infiltration coefficient to add to the model in the form of an injection well. Therefore, the grids where the injection well was located were refined twice, and the grid size after refinement was 25 × 25 m. A total of 136,721 grids existed in the model.

The simulation period was from 1 January 2019 to 31 December 2019. During the simulation period, we allocated the ecological water supplement from April to July to the "stress period" of days, and the rest of the non-refill months to each natural month, as a stress period. Each stress period was processed by the corresponding software package according to the actual source and sink data, and then assigned to each time step. The initial values of hydrogeological parameters referred to the parameter partition values generated by previous work experience and fine-tuned through fitting correction.

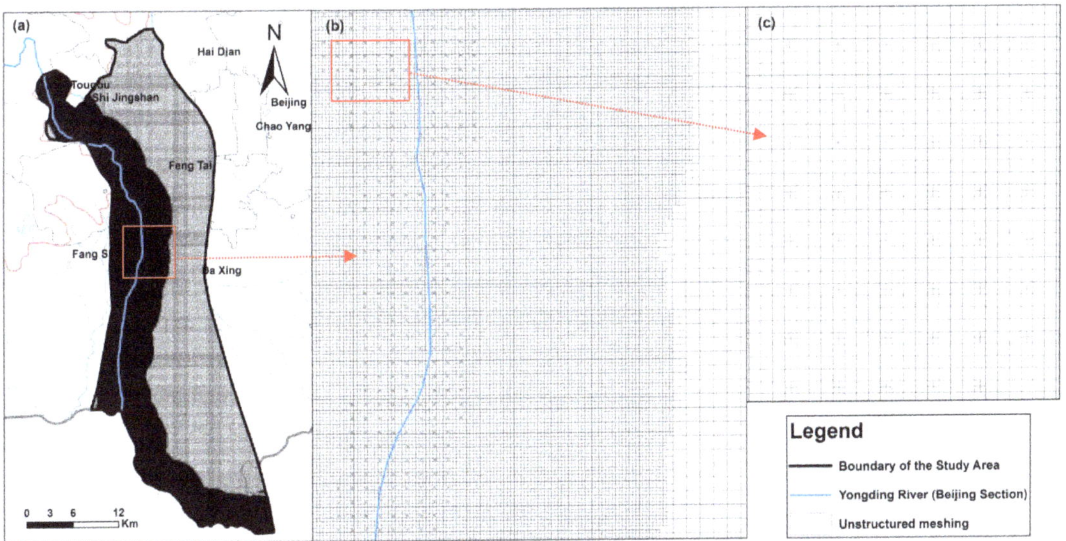

Figure 5. (**a**) Model quadtree grids division diagram. (**b**) Schematic diagram of a partial enlargement of the first-level encryption in a 3 km range along the Yongding River. (**c**) Schematic diagram of partial enlargement of secondary intensification of ecological water injection wells.

2.4. Model Calibration and Validation

The measured groundwater flow field on 31 December 2019, was used as the identification and verification flow field, and the model was identified and verified by trial-and-error calibration. The results showed that the simulated flow field and the measured flow field were (basically) the same in trend and flow patterns. Except for the piedmont zone on the north side and the Pang Gezhuang area in the Daxing district, which had a poor simulation effect, the other parts all reflected the actual groundwater flow trend (Figure 6a). The simulated groundwater level process line of the typical observation wells was consistent with the measured groundwater level, which accurately reflected the change process of groundwater before and after water replenishment (Figure 6b). The parameter partition after identification and correction is shown in Figure 6c. The hydraulic conductivity changes showed a clear trend of gradually decreasing from the upper part of the alluvial fan to the downstream plain region, and the parameter was not much different from the initial value. Therefore, the established numerical model could reflect the variation characteristics of the groundwater after the Yongding River ecological water supplement. We could use it to simulate and predict the impact of groundwater restoration after the Yongding River ecological water supplement.

Through the calibration and validation of the model, we obtained the results of the groundwater zone budget in the study area in 2019 (Table 2). During the simulation period, the total recharge of groundwater in the study area was 359.79×10^6 m^3, the total discharge was 238.17×10^6 m^3, and the recharge difference was 121.62×10^6 m^3. The groundwater was in a positive equilibrium state as a whole. Among the recharge items, the rainfall infiltration was 143.28×10^6 m^3, followed by the Yongding River ecological water supplement infiltration, 101.28×10^6 m^3. They accounted for 67.97% of the total groundwater recharge in the study area. The discharge was mainly assembled by artificial mining, which was 188.97×10^6 m^3, accounting for 79.34% of the total discharge.

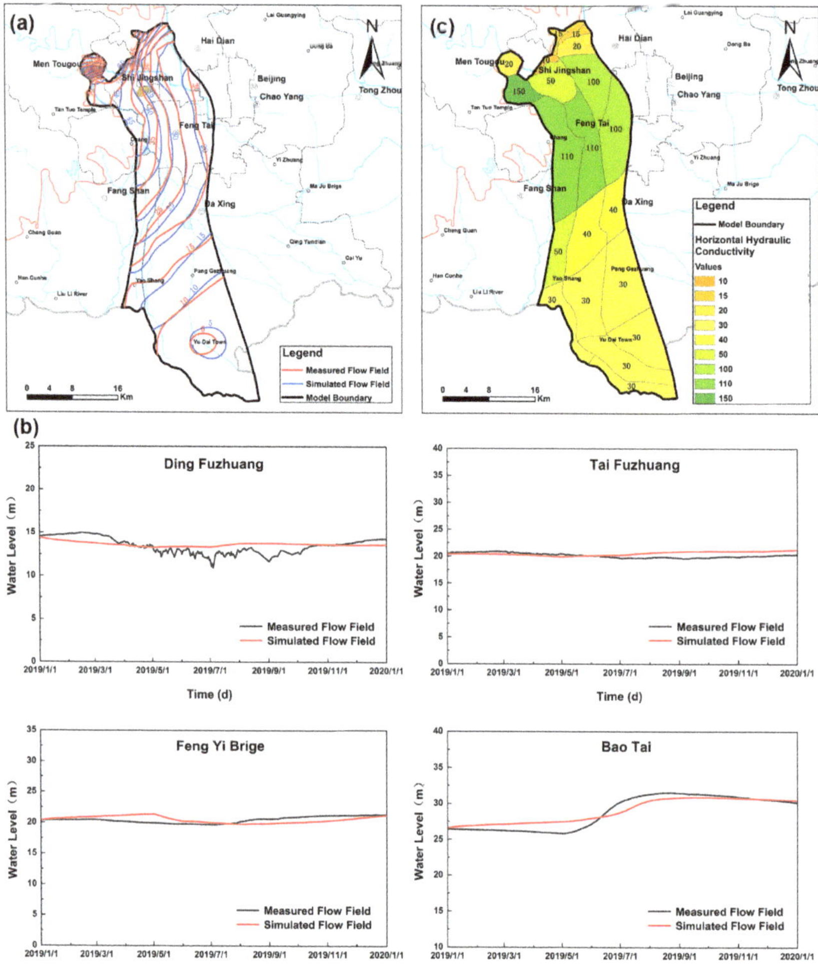

Figure 6. (**a**) Flow-field fitting diagram at the end of the simulation period. (**b**) The fitting curve of typical observation well in the ecological water supplement region. (**c**) Identification verified the aquifer parameter-zoning diagram (m/d).

Table 2. Groundwater zone budget in the study area in 2019.

	Zone Budget Terms	Volume ($\times 10^6$ m^3)	Percentage
	Rainfall infiltration	143.28	39.82%
	Lateral inflow	92.82	25.80%
Recharge	Pipe network infiltration	22.41	6.23%
	Ecological water supplement infiltration	101.28	28.15%
	Total recharge	359.79	100%
	Artificial mining	−188.97	79.34%
Discharge	Lateral outflow	−49.20	20.66%
	Total discharge	−238.17	100%
Recharge and discharge difference		121.62	

3. Results and Discussion

3.1. Identification of Infiltration Volume and Leakage Coefficient in the Yongding River Channel

The artificially ecological water supplement infiltration volume in each section of the Yongding River was calibrated and validated by the corrected model. Therefore, we also obtained the accurate leakage coefficients to provide calculation parameters for the groundwater flow prediction model of the Yongding River ecological water supplement in 2020. The ecological water supplement "head" of the Yongding River, in 2019, only reached 12 km downstream of the Lugouqiao barrage. The calculation only included the section of the river from below Wanping Lake to the point where the final hydraulic head reached. Due to the runoff interception effect of the Lugouqiao barrage causing the water level to rise, the infiltration capacity of the sandy gravel river channel had tremendously improved. Therefore, the section from the Yuan Bo lake to the Lugouqiao barrage was significantly larger than other river sections. Finally, we list the calculated infiltration amount and leakage coefficient of the Yongding River ecological water supplement in Table 3. About 21% of the total ecological water supplement volume was lost in 2019, and only 79% of the ecological water supplement volume successfully infiltrated and recharged groundwater. The loss of ecological supplement water could be explained as the water loss due to wetting of the river and intercepted by the soil during the infiltration of the ecological water supplement head through the river. The daily evaporation and infiltration of urban rivers and lakes could also occupy some ecological water supplement flow. Moreover, the lost volume, because the water surface of the entire river course evaporated, and was lost in the process of the ecological water supplement, was also included.

Table 3. Ecological water supplement infiltration volume and leakage coefficient for each section of Yongding River in 2019.

Section	Ecological Water Supplement Volume of Each Section ($\times 10^6$ m^3)	Ecological Water Supplement Infiltration Volume of Each Section ($\times 10^6$ m^3)	Leakage Coefficient
Seepage reduction channel section from the San Jiadian barrage to the Yuan Bo lake	68.64	5.49	0.08
Section from the Yuan Bo lake to the Lugouqiao barrage		51.13	0.72
Seepage reduction channel section from the Lugouqiao barrage to the Wanping Lake	59.54	8.93	0.15
Section of natural river course from below Wanping Lake to the point where the water head reaches		35.73	0.60
Total	128.18	101.28	0.79

3.2. Impact Range Analysis of Ecological Water Supplement in Yongding River in 2019

To obtain the impact range under the single factor of the ecological water supplement infiltration, the simulation results of the model with ecological water supplement conditions and the model without water supplement conditions were compared. We also used the groundwater level data on 31 December 2019, simulated by the two models to map the uplift variation of the groundwater level in the ecological water supplement region (Figure 7). It was shown that, by the end of 2019, the ecological water supplement head would gradually spread outward. Considering the model interpolation errors and the small fluctuations in the groundwater level, we finally selected the groundwater level rise value greater than 0.25 m as the standard to quantitatively calculate the impact range. The calculated affected area was about 265.19 km^2, accounting for 29.86% of the study area. It was distributed within about 6 km on both sides of the Yongding River channel; the widest impact range was located near the Lugouqiao barrage, reaching 9.7 km. Due to the impact of seepage reduction in the upstream channel below the San Jiadian barrage, the groundwater impact range was relatively small, with a maximum width of only 6–7 km. Owing to the ecological water supplement head only reaching 12 km downstream of the Lugouqiao barrage, the ecological water supplement had no impact on the groundwater level of the downstream region, such as Daxing District. Finally, within the range of the ecological water supplement, the water level rise at the center of the river near the Lugouqiao barrage was the largest, which could reach about 7–8 m. While the average groundwater level rise within the impact range was calculated to be only 2.10 m.

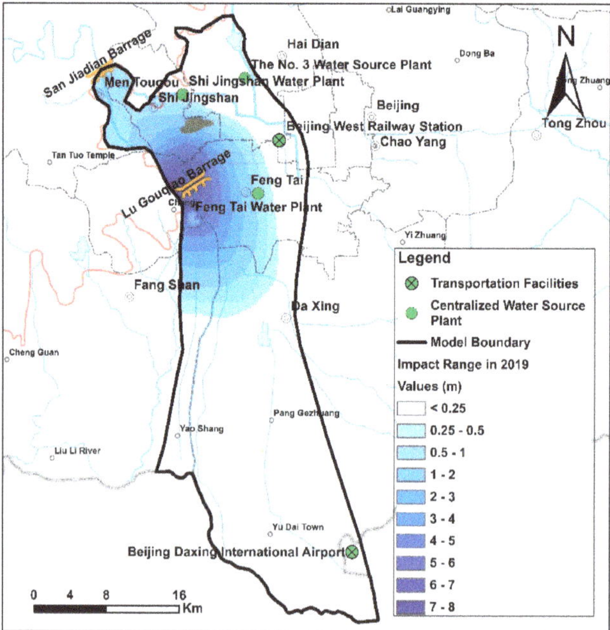

Figure 7. Uplift variation map of the groundwater level in the ecological water supplement region.

Combined with the obtained groundwater level rise variation map and the delineated groundwater level impact range, we analyzed the impact of the main typical buildings within the study area, such as the centralized water source plant, Beijing West Railway Station (BWRS), and Beijing Daxing International Airport (BDIA). The analysis was only limited to the analysis of the groundwater level around the typical buildings and did not involve the analysis of the impact on the structure of the engineering buildings. The same was true of the analysis in the following forecast for 2020. While the Fengtai water plant is within the delineated impact area, the impact of the ecological water supplement on it only caused the surrounding groundwater level, rising by 0.5 m. Such a small increase will give the centralized water source plant a few positive impacts. In addition, several typical buildings, including the Shi Jingshan water plant, the no. 3 water source plant, BWRS, and BDIA were excluded in the impact range of this ecological water supplement. Thus, the ecological water supplement would not cause a significant impact on them under this condition.

3.3. Prediction of Impact Range of Groundwater Level Restoration in Yongding River in 2020

The simulation period of the numerical model was extended to 31 December, 2020, to predict the groundwater level restoration of Yongding River under the ecological water supplement scenario in 2020. Moreover, the ecological water supplement data in each section of the river in 2020 were substituted into the model. Through the comparison of the simulated groundwater level data on 31 December, 2020, with or without ecological supplement conditions, the groundwater level restoration impact range of the ecological water supplement in 2020 was obtained, as shown in Figure 8. The overall distribution concentrated in the range of about 6–7 km on both sides of the Yongding River. The affected area was about 506.88 km², which accounted for 57.01% of the study area. Obviously, only the Fengtai water plant was within the delineated area of impact range, and the remaining typical buildings, including the no. 3 water source plant, Shi Jingshan water plant, BWRS, and BDIA, were not included in it.

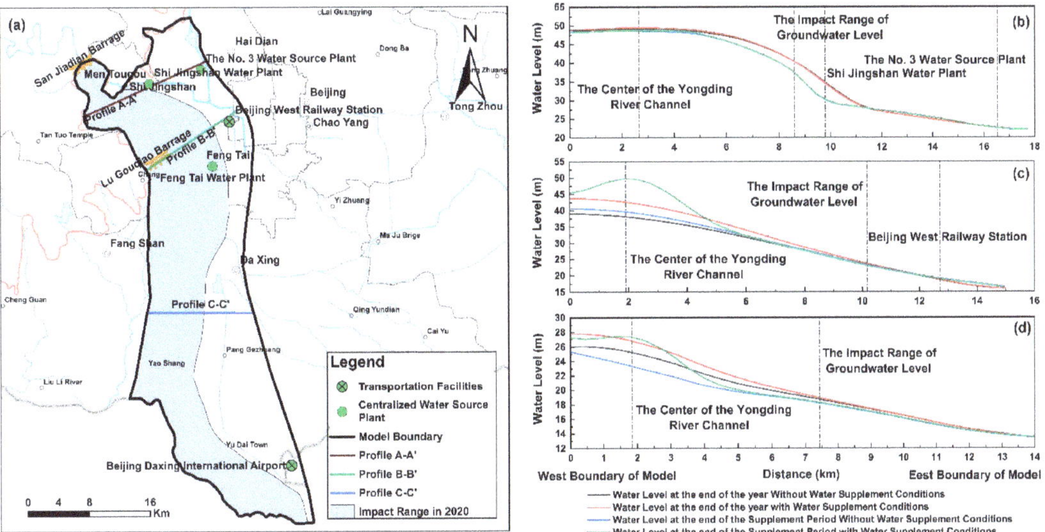

Figure 8. (**a**) Impact range of groundwater level restoration in the Yongding River supplement region in 2020. (**b**) Profile A–A' water level change map for the upper reaches of the Yongding River. (**c**) Profile B–B' water level change map for the middle reaches of the Yongding River. (**d**) Profile C–C' water level change map for the lower reaches of the Yongding River.

We set three calculation sections perpendicular to the Yongding River, along the upper, middle, and lower reaches of the Yongding River, to analyze the changes in the width of the impact range affected by the ecological water supplement. The position of the calculation section is shown in Figure 8a, and the groundwater level process lines of each profile are shown in Figure 8b–d. In regard to the upper reaches of the Yongding River, most of the river channels had seepage reduction effects, and the impact range of groundwater level rise was only about 7–9 km. In the middle reaches of the area, affected by the runoff interception effect of the Lugouqiao barrage, the impact range could be up to 9–11 km. In the downstream area, due to the decrease of the ecological water supplement, the width of its impact range was merely about 6–7 km.

3.4. Prediction of the Restoration Degree of the Groundwater Level in Yongding River in 2020

The maximum restoration degree of the groundwater level under the influence of the ecological water supplement infiltration in 2020 was obtained by the data at the end of the ecological water supplement period (12 June 2020) (Figure 9a). Identically, the restoration degree of the groundwater level at the end of the year, was obtained by the data on 31 December 2020 (Figure 9b). Combined with the calculated impact range of the Yongding River ecological water supplement in 2020, we also got the average groundwater level restoration degree.

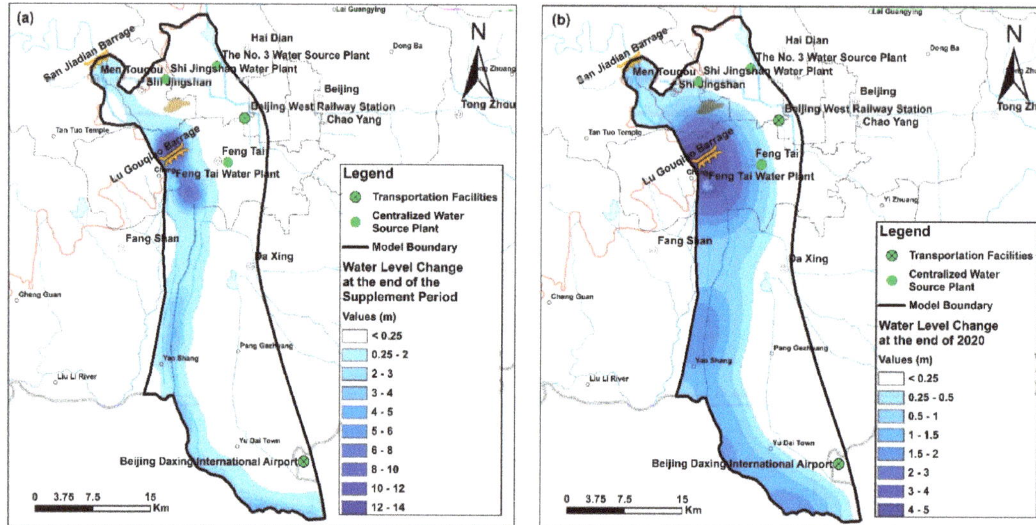

Figure 9. (**a**) Variation map of groundwater level in the study area at the end of the supplement period in 2020. (**b**) Variation map of groundwater level in the study area at the end of 2020.

The largest restoration degree of the groundwater level rise at the end of the ecological water supplement period in 2020 occurred at the center of the channel near the Lugouqiao barrage, which was about 12–14 m. We explained this by the runoff interception effect of the Lugouqiao barrage, which caused the water level to rise and then led to the increased infiltration of the sandy gravel channel. Similarly, because of the general seepage reduction effect in the upper reaches of the Yongding River, the water level rise was generally smaller than the middle and lower natural river sections. Until the end of 2020, the water level gradually dissipated outwards and the water level at the center of the river channel gradually decreased. The change range of the high water level region near the Lugouqiao barrage dropped to 4–5 m, with the rest of the river channel generally declining to below 2.0 m. Moreover, the average groundwater level restoration degree of the impact range was 1.25 m.

For typical buildings, they were not affected by the ecological water supplement head at the end of the ecological water supplement period in 2020. However, with the gradual spread of the ecological water supplement head, only the Fengtai water plant was not significantly affected at the end of 2020. The groundwater level rose about 0.5 m. Outside of the range of the groundwater level restoration, although the Shi Jingshan water plant was unaffected, it was already at the edge of the groundwater level restoration area. When the next large-scale ecological water supplement occurs, it will be affected to a certain extent. However, the no. 3 water source plant, BWRS, and BDIA, were still more than 2–3 km away from the groundwater level restoration area. In the future, the ecological water supplement process would be relatively less susceptible to the impact of groundwater level restoration.

3.5. Analysis on the Ecological Water Supplement Infiltration Volume of Yongding River in 2020

By analyzing the simulation results, the ecological water supplement infiltration volume of each section of the Yongding River in 2020 is given in Table 4. Compared with 2019 (Table 3), in order to accomplish the ecological water supplement demand of the downstream reaches, the ecological water supplement of the two river sections above the Lugouqiao barrage was reduced by 32%. For the river channel below the Lugouqiao barrage, there was no runoff interception effect, and the amount of the ecological water supplement was inversely related to the distance from the section to the Lugouqiao barrage.

The leakage coefficients of each channel segment were selected from the leakage coefficients after calibration and validation of the ecological water supplement's numerical model in 2019. The ecological water supplement infiltration volume of Yongding River in 2020 was about 102.64×10^6 m^3, accounting for about 80% of the total ecological water supplement volume, which was roughly equivalent to the ecological water supplement infiltration volume in 2019.

Table 4. Ecological water supplement infiltration volume for each section of Yongding River in 2020.

Section	Ecological Water Supplement Volume of Each Section ($\times 10^6$ m^3)	Ecological Water Supplement Infiltration Volume of Each Section ($\times 10^6$ m^3)
Seepage reduction channel section from the San Jiadian barrage to the Yuan Bo lake	46.69	3.74
Section from the Yuan Bo lake to the Lugouqiao barrage		33.62
Seepage reduction channel section from the Lugouqiao barrage to the Wanping Lake	38.26	3.06
Section from the Wanping Lake to the Ethylene Pipe Bridge, Jing Liang Road		27.55
Section from the Ethylene Pipe Bridge, Jing Liang Road to the Sixth Ring Road	10.00	8.00
Section from the Sixth Ring Road to the Gu An monitoring station	21.82	17.45
Section from the Gu An monitoring station to the Cui Zhihuiying village (the Beijing administrative boundary)	11.53	9.22
Total	128.30	102.64

4. Conclusions

By establishing a groundwater flow model, this article studied the ecological water supplement infiltration volume, groundwater level restoration, and the impact of water level rise on typical buildings under different water supplement scenarios in the Yongding River in 2019 and 2020. The numerical model could accurately reflect the dynamic change characteristics of groundwater and predict the effect of the ecological water supplement. The conclusions were as follows:

(1) The total recharge of groundwater in the study area was 359.79×10^6 m^3, the total discharge was 238.17×10^6 m^3, and the recharge difference was 121.62×10^6 m^3. The groundwater was in a positive equilibrium state as a whole. The ecological water supplement infiltration volume of the Yongding River reached a 28.15% proportion in the recharge items, which was 101.28×10^6 m^3 in 2019 and 102.64×10^6 m^3 in 2020, accounting for 80% of the total ecological water supplement.

(2) The total impact range of the ecological water supplement of the Yongding River in 2019 was 265.19 km^2, which was concentrated in the 6 km range on both sides of the Yongding River. The maximum groundwater level restoration was located at the center of the river channel near the Lugouqiao barrage, in which the groundwater level rose up to 7–8 m. Moreover, the average groundwater level rise was 2.10 m in the affected area.

(3) The predicted area affected by the ecological water supplement of the Yongding River in 2020 was 506.88 km^2, and the maximum groundwater level restoration range was up to 9–11 km. The upstream section of the river had a smaller impact range due to the seepage reduction effect, merely about 7–9 km. The largest groundwater level rise rose at the end of the ecological water supplement period, which was up to 12–14 m. Moreover, the average groundwater level rise was 1.25 m at the end of 2020.

(4) For the groundwater level around the typical buildings, such as Beijing West Railway Station, Beijing Daxing International Airport, and the centralized water source plant within the study area, only the Fengtai water plant was slightly affected by the two ecological water supplements in 2019 and 2020. The rest were not significantly affected.

Author Contributions: Formal analysis, Z.J. and Y.C.; investigation, Z.J., S.Z. and J.S.; data curation, Z.J., Y.C., S.Z., W.C. and J.S.; writing—original draft preparation, Z.J.; writing—review and editing, Y.C. and J.S.; supervision, Y.C, S.Z., W.C. and J.S. All authors have read and agreed to the published version of the manuscript.

Funding: This research was funded by the Beijing Science and Technology Plan, China, grant number (Z191100006919001), and the Special Foundation for the Basic Resources Survey Program of the Ministry of Science and Technology, China, grant number (2017FY100405).

Institutional Review Board Statement: Not applicable.

Informed Consent Statement: Not applicable.

Conflicts of Interest: The authors do not have any commercial or associative interests that represent conflicts of interest in connection with the submitted work.

References

1. Lin, Z.; Han, L.; Lianwei, Z. Negative environmental effects of water cons-ervancy projects on Yongding River in Beijin-g from 1950 to 1990. *Water Resour. Prot.* **2016**, *32*, 130–135.
2. Yixuan, W.; Yanjun, S.; Ya, G.; Hang, L. Research progress on the changes of environmental and water resources in the upper Yongding River Basin. *South-North Water Transf. Water Sci. Technol.* **2021**, *4*, 656–668.
3. Miao, Y.; Yuansong, W.; Junguo, L.; Peibin, L. Impact of socioeconomic development on water resource and water environment of Yongding River in Beijing. *Acta Sci. Circumstantiae* **2011**, *31*, 1817–1825.
4. Dan, D.; Xiangqin, X.; Mingdong, S.; Chenlin, H.; Xubo, L.; Kun, L. Decrease of both river flow and quality aggravates water crisis in North China: A typical example of the upper Yongding River watershed. *Environ. Monit. Assess.* **2020**, *192*, 1–13.
5. Peng, W. Study on River Health Assessment and Restoration. Master's Thesis, Hefei University of Technology, Hefei, China, 2007.
6. Piégay, H.; Cottet, M.; Lamouroux, N. Innovative approaches in river management and restoration. *River Res. Appl.* **2020**, *36*, 875–879. [CrossRef]
7. Md Anawar, H.; Chowdhury, R. Remediation of Polluted River Water by Biological, Chemical, Ecological and Engineering Processes. *Sustainability* **2020**, *12*, 7017. [CrossRef]
8. Ge, P.; Chen, M.; Zhang, L.; Song, Y.; Mo, M.; Wang, L. Study on Water Ecological Restoration Technology of River. *IOP Conf. Ser. Earth Environ. Sci.* **2019**, *371*, 1–4.
9. Zhaozhong, F.; Shuo, L.; Pin, L. Research progress in eco-environmental issues and eco-restoration strategy in Yongding River basin. *J. Univ. Chin. Acad. Sci.* **2019**, *36*, 510–520.
10. Tao, P.; Zhenming, Z.; Junguo, L. Discussion on the Ecological Restoration Goals of the Yongding River in Beijing Based on Ecosystem Service Functions Analysis. *Chin. Agric. Sci. Bull.* **2010**, *26*, 287–292.
11. Jing, M.; Qingrui, Y.; Wenqi, P.; Huihuang, L.; Yuling, H. Habitat Quality Evaluation of Yongding River in Beijing Mountain Section after Ecological Rehydration. *China Rural Water Hydropower* **2021**, 30–36. Available online: https://global.cnki.net/kcms/detail/detail.aspx?filename=ZNSD202102006&dbcode=CJFQ&dbname=CJFDTEMP&v= (accessed on 29 September 2021).
12. Jian, W.; Xingyao, P.; Gang, K.; Tao, B.; Qian, H. Study on ecological restoration of water-deficient rivers based on ecological water supplement method. *J. Water Resour. Water Eng.* **2020**, *31*, 64–69.
13. Yong, D.; Chao, W. Study on water demand for water supply of whole Yongding river and its security plan. *Water Resour. Plan. Des.* **2020**, 14–17. Available online: https://global.cnki.net/KCMS/detail/detail.aspx?dbcode=CJFQ&dbname=CJFDLAST2020&filename=SLGH202007004&v=MDAwNDFyQ1VSN3VlWnVadEZ5amhWcnpOTmlITVpyRzRITkhNcUk5RllJUjhlWDFMdXhZUzdEaDFUM3FUcldNMUY= (accessed on 29 September 2021).
14. Hu, H.; Mao, X.; Yang, Q. Impacts of Yongding River Ecological Restoration on the Groundwater Environment: Scenario Prediction. *Vadose Zone J.* **2018**, *17*, 1–15. [CrossRef]
15. Luo, Z.; Zhao, S.; Wu, J.; Zhang, Y.; Liu, P.; Jia, R. The influence of ecological restoration projects on groundwater in Yongding River Basin in Beijing, China. *Water Supply* **2019**, *19*, 2391–2399. [CrossRef]
16. Litang, H.; Jianli, G.; Shouquan, Z.; Kangning, S.; Zhengqiu, Y. Response of groundwater regime to ecological water replenishment of the Yongding River. *Hydrogeol. Eng. Geol.* **2020**, *47*, 5–11.
17. Yao, M.; Yong, Y.; Guojin, H. Analysis of groundwater replenishment caused by water supplement in Yongding River (Beijing section). *Beijing Water* **2020**, *4*, 22–27.
18. Kangning, S.; Litang, H.; Jianli, G.; Zhengqiu, Y.; Yuanzheng, Z.; Shouquan, Z. Enhancing the understanding of hydrological responses induced by ecological water replenishment using improved machine learning models: A case study in Yongding River. *Sci. Total Environ.* **2021**, *768*, 1–12.

19. El-Zehairy, A.A.; Lubczynski, M.W.; Gurwin, J. Interactions of artificial lakes with groundwater applying an integrated MODFLOW solution. *Hydrogeol. J.* **2018**, *26*, 109. [CrossRef]
20. Ghysels, G.; Mutua, S.; Veliz, G.B.; Huysmans, M. A modified approach for modelling river–aquifer interaction of gaining rivers in MODFLOW, including riverbed heterogeneity and river bank seepage. *Hydrogeol. J.* **2019**, *27*, 1851–1863. [CrossRef]
21. Chakraborty, S.; Maity, P.K.; Das, S. Investigation, simulation, identification and prediction of groundwater levels in coastal areas of Purba Midnapur, India, using MODFLOW. *Environ. Dev. Sustain.* **2020**, *22*, 3805–3837. [CrossRef]
22. Fadoua, H.; Mounira, Z.; Meriem, A.; Mohamedou, B.S.; Moncef, G.; Rachida, B. Hydrogeochemical modeling for groundwater management in arid and semiarid regions using MODFLOW and MT3DMS: A case study of the Jeffara of Medenine coastal aquifer, South-Eastern Tunisia. *Nat. Resour. Model.* **2020**, *33*, 1–23.
23. Lea, D.; Laurence, G. Modeling spring flow of an Irish karst catchment using Modflow-USG with CLN. *J. Hydrol.* **2021**, *597*, 1–14.
24. Jiang, B.; Wong, C.P.; Lu, F.; Ouyang, Z.; Wang, Y. Drivers of drying on the Yongding River in Beijing. *J. Hydrol.* **2014**, *519*, 69–79. [CrossRef]
25. Wang, L.; Wang, Z.; Koike, T.; Yin, H.; Yang, D.; He, S. The assessment of surface water resources for the semi-arid Yongding River Basin from 1956 to 2000 and the impact of land use change. *Hydrol. Process.* **2010**, *24*, 1123–1132. [CrossRef]
26. Xiaofang, Y.; Mingyu, W.; Liya, W.; Jianhui, Z. Investigation of key controlling factors and numerical simulation uncertainty of the groundwater level companying with Yongding river ecological restoration. *J. Univ. Chin. Acad. Sci.* **2015**, *32*, 192–199.
27. Qi-chen, H.; Jing-li, S.; Yu, L.; Zhenhua, X.; Yali, C. Optimization of artificial recharge of groundwater system based on parallel genetic algorithm—A case study in the alluvial fan of Yongding River in Beijing. *South-North Water Transf. Water Sci. Technol.* **2015**, *13*, 67–71.
28. Huan, H.; Jin-sheng, W.; Yuan-zheng, Z.; Jie-qiong, Z. Chemical Characteristics and Evolution of Groundwater in the Yongding River Alluvial Fan of Beijing Plain. *Acta Geosci. Sin.* **2011**, *32*, 357–366.
29. Ezzeldin, M.; El-Alfy, K.; Abdel-Gawad, H.; Abd-Elmaboud, M. Comparison between Structured and Unstructured MODFLOW for Simulating Groundwater Flow in Three-Dimensional Multilayer Quaternary Aquifer of East Nile Delta, Egypt. *Hydrol. Curr. Res.* **2018**, *9*, 1–12. [CrossRef]
30. Shaoqing, T.; Yanhui, D.; Liheng, W. Methods and applications of an unstructured grid version of MODFLOW: MODFLOW-USG. *Hydrogeol. Eng. Geol.* **2016**, *43*, 9–16.
31. Wei-zhe, C.; Qichen, H.; Kang, C.; Fei, C.; Shinan, T. Comparison of Different Grid Refinement Methods in Simulations of Groundwater Recharge by Rivers. *J. Earth Sci. Environ.* **2020**, *42*, 394–404.

Article

Research on the Application of Typical Biological Chain for Algal Control in Lake Ecological Restoration—A Case Study of Lianshi Lake in Yongding River

Pengfei Zhang [1,2], Xiaoyu Cui [2], Huihuang Luo [2,*], Wenqi Peng [2] and Yunxia Gao [1]

[1] School of Municipal and Environmental Engineering, Hebei University of Architecture, Zhangjiakou 075000, China; pfzhang1001@163.com (P.Z.); wljgyx@163.com (Y.G.)
[2] Department of Water Ecological Environment, China Institute of Water Resources and Hydropower Research, Beijing 100038, China; cuixy@iwhr.com (X.C.); pwq@iwhr.com (W.P.)
*   Correspondence: luohh@iwhr.com or luohuihuang@sina.com; Tel.: +86-138-1089-8180

Abstract: Maintaining the health of lake ecosystems is an urgent issue. However, eutrophication seriously affects lakes' ecological functions. Eutrophication is also the main target of lake ecological restoration. It is vital to carry out research on lake eutrophication control and energy flow evaluation in ecosystems scientifically. Based on in situ survey results for the aquatic life data for Lianshi Lake from 2018 to 2019, the Ecopath model was used to establish an evaluation index system for the typical biological chain to screen out the key species in the water ecosystem, and the fuzzy comprehensive evaluation (FCE) method was used to screen all the biological chains controlling algae. A combination of the FCE coupled with the Ecopath screening method for typical biological chains for algal control was applied to the Lianshi Lake area; the results show that the typical biological chain for algal control is phytoplankton (Phyt)–zooplankton (Zoop)–macrocrustaceans (Macc)–other piscivorous (OthP). Upon adjusting the biomass of Zoop and Macc in the typical biological chain for algal control to three times that of the current status, the ecological nutrition efficiency of Phyt was increased from 0.308 to 0.906. The material flow into the second trophic level from primary producers increased from 3043 to 8283 t/km^2/year. The amount of detritus flowing into primary producers for sedimentation decreased from 7618 to 2378 t/km^2/year. Finally, the total primary production/total respiratory volume (TPP/TR) decreased from 9.224 to 3.403, the Finn's cycle index (FCI) increased from 13.6% to 17.5%, and the Finn's average energy flow path length (FCL) increased from 2.854 to 3.410. The results suggest that the problem of eutrophication can be solved by introducing Zoop (an algal predator) and Macc to a large extent, resulting in improved ecosystem maturity. The research results can facilitate decision making for the restoration of urban lake water ecosystems.

Keywords: alga control; typical biological chain; Ecopath model; ecological restoration; Lianshi Lake

Citation: Zhang, P.; Cui, X.; Luo, H.; Peng, W.; Gao, Y. Research on the Application of Typical Biological Chain for Algal Control in Lake Ecological Restoration—A Case Study of Lianshi Lake in Yongding River. *Water* **2021**, *13*, 3079. https://doi.org/10.3390/w13213079

Academic Editor: Antonio Zuorro

Received: 4 September 2021
Accepted: 24 October 2021
Published: 2 November 2021

Publisher's Note: MDPI stays neutral with regard to jurisdictional claims in published maps and institutional affiliations.

1. Introduction

The rapid expansion of the world's population has exacerbated the degradation of global lake ecosystems [1]. According to statistics, more than 60% of the lakes in the world are in different degrees of eutrophication [2]. The increasing eutrophication of lakes has become a global water environment problem [3] (e.g., for Lake Canyon in the United States [4], Lake Geneva in Switzerland [5], and Sugarloaf Lake in Australia [6]).

The interaction between lake water environments and water ecology is complex. It is important to ensure the integrity of the ecosystem while improving the quality of the water environment [7–9]. Therefore, how to deal with lake eutrophication and restore aquatic ecosystems has become an urgent problem to be solved in current limnology, environmental science, freshwater ecology and other disciplines, and has been closely followed by international scholars [10,11].

At present, the recognized theories of lake ecological restoration mainly include the nutrient salt concentration limit theory by Scheffer [12], multi-steady state theory by Lewontin [13] and biological manipulation theory by Hrbacek [14]. Among them, the biological manipulation theory has the most promising application prospects and has led to many successful cases of lake ecological restoration. For example, the biological manipulation process implemented by putting silver carp (Silc) and bighead carp (Bigc) in Donghu Lake in Wuhan is one of the main reasons for the disappearance of cyanobacteria blooms in the lake [15]. Olin et al. [16] reported that the removal of common carp (Comc) from 10 lakes in southern Finland effectively reduced the biomass of cyanobacteria and the degree of algal outbreaks. Shapiro et al. [17] adjusted the ratio of other piscivarious (Othp) and plankton-eating fish from 1:1.65 to 1:2.2 in the Round Lake, and the concentrations of total nitrogen (TN), total phosphorus (TP) and chlorophyl-a (Chl-*a*) in the lake were reduced to varying degrees. However, biological manipulation cannot enable all ecosystems to achieve the expected ecological functions and may lead to changes in the species diversity of the ecosystem, the decline of the average trophic level in the system, and the destruction of habitats [18,19]. For example, Razlutskij et al. [20], through 72 days of outdoor experiments, found that introducing *Carassius auratus* increased the biomass of Phyt. Recent ecological studies have revealed significant negative effects of crucian carp (Cruc) on the water quality and ecological states of shallow lakes, e.g., increasing nutrient levels, leading to reduced water clarity [21,22]. The current international guidance on the application of biological manipulation technology to the process of lake ecological restoration remains far from sufficient [23].

In light of this, this study intended to establish a screening method for typical biological chains of ecosystems and propose a biomass control strategy for the typical biological chain of alga control. The purpose of this study was to provide strong theoretical support for the management of urban lake eutrophication and water ecological restoration, and it consisted of the following: (1) Key species and target biological chains were screened, based on the Ecopath model; the key species of the ecosystem were analyzed, and the biological chain with phytoplankton (Phyt) as the primary producer and including the key species was selected. (2) Typical biological chains were screened, and an evaluation index that could reflect the efficiency and stability of the biological chain was constructed. The index weight was used to select typical biological chains based on the fuzzy comprehensive evaluation method. (3) The regulatory impact was analyzed, the biomass of key species in the typical biological chain was regulated, and then the impacts on the regulation target and the ecosystem were judged.

2. Materials and Methods

2.1. Overview of the Study Area and Data Sources

2.1.1. Overview of the Study Area

As a river-type lake in the urban section of the plain of the Yongding River, Lianshi Lake is located at the junction of Mentougou and Fengtai District in the southwest of Shijingshan District (116°6′58″~116°9′34″ E, 39°56′3″~39°53′15″ N), which belongs to the temperate continental monsoon climate, with high temperatures and rain in summer and cold and dry conditions in winter. The total length of Lianshi Lake is 5.8 km, the average width of the lake is 376 m, the widest point is about 500 m, the total water area is 106 hm^2, and the average depth of the water body is 1.6 m. The phytoplankton in Lianshi Lake are dominated by *Pseudanabaena moniliformis* (Cyanophinyta) and *Limnothrix*. Lianshi Lake belongs to the northern freshwater city lake. The distribution of the study area and monitoring points are shown in Figure 1.

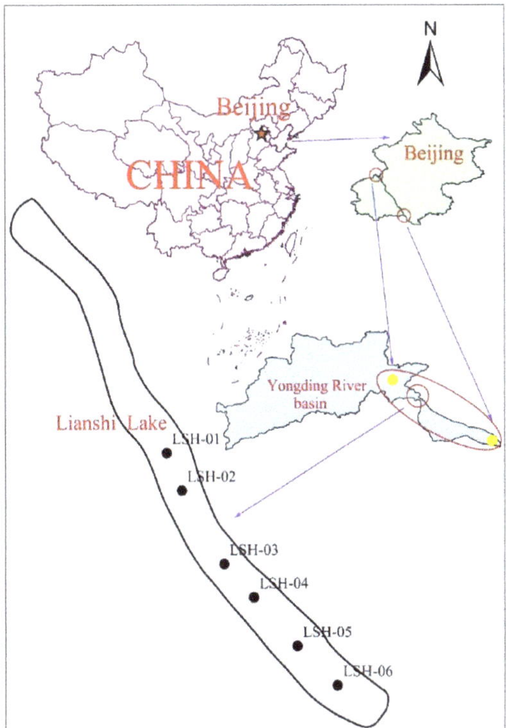

Figure 1. The location of Lianshi Lake and the distribution of monitoring points.

2.1.2. Data Sources

The data sources included long-term field sampling survey data, formula calculations and reference comparisons. The main source of the biomass B_i of each function group was a field survey of the biomass at six points in Lianshi Lake from 2018 to 2019. The macroinvertebrates were collected using a Surber sampler with a mesh diameter of 40 mesh (500 μm) and a sampling area of 0.09 m², and 70% alcohol was added to the test tube bottle to be stored for inspection. The zooplankton were collected using a No. 25 plankton net (200 mesh) and immediately fixed with 4% formaldehyde. For the collection of phytoplankton, a 1 L water sample (0.5 m below the surface of the water body) was collected in a plexiglass water collector, and 15 mL of Lugol's reagent was added for fixation. Sampling areas of 0.5 m × 0.5 m and 1 m × 1 m were used for vascular plants and submerged plants, respectively. In the field, the plant species were distinguished and weighed (wet weight). Fish resources were collected mainly using trawl nets, and the travel distance for each sampling point was no more than 100 m. The biomass of organic detritus was calculated based on the relationship between the primary productivity and water transparency [24] (Equation (1)).

$$logD = 0.954logPP + 0.863logE - 2.41 \qquad (1)$$

where D is the detrital biomass (g C·m⁻²), PP is the primary production volume (g C·m⁻²·a⁻¹), and E is the average transparency (m).

2.2. Ecopath Model

2.2.1. Principle of the Ecopath Model

The Ecopath is an ecological model that can directly determine the structure of an ecological system and describe its energy flow and mass transfer through the principle of nutrition dynamics. The Ecopath model stipulates that each function group (i) energy input and output in the ecosystem are balanced. The model uses a set of simultaneous linear equations to define an ecosystem, and each function group is represented by a linear equation [25] (Equation (2)).

$$B_i \times \left(\frac{P}{B}\right)_i \times EE_i = \sum_{j=1}^{n} B_j \times \left(\frac{Q}{B}\right)_i \times DC_{ij} + EX_i \qquad (2)$$

In the formula, B_i is the biomass of the i-th function group. $\left(\frac{P}{B}\right)_i$ is the ratio of the annual average production of the i-th function group to the annual average biomass, which is the biomass turnover rate. EE_i is the ecological nutrient conversion efficiency of the i-th function group and is usually obtained by the calculation of the model [26]. $\left(\frac{Q}{B}\right)_i$ is the ratio of the consumption of the i-th function group to the biomass. DC_{ij} is the ratio of the i-th the prey group to the total predation of the predator group j. EX_i is the output of the i-th function group.

2.2.2. Division of Function Groups

The purpose of the establishment of function groups is to merge populations with highly overlapping niches to simplify the food web. Based on the survey results for the Lianshi Lake ecosystem, characteristics of the community, and survival habits of each species, this work identified organisms with similar ecological characteristics, which were grouped into function groups [24]. A total of 14 function groups were set up (Table 1).

Table 1. Ecosystem function groups and main types for Lianshi Lake.

No.	Function Group	Abbreviation for Composition	Included Types
1	Other piscivorous	OthP	Horsemouth, yellow catfish
2	Common carp	Comc	Common carp
3	Crucian carp	Cruc	Crucian carp
4	Bighead carp	Bigc	Bighead carp
5	Silver carp	Silc	Silver carp
6	Herbivorous fish	HerF	Grass carp, bream
7	Other fish	OthF	Wheat ear fish, tortoisefish
8	Macrocrustaceans	Macc	Green prawns, prawns, Chinese mitten crabs, etc.
9	Other benthos	OthB	*Hydrophilia, Ceratobranchus, Longbrachium, Fanchus, Chironomus*, Chironomidae, etc.
10	Zooplankton	Zoop	Protozoa, rotifers, cladocerans, copepods, etc.
11	Phytoplankton	Phyt	Cyanobacteria, green algae, euglena, dinoflagellate, cryptophytes, golden algae, etc.
12	Submerged macrophytes	SubM	*Potamogeton, Myriophyllum, Hydrilla verticillata, Ceratophyllum*, etc.
13	Other macrophytes	OthM	Reeds, cattails, etc.
14	Detritus	Detr	Organic detritus

2.2.3. Parameter Setting

Some of the parameters refer to lakes with water environmental conditions similar to those of Lianshi Lake, such as the production/biomass (P/B) and consumption/biomass (Q/B) parameters of fish resources that were obtained by querying (Available online: http://www.fishbase.org) (accessed on 20 May 2021). The plant P/B coefficient refers to related research on Taihu Lake [27]. The P/B coefficients of zooplankton (Zoop), other benthos (Othb) and macrocrustaceans (Macc) were estimated based on the measured data by referring to the research results for Taihu Lake [27], Qiandao Lake [28], Dianshan Lake [29] and Zhushan Bay [30]. The P/Q coefficient refers to recognized data. The values for Zoop, OthB and Macc were 0.05 [31], 0.02 [32] and 0.075 [33], respectively. The food composition (DC_i) is shown in Table 2. In addition to ecological investigation and research, we also referred to related research results [27,34], such as those for Zhushan Lake [30] and Qiandao Lake [35,36]. According to the cited literature, the *GS* values for general OthP and herbivorous fish (HerF) were set to 0.2 and 0.41, and the *GS* values for Zoop, OthB and Macc were set as 0.65, 0.94 and 0.7, respectively [31–33,37].

Table 2. Food composition data entered into the model.

No.	Prey/Predator	1	2	3	4	5	6	7	8	9	10
1	OthP	0.007									
2	Comc	0.15									
3	Cruc	0.27									
4	Bigc										
5	Silc										
6	HerF	0.07						0.006			
7	OthF										
8	Macc	0.38									
9	OthB	0.073	0.13					0.230			
10	Zoop		0.82	0.24	0.501	0.213		0.620	0.350	0.005	0.009
11	Phyt		0.048		0.361	0.620			0.300	0.022	0.801
12	SubM		0.002	0.09			0.997	0.003		0.101	
13	OthM	0.01					0.003				
14	Detr	0.04		0.67	0.138	0.167		0.141	0.350	0.872	0.190
15	Sum	1	1	1	1	1	1	1	1	1	1

2.2.4. Model Balance Calculation

The Ecopath model requires the input of six basic parameters: B_i, $\left(\frac{P}{B}\right)_i$, EE_i, $\left(\frac{Q}{B}\right)_i$, DC_{ij} and EX_i. Model balancing was executed, which was established on the basis of conforming to objective laws. The model parameters could be slightly modified to meet the requirements of the model operation, but it was necessary to avoid changing reliable data sources [25]. The nutritional conversion efficiency (EE_i) ranged from 0 to 1 [38], the group *EE* was close to 1 in the face of considerable predation pressure, and the underutilized function group had a lower *EE* value. The value range of $GE\left(\frac{P}{Q}\right)$ was generally 0.1–0.3. If the balance test had one or more $EE > 1$, it was necessary to locate which predators caused the problem for specific prey groups in the predation mortality. The level of model confidence is mainly related to the quality and reliability of the acquired data. The accuracy of the Ecopath model was measured using the Pedigree index. The higher the index, the higher the quality of the model. The input and output parameters of the balanced model are shown in Table 3.

Table 3. Parameters of Lianshi Lake ecosystem construction.

Function Group	Biomass (t/km^2)	Production/ Biomass	Consumption/Biomass	Eco-Nutrition Efficiency	Production/ Consumption	Proportion of Unassimilated Food
OthP	0.13	1.670	6.1	0.026	0.274	0.200
Comc	0.5	0.960	10.7	0.248	0.090	0.200
Cruc	0.5	1.130	12.3	0.379	0.092	0.200
Bigc	1.8	0.990	6.9	0.001	0.143	0.200
Silc	1.2	1.100	8.0	0.001	0.138	0.200

Table 3. *Cont.*

Function Group	Biomass (t/km²)	Production/ Biomass	Consumption/Biomass	Eco-Nutrition Efficiency	Production/ Consumption	Proportion of Unassimilated Food
HerF	0.27	0.987	7.1	0.778	0.139	0.410
OthF	2.3	2.155	11.0	0.001	0.196	0.410
Macc	1.58	3.090	41.0	0.062	0.075	0.700
OthB	16.141	4.130	206.5	0.099	0.020	0.940
Zoop	7.85	20.680	413.6	0.606	0.050	0.650
Phyt	47.42	185.000		0.308		
SubM	1460	1.250		0.186		
OthM	64	1		0.001		
Detr	3.230			0.272		

2.3. Fuzzy Comprehensive Evaluation Method

The fuzzy comprehensive evaluation method is based on fuzzy mathematics. There is a certain characteristic of n things to be evaluated; these n things comprise the object set $X = \{x_1, x_2, \ldots, x_n\}$, the factor set $U = \{u_1, u_2, \ldots, u_n\}$ and the evaluation set $V = \{v_1, v_2, \ldots v_m\}$. Suppose that the weight distribution of the factors is the fuzzy subset A on V, denoted as $A = \{a_1, a_2, \ldots, a_n\}$. In the formula, a_i is the weight corresponding to the i-th factor u_i, and the general rule is $\sum_{i=1}^{n} a_i = 1$.

The evaluation steps for the fuzzy comprehensive evaluation method are as follows:

(1) Establish a factor set.

Assuming there are a total of i factors of the judged object, the factor universe of the evaluated object U is $U = \{u_1, u_2 \ldots, u_i\}$.

(2) Determine the membership function.

Assuming that the evaluation level is divided into i levels, the set is $V = \{v_1, v_2, v_3, \ldots v_i\}$.

(3) Establish a fuzzy relationship evaluation matrix.
(4) Establish a weight vector.

Determine the weight vector of the judgement factor $A = (a_1, a_2, \ldots, a_n)$. A is the subordination relationship of each factor in U to the thing being judged; it depends on the focus of people when making fuzzy comprehensive judgments, and it is equivalent to assigning weights according to the importance of each factor in the judgment.

(5) Establish a fuzzy comprehensive evaluation matrix.

According to the calculated maximum membership value, the program selection or category evaluation is performed.

3. Results

3.1. Screening of Typical Biological Chains of Fuzzy Comprehensive Evaluation Coupled with Ecopath

The Ecopath model was used to screen the key species in the Lianshi Lake ecosystem, identify the types of algae-controlling biological chains containing the key species, and establish a biological chain evaluation index system that could simultaneously express the delivery efficiency and stability. Finally, the fuzzy comprehensive evaluation method was used to screen out the biological chain with the largest weight value.

See Flowchart 2 for specific screening ideas (Figure 2).

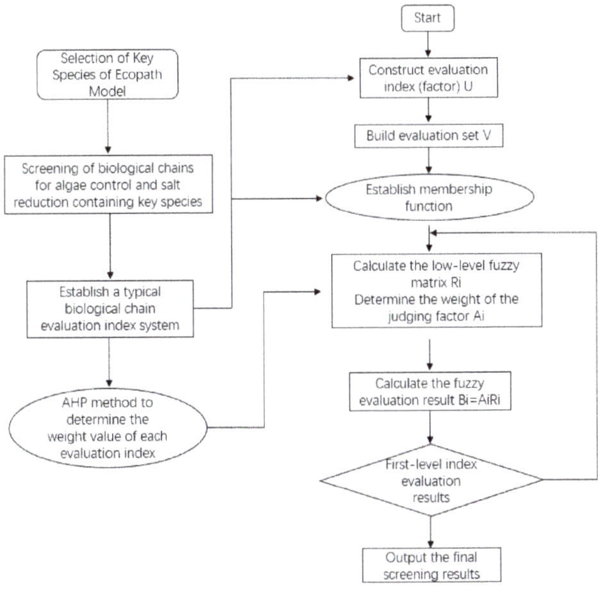

Figure 2. Screening flowchart of a typical biological chain.

(1) Analysis of key species in the ecosystem

The key function group plays an important role in the function of the Lianshi Lake ecosystem. This study used the method developed by Libralato et al. [39] to calculate the keystoneness (KS) index of each function group. The one with the largest KS value is regarded as the key function group. Compared to the calculation method for the KS value proposed by Valls et al. [40] and Power et al. [41], this calculation is simpler and comprehensively considers the effects of biomass and capacity. The equation is $KS_i = log[\varepsilon_i(1 - p_i)]$, $\varepsilon_i = \sqrt{\sum_{j \neq 1}^{n} m_{ij}^2} \varepsilon$, $p_i = \frac{B_i}{\sum_i^n B_K}$, where KS_I is the criticality index of function group i, ε_i is the total impacts of function group i in the ecosystem (total impacts), and m_{ij} is the value of the mixed nutrition effect of function group i on function group j, indicating the mutual relationship between each other's strengths, and P_i is the ratio of the biomass B_i of function group i to the biomass of the entire ecosystem $\sum_k^n B_k$. Since KS_i and p_i are negatively correlated, the criticality index will not be too high due to the high biomass of the function groups (Table 4).

Table 4. Criticality data table for each function group of Lianshi Lake.

No.	Function Group	Biomass	P_i	ε_i	KS_i	Relative Total Impact
1	OthP	0.13	0.0000809	0.899516	−0.04603	0.622
2	Comc	0.50	0.000311	0.147707	−0.83073	0.122
3	Cruc	0.50	0.000311	0.197666	−0.7042	0.17
4	Bigc	1.80	0.00112	0.054332	−1.26543	0.0531
5	Silc	1.20	0.000747	0.017978	−1.74559	0.0175
6	HerF	0.27	0.000168	0.429709	−0.3669	0.425
7	OthF	2.30	0.001431	0.845322	−0.0736	0.749
8	Macc	1.58	0.000983	0.361432	−0.4424	0.328
9	OthB	16.14	0.010045	1.077714	0.028119	0.867
10	Zoop	7.85	0.004885	0.882764	−0.05628	0.701
11	Phyt	47.42	0.02951	0.914433	−0.05186	0.844
12	SubM	1460	0.90857	1.014112	−1.03283	1
13	OthM	64	0.039828	0.007863	−2.12205	0.00724

As shown in Figure 3, the first key function groups in the Lianshi Lake ecosystem are submerged macrophytes (SubM). However, other macrophytes (OthM) are redundant in the food web of the entire ecosystem. Other benthos (OthB), Phyt, other fish (OthF) and Zoop have a relative total impact second only to SubM, and their criticality index ranks in the top four.

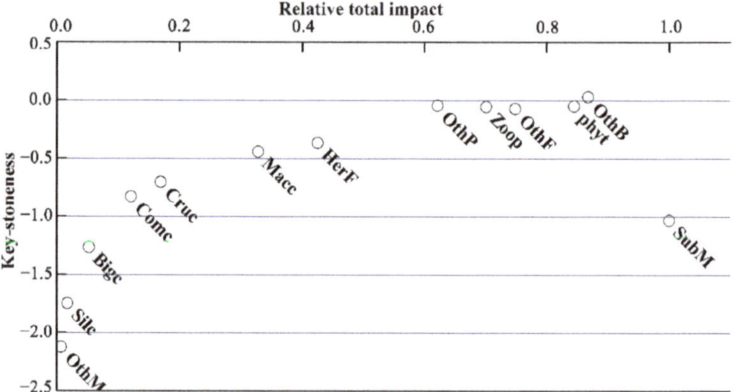

Figure 3. Key species in the ecosystem of Lianshi Lake.

(2) Screening of target biological chains for alga control and salt reduction

With the goal of alga control, we identified the biological chain with Phyt and SubM as primary producers, and then determined the biological chain containing key species such as Zoop, OthB and OthF. The results are shown in Figure 4.

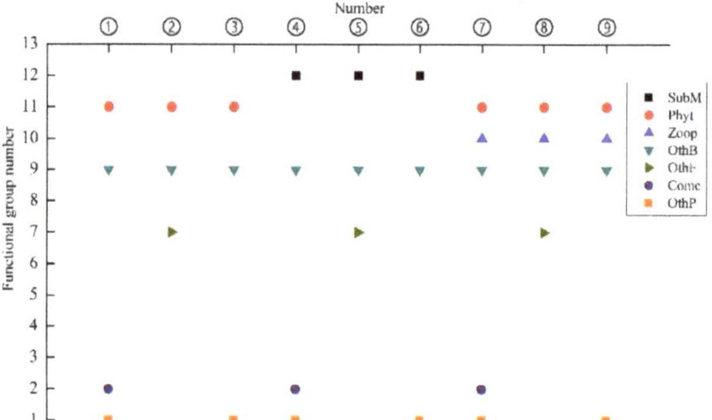

(**a**) Alga control and salt–cutting biological chain containing key species of OthB.

Figure 4. *Cont.*

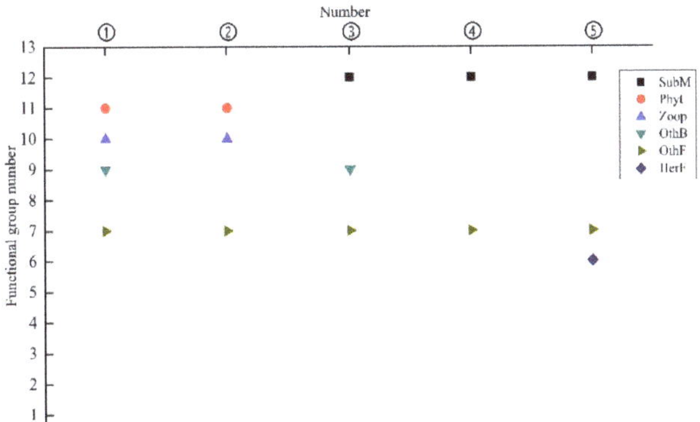

(**b**) Alga control and salt–cutting biological chain containing key species of OthF.

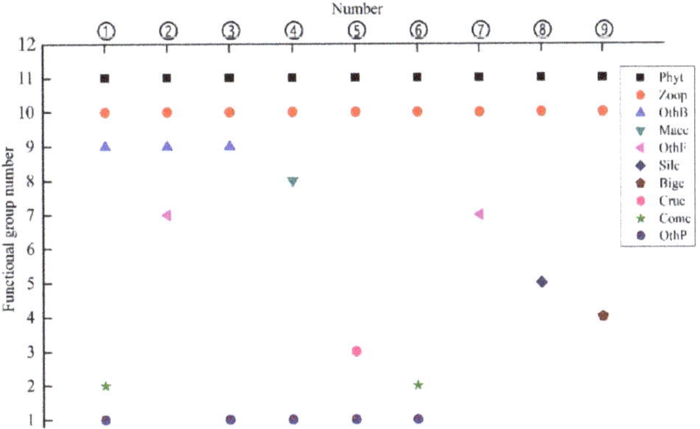

(**c**) Alga control and salt–cutting biological chain containing key species of Zoop.

Figure 4. The biological chain of algal control in Lianshi Lake containing key species.

(3) Establishment of evaluation indicators for typical biological chains

In order to express the characteristics of the Lianshi Lake ecosystem, the ecotrophic efficiency (EE), production/respiration (P/R), relative total impact (RTI), omnivory index (OI) and trophic level (TL) are listed as typical indicators for biological chain evaluation (Table 5). The evaluation indicators have the following characteristics: (1) EE stands for the utilization and conversion efficiency for the energy contained in the previous trophic organism. The value of EE ranges from 0 to 1, and it is close to 1 in a group with considerable predation pressure. (2) P/R represents an important indicator of the maturity of the biological chain, which is close to 1 in a mature ecosystem. (3) RTI represents the relative total impact of a single function group on the entire ecosystem. (4) OI indicates that, when the omnivorous index is zero, the corresponding prey is oriented. (5) TL refers to the level of the organism in the food chain of the ecosystem. The higher the trophic level of the organism, the greater the contribution of the food chain to the stability of the ecosystem.

Table 5. Basic data of evaluation indicators for each function group of Lianshi Lake.

Serial Number	Function Group	EE	P/R	OI	TL	RTI
1	OthP	0.026	0.520	0.171	3.303	0.622
2	Comc	0.248	0.126	0.048	2.958	0.122
3	Cruc	0.379	0.130	0.186	2.242	0.17
4	Bigc	0.000	0.219	0.255	2.506	0.053
5	Silc	0.000	0.208	0.171	2.215	0.018
6	HerF	0.778	0.308	0.000	2.000	0.425
7	OthF	0.000	0.497	0.125	2.863	0.749
8	Macc	0.062	0.336	0.232	2.353	0.328
9	OthB	0.099	0.500	0.005	2.005	0.867
10	Zoop	0.606	0.167	0.009	2.009	0.701
11	phyt	0.308	-	0.000	1.000	0.844
12	SubM	0.186	-	0.000	1.000	1
13	OthM	0.000	-	0.000	1.000	0.007
14	Detritus	0.272	-	0.251	1.000	-

(4) Fuzzy comprehensive evaluation method for screening typical biological chains

The research did not consider the comment-level domain in the fuzzy comprehensive evaluation, and selected the typical biological chain according to the maximum weight (from Equation (3) to Equation (6)):

① The factor domain U, $U = (EE, P/R, RTI, OI, TL)$ of the judged object was determined.

② The membership function was established.

The higher the value, the higher the membership function:

$$r_i = \begin{cases} 0 & (x \leq S_1) \\ \frac{x-S_1}{S_2-S_1} & (S_1 < x < S_2) \\ 1 & (x \geq S_2) \end{cases} \tag{3}$$

(I) The determination of the membership functions of EE, P/R, RTI, and OI As the variation range of EE, P/R, RTI, and OI was between 0 and 1, the membership function was:

$$U_A(x) = x \tag{4}$$

(II) The determination of the TL membership function

According to the literature, the lowest trophic level of primary producers in lake ecosystems is 1, and the trophic level of the function group that is generally at the top is 4, so the membership function is:

$$U_A(x) = \frac{x-1}{3} \tag{5}$$

③ Single factor evaluation was performed, and a fuzzy relationship matrix was established.

The fuzzy evaluation matrix was established according to the following formula:

$$R = [r_{ij}] = \begin{bmatrix} r_{11} & \cdots & r_{15} \\ \vdots & \ddots & \vdots \\ r_{51} & \cdots & r_{55} \end{bmatrix} \tag{6}$$

In the formula, in the i-th row $R_i = (r_{i1}, r_{i2}, \ldots r_{im})$, $i = 1, \ldots m$, i is the degree of the membership of the i-th evaluation factor to the evaluation standards at all levels; for the j-th column $R_j = (r_{1j}, r_{2j}, \ldots, r_{nj})$, j is the degree of the membership of each evaluation factor to the j-th evaluation standard. Taking the nine biological chains containing the Zoop group as an example, the membership degrees corresponding to different evaluation indicators of different biological chains were calculated and are shown in Table 6 below.

Table 6. The membership degrees corresponding to different evaluation indicators of different biological chains with the Zoop group.

Zoop Group Project Evaluation Index	①	②	③	④	⑤	⑥	⑦	⑧	⑨
EE	0.257	0.338	0.260	0.251	0.330	0.297	0.457	0.457	0.457
P/R	0.328	0.388	0.396	0.341	0.272	0.271	0.332	0.188	0.193
RTI	0.631	0.79	0.759	0.624	0.584	0.572	0.765	0.521	0.533
OI	0.058	0.046	0.062	0.137	0.122	0.076	0.067	0.09	0.132
TL	0.418	0.323	0.360	0.389	0.380	0.439	0.319	0.247	0.279

The fuzzy relation matrix with the Zoop group:

$$R = \begin{bmatrix} 0.257 & 0.338 & 0.260 & 0.251 & 0.330 & 0.297 & 0.457 & 0.457 & 0.457 \\ 0.328 & 0.388 & 0.396 & 0.341 & 0.272 & 0.271 & 0.332 & 0.188 & 0.193 \\ 0.631 & 0.790 & 0.759 & 0.624 & 0.584 & 0.572 & 0.765 & 0.521 & 0.533 \\ 0.058 & 0.046 & 0.062 & 0.137 & 0.122 & 0.076 & 0.067 & 0.090 & 0.132 \\ 0.418 & 0.323 & 0.360 & 0.389 & 0.380 & 0.439 & 0.319 & 0.247 & 0.279 \end{bmatrix}$$

④ A weight vector was established.

The weight vector $A = (a_1, a_2, \ldots, a_n)$ of the judgement factor was determined. A is the subordination relationship of each factor in U to the thing being judged. Due to the importance of the distribution of the weight, this study used the analytic hierarchy process to determine the weight of each evaluation index; the hierarchical structure is shown in Figure 5.

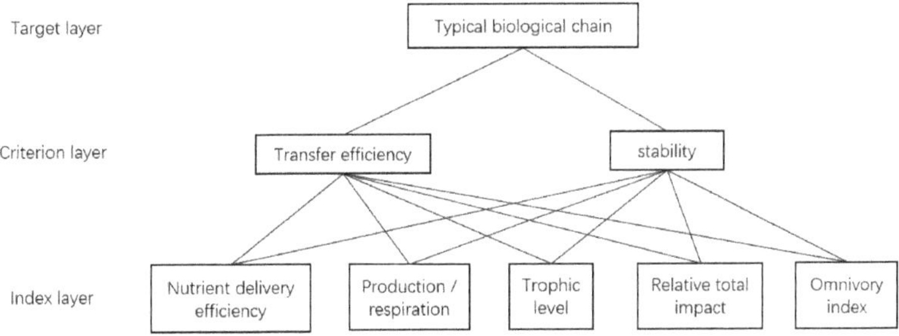

Figure 5. Hierarchical structure model.

$$\text{The judgement matrix } T = \begin{bmatrix} 1 & 4 & 0.33 & 0.143 & 0.25 \\ 0.25 & 1 & 0.25 & 0.143 & 0.2 \\ 3 & 4 & 1 & 0.5 & 3 \\ 7 & 7 & 2 & 1 & 8 \\ 4 & 5 & 0.33 & 0.125 & 1 \end{bmatrix}. \text{ The calculated maxi-}$$

mum characteristic value is $\lambda_{max} = 5.42$, the random consensus ratio is $0.0944 < 0.1$, and the consistency test shows that the weight distribution is reasonable, Therefore, the weight of each evaluation index of a typical biological chain is $A = (0.0712, 0.0388, 0.2224, 0.0534, 0.1336)$.

⑤ The comprehensive evaluation results were analyzed.

$$B = AR = (0.2582, 0.2825, 0.2839, 0.2950, 0.2798, 0.2581, 0.2940, 0.2368, 0.2663)$$

It can be seen that the typical biological chain containing the Zoop group as the key species is ④, equal to phytoplankton (Phyt)–zooplankton (Zoop)–macrocrustaceans (Macc)–other piscivorous (OthP).

According to the same calculation principle, the typical algal control biological chain containing OthB and OthF as the key species does not exist.

3.2. Analysis of Typical Algal Control Biological Chain and Nutritional Structure and Function Response

Through modeling, it was found that the production of primary producers in the Lianshi Lake ecosystem was 10,661.7 t/km^2/year, of which the amount of food consumed was 3043 t/km^2/year, and the remaining 71.5% was not consumed by other predators but flowed into the debris, entered recirculation, and became deposited through mineralization. Therefore, the effective removal of plant organisms from the water body plays an important role in controlling the total amount of nutrients, dissolved oxygen and transparency of the water body.

In the Lianshi Lake ecosystem, there are 16 biological chains with Phyt as the primary producer. Among them, the number of biological chains where Phyt is directly grazed by Zoop is the largest; it has nine chains. The transfer efficiency of Phyt is 0.308. The transfer efficiency of Zoop reached 0.606. It can be seen that the types of biological chains that impose significant biological density constraints on algae are mainly Zoop predation on Phyt. Combined with the screening of typical biological chains, a typical biological chain of algal control in the Lianshi Lake ecosystem is phytoplankton (Phyt)–zooplankton (Zoop)–macrocrustaceans (Macc)–other piscivorous (OthP).

This study improved the efficiency of nutrient delivery for primary producers by simulating an increase in biomass of Zoop and Macc. Under the guidance of biological manipulation theory, the key factor in the regulation of the algal control food chain is Zoop. Through the introduction of *Daphnia magna*, the food chain is opened up through insects eating algae and fish-eating insects. A symbiotic system of "*Daphnia magna*–underwater forest–aquatic animals–microbes" was constructed, and the "grass-type clean water state" self-purification system was restored.

The proliferation of ecological capacity continuously increases the biomass of the target species. By observing the changes in other function groups such as the eaten organisms in the model, when the ecological nutrition transfer efficiency of a function group in the model $EE > 1$, the model will become unbalanced; the biomass value of the released species before the imbalance of the model is the ecological capacity. The biomass of Zoop and Macc have been increased to 1.5, 2, 3 and 4 times, respectively. It can be seen from Table 7 that, within the ecological capacity of Zoop and Macc, with increases in the biomasses of both, the ecological nutrition transfer efficiency of Phyt gradually increased, from 0.308 (the current situation) to 0.458 (1.5 times), 0.607 (2 times) and 0.906 (3 times). The flow of material flowing into the second trophic level from primary producers also increased from 3043 (the current situation) to 4353 (1.5 times), 5663 (2 times) and 8283 t/km^2/year (3 times).

Table 7. Comparison table for overall characteristics of the Lianshi Lake ecosystem.

Index	Current State	1.5 Times	2 Times	3 Times	4 Times	Unit
Zoop biomass	7.85	11.775	15.7	23.55	31.4	t/km^2/year
Macc biomass	1.58	2.37	3.16	4.74	6.32	t/km^2/year
Phyt Eco-nutrition efficiency	0.308	0.458	0.607	0.906	1.204	-
SubM Eco-nutrition efficiency	0.186	0.186	0.186	0.186	0.186	-
The amount of material flowing into the second trophic level from primary producers	3043	4353	5663	8283	10904	t/km^2/year
The amount of debris flowing into the primary producer	7618	6308	4998	2378	−241.9	t/km^2/year
Total primary production/total respiratory volume (TPP/TR)	9.224	6.461	4.972	3.403	2.587	-
Finn's cycle index (FCI)	13.59%	14.33%	15.23%	17.51%	20.38%	-
Finn's average energy flow path length (FCL)	2.854	2.993	3.132	3.410	3.688	-
Connection coefficient (CI)	0.225	0.226	0.226	0.226	0.226	-
System omnivorous degree (SOI)	0.092	0.092	0.093	0.093	0.094	-

The amount of debris flowing into the primary producers for deposition continuously decreased, from 7618 (the current situation) to 6308 (1.5 times), 4998 (two times) and

2378 t/km^2/year (three times). The total primary production/total respiratory volume (TPP/TR) continued to decrease from 9.224 (the current situation) to 6.461 (1.5 times), 4.972 (two times) and 3.403 (three times). The Finn's circulation index (FCI) continued to rise from 13.59% (the current situation) to 14.33% (1.5 times), 15.23% (two times) and 17.51% (three times), whereas the Finn's average energy flow path length (FCL) rose from 2.854 (the current situation) to 2.993 (1.5 times), 3.132 (two times) and 3.410 (three times).

Based on the analysis of the nutritional structure of the Lianshi Lake ecosystem, the artificial introduction of Zoop (*Daphnia magna*) and Macc can increase the transfer efficiency and the maturity of the ecosystem. To a certain extent, it can solve the problem of excessive primary production.

4. Discussion

4.1. Development Characteristics of Lianshi Lake Ecosystem

The total primary production/total respiratory volume (TPP/TR) is an important evaluation index, which is close to 1 in mature ecosystems, far greater than 1 in developing ecosystems, and less than 1 in polluted ecosystems. The Finn's cycling index (FCI) is the ratio of the circulation flow to the total flow in the system, and the Finn's mean path length (FMPL) is the average length of each circulation through the food chain. The higher the ratio of material recycling, the longer the food chain through which the nutrient flows, and the FCI value of a mature ecosystem is close to 1. The current Lianshi Lake TPP/TR is 9.224, the FCI is 13.59%, and the FMPL is 2.854, indicating that the Lianshi Lake ecosystem is immature.

In the typical biological chain of algal control, with an increase in the biomass of Zoop (*Daphnia magna*) and Macc within the ecological capacity, the amount of phytoplankton as a primary producer flowing into the next trophic level gradually increases, the transfer efficiency gradually increases, the FCI and FMPL gradually increase, and the TPP/TR gradually approaches 1, which reduces the risk of lake eutrophication and increases ecosystem maturity to a certain extent.

Therefore, there may be two reasons for the low maturity of the Lianshi Lake ecosystem: first, the biomass of the function group that plays a key role in the Lianshi Lake ecosystem is much lower than the ecological capacity, resulting in insufficient driving force for the ecosystem to develop to a mature state. Second, the biodiversity of Lianshi Lake is low, and the flow of energy to higher levels is hindered. It is recommended to introduce Zoop (*Daphnia magna*) and indigenous herbivorous fishes to build a food chain in order to promote material and energy cycles.

4.2. Prospects of Ecopath Model in Lake Ecological Restoration

As early as 1975, Shapiro et al. [17] proposed the biological manipulation theory. Biomanipulation methods have been developed for nearly 50 years; there have been many reports in Western European countries, but this technology has not been widely promoted. This may be because biological manipulation methods involve complex biological networks, and too many factors are affected. Traditional research methods can only study the behavior of individual organisms in simple habitats or competitive environments, and there is very little research on the biological chain that significantly affects the regulation goals and ecosystem conditions. Therefore, it is generally believed that research on complex ecosystems must rely on the guidance of mathematical models or theories [42].

The Ecopath with Ecosim (EWE) model, also known as the ecological channel model, is an ecological model that can assess the true structure of the ecosystem and describe its energy flow and mass balance. The Ecopath model was originally created in 1984 [43]. After years of development, the Ecopath model has become a key tool for ecosystem research. Fuzzy comprehensive evaluation (FCE) is based on fuzzy mathematics [44], which can express the fuzzy relationship between various factors and solve the problem of ambiguity between multi-factor evaluations that cannot be solved by traditional methods, and it has been widely used in the field of policy evaluation and risk assessment. A typical biological chain of alga control combines the advantages of the Ecopath model and FCE method, and the results show that the biomass of phytoplankton in Lianshi Lake has been effectively controlled and that the ecosystem's maturity has been improved.

Ecopath is a powerful model but is mostly used to assess the condition of the ecosystem and provide scientific management and control solutions. In the future, the Ecopath model will continue to be developed and coupled with other models, such as pollutant diffusion models and ecotoxicology models, which will have important scientific research significance for exploring the restoration of lake ecosystems.

5. Conclusions

Based on the survey results for aquatic organisms in Lianshi Lake from 2018 to 2019, this study established, for the first time, a typical biological chain screening method with fuzzy comprehensive evaluation coupled with Ecopath. Among the organisms, Phyt is the primary producer, and the typical biological chain of alga control with Zoop as the key species is phytoplankton (Phyt)–zooplankton (Zoop)–macrocrustaceans (Macc)–other piscivorous (OthP).

In a typical biological chain with significant biological density constraints on algae, when the biomass of Zoop was increased from 7.85 (the current situation) to 23.55 t/km^2/year (three times) and Macc was increased from 1.58 (the current situation) to 4.74 t/km^2/year (three times), the results show that the ecological nutrition efficiency of Phyt increased from 0.308 (the current situation) to 0.906 (three times), the material flow into the second trophic level from primary producers increased from 3043 (the current situation) to 8283 t/km^2/year (three times), the amount of debris flowing into primary producers for sedimentation decreased from 7618 (the current situation) to 2378 t/km^2/year (three times), the total primary production/total respiratory volume (TPP/TR) decreased from 9.224 (the current situation) to 3.403 (three times), the Finn's cycle index (FCI) increased from 13.59% (the current situation) to 17.51% (three times), and the Finn's average energy flow path length (FCL) increased from 2.854 (the current situation) to 3.410 (three times). In the typical biological chain of alga control, the artificial release of Zoop (*Daphnia magna*) and Macc can improve the transfer efficiency of phytoplankton as primary producers to a certain extent, reduce the harm caused by eutrophication to lake ecosystems, and improve the maturity of the lake ecosystem.

Author Contributions: P.Z. and X.C. designed the research content; P.Z., H.L. and W.P. processed the data and analyzed the results. All the authors participated in the writing of the manuscript. Y.G. and X.C. revised the manuscript. All authors have read and agreed to the published version of the manuscript.

Funding: This research was supported by the Major Science and Technology Program for Water Pollution Control and Treatment (No.2018ZX07101005).

Acknowledgments: We thank Min Zhang for providing hydrobiological data and Su Wei for assistance with the fieldwork. We would like to thank Guanglei Qiu for his guidance on the translation of the manuscript.

Conflicts of Interest: The authors declare no conflict of interest.

References

1. Zhai, Y.; Xia, X.; Yang, G.; Lu, H.; Ma, G.; Wang, G.; Teng, Y.; Yuan, W.; Shrestha, S. Trend, seasonality and relationships of aquatic environmental quality indicators and implications: An experience from Songhua River, NE China. *Ecol. Eng.* **2020**, *145*, 105706. [CrossRef]
2. Wang, S.; Li, J.; Zhang, B.; Spyrakos, E.; Tyler, A.N.; Shen, Q.; Zhang, F.; Kuster, T.; Lehmann, M.K.; Wu, Y.; et al. Trophic state assessment of global inland waters using a MODIS-derived Forel-Ule index. *Remote Sens. Environ.* **2018**, *217*, 444–460. [CrossRef]
3. Zhao, Y.; Deng, X.; Zhan, J. Research progress on prevention and control strategies of lake eutrophication. *Environ. Sci. Technol.* **2010**, *33*, 92 98. (In Chinese)
4. Davis, K.; Anderson, M.A.; Yates, M.V. Distribution of indicator bacteria in Canyon Lake, California. *Water Res.* **2005**, *39*, 1277–1288. [CrossRef]

5. Razmi, A.M.; Barry, D.A.; Lemmin, U.; Bonvin, F.; Kohn, T.; Bakhtyar, R. Direct effects of dominant winds on residence and travel times in the wide and open lacustrine embayment: Vidy Bay (Lake Geneva, Switzerland). *Aquat. Sci.* **2014**, *76*, 59–71. [CrossRef]

6. Hipsey, M.R.; Antenucci, J.P.; Brookes, J. A generic, process-based model of microbial pollution in aquatic systems. *Water Resour. Res.* **2008**, *44*, 76–87. [CrossRef]

7. Su, W.; Wu, J.; Zhu, B.; Chen, K.; Peng, W.; Hu, B. Health Evaluation and Risk Factor Identification of Urban Lakes—A Case Study of Lianshi Lake. *Water* **2020**, *12*, 1428. [CrossRef]

8. Shan, B.; Wang, C.; Li, X. The formulation method and case study of river governance plan based on water quality objective management. *Acta Sci. Circumstantiae* **2015**, *35*, 2314–2323. (In Chinese) [CrossRef]

9. Birk, S.; Bonne, W.; Borja, A.; Brucet, S.; Courrat, A.; Poikane, S.; Solimini, A.; van de Bund, W.; Zampoukas, N.; Hering, D. Three hundred ways to assess Europe's surface waters: An almost complete overview of biological methods to implement the Water Framework Directive. *Ecol. Indic.* **2012**, *18*, 31–41. [CrossRef]

10. Jiang, Y.; Zhang, J.; Zhu, Y.; Du, Q.; Teng, Y.; Zhai, Y. Design and Optimization of a Fully-Penetrating Riverbank Filtration Well Scheme at a Fully-Penetrating River Based on Analytical Methods. *Water* **2019**, *11*, 418. [CrossRef]

11. Li, Y.; Zhang, Q.; Yao, J. Investigation of Residence and Travel Times in a Large Floodplain Lake with Complex Lake-River Interactions: Poyang Lake (China). *Water* **2015**, *7*, 1991–2012. [CrossRef]

12. Scheffer, M.; Straile, D.; van Nes, E.; Hosper, S.H. Climatic warming causes regime shifts in lake food webs. *Limnol. Oceanogr.* **2001**, *46*, 1780–1783. [CrossRef]

13. Lewonin, R.C. The meaning of stability: Diversity and stability in ecological systems. *Brookhaven Symp. Biol.* **1969**, *22*, 13–24.

14. Hrbáčke, J.; Dvořakova, M.; Kořínek, V.; Procházková, L. Demonstration of the effect of the fish stock on the species composition of zooplankton and the intensity of metabolism of the whole plankton association. *Verh. Int. Ver. Limnol.* **1961**, *14*, 192–195. [CrossRef]

15. Liu, J.; Xie, P. Uncover the mystery of the disappearance of cyanobacteria blooms in East Lake, Wuhan. *Resour. Environ. Yangtze Basin* **1999**, *3*, 85–92. (In Chinese)

16. Olin, M.; Rask, M.; Ruuhijärvi, J.; Keskitalo, J.; Horppila, J.; Tallberg, P.; Taponen, T.; Lehtovaara, A.; Sammalkorpi, I. Effects of Biomanipulation on Fish and Plankton Communities in Ten Eutrophic Lakes of Southern Finland. *Hydrobiologia* **2006**, *553*, 67–88. [CrossRef]

17. With, J.S.; Wright, D.I. Lake restoration by biomanipulation: Round Lake, Minnesota, the first two years. *Freshw. Biol.* **1984**, *14*, 371–383. [CrossRef]

18. He, H.; Han, Y.; Li, Q.; Jeppesen, E.; Li, K.; Yu, J.; Liu, Z. Crucian Carp (*Carassius carassius*) Strongly Affect C/N/P Stoichiometry of Suspended Particulate Matter in Shallow Warm Water Eutrophic Lakes. *Water* **2019**, *11*, 524. [CrossRef]

19. Rakshit, N.; Banerjee, A.; Mukherjee, J.; Chakrabarty, M.; Borrett, S.R.; Ray, S. Comparative study of food webs from two different time periods of Hooghly Matla estuarine system, India through network analysis. *Ecol. Model.* **2017**, *356*, 25–37. [CrossRef]

20. Vladimir, R.; Xueying, M.; Natallia, M.; Elena, S.; Dzmitry, L.; Andrei, M.; Erik, J.; Xiufeng, Z. Omnivorous Carp (*Carassius gibelio*) Increase Eutrophication in Part by Preventing Development of Large-Bodied Zooplankton and Submerged Macrophytes. *Water* **2021**, *13*, 1497.

21. Gu, J.; He, H.; Jin, H.; Yu, J.; Jeppesen, E.; Nairn, R.W.; Li, K. Synergistic negative effects of small-sized benthivorous fish and nitrogen loading on the growth of submerged macrophytes—Relevance for shallow lake restoration. *Sci. Total Environ.* **2018**, *610–611*, 1572–1580. [CrossRef]

22. He, H.; Hu, E.; Yu, J.; Luo, X.; Li, K.; Jeppesen, E.; Liu, Z. Does turbidity induced by *Carassius carassius* limit phytoplankton growth? A mesocosm study. *Environ. Sci. Pollut. Res.* **2016**, *24*, 5012–5018. [CrossRef] [PubMed]

23. Li, C.; Xian, Y.; Ye, C.; Wang, Y.; Wei, W.; Xi, H.; Zheng, B. Wetland ecosystem status and restoration using the Ecopath with Ecosim (EWE) model. *Sci. Total Environ.* **2019**, *658*, 305–314. [CrossRef] [PubMed]

24. Christensen, V.; Pauly, D. ECOPATH II—A software for balancing steady-state ecosystem models and calculating network characteristics. *Ecol. Model.* **1992**, *61*, 169–185. [CrossRef]

25. Christensen, V.; Walters, C.J.; Pauly, D. *Ecopath with Ecosim: A User's Guide*; Fisheries Centre, University of British Columbia: Penang, Malaysia, 2005; Available online: https://www.researchgate.net/publication/267193103_Ecopath_with_Ecosim_A_User\T1\textquoterights_Guide (accessed on 15 April 2021).

26. Huang, X.; Bing, X.; Zhang, X. Application of EwE model in evaluating fishery water ecosystem. *J. Hydroecology* **2011**, *32*, 125–129. (In Chinese) [CrossRef]

27. Song, B. Research on the Ecosystem Model of Taihu Lake Fishery and Environment. Ph.D. Thesis, East China Normal University, Shanghai, China. Available online: https://kns.cnki.net/kcms/detail/detail.aspx?FileName=2004087212.nh&DbName=CDFD2004 (accessed on 2 June 2021). (In Chinese).

28. Yu, J.; Liu, J.; Wang, L. Analysis of Structure and Function of Qiandao Lake Ecosystem Based on Ecopath Model. *Acta Hydrobiol. Sin.* **2021**, *45*, 308–317. (In Chinese)

29. Feng, D. Research on ECOPATH Model of Dianshan Lake Ecosystem Structure and Energy Flow Characteristics. Master's Thesis, East China Normal University, Shanghai, China. Available online: https://kns.cnki.net/kcms/detail/detail.aspx?FileName=1011131168.nh&DbName=CMFD2011 (accessed on 18 June 2021). (In Chinese).

30. Xian, Y.; Ye, C.; Li, C. Construction and Analysis of EWE Model of Lake Buffer Zone Wetland Ecosystem in Zhushan Bay. *Chin. J. Appl. Ecol.* **2016**, *27*, 2101–2110. (In Chinese) [CrossRef]

31. Park, R.A. A generalized model for simulating lake ecosystems. *Simulation* **1974**, *23*, 33–50. [CrossRef]
32. Yan, Y.; Liang, Y. Energy flow of benthic fauna in Biandan Pond. *Acta Ecol. Sin.* **2003**, *3*, 527–538. (In Chinese)
33. Halfon, E.; Schito, N.; Ulanowicz, R.E. Energy flow through the Lake Ontario food web: Conceptual model and an attempt at mass balance. *Ecol. Model.* **1996**, *86*, 1–36. [CrossRef]
34. Li, Y.; Liu, E.; Wang, H. Analysis of the Structure and Function of Taihu Lake Ecosystem Based on Ecopath Model. *Chin. J. Appl. Ecol.* **2014**, *25*, 2033–2040. (In Chinese) [CrossRef]
35. Song, G.; Hu, M.; Liu, Q. Using Stable Isotope Technology to Study the Feeding Habits and Nutritional Levels of Gillnet Catch in Qiandao Lake in Autumn. *J. Shanghai Ocean. Univ.* **2014**, *23*, 117–122. (In Chinese)
36. Liu, Q. Qiandao Lake Water Conservation Fishery and Its Impact on Lake Ecosystem. Ph.D. Thesis, East China Normal University, Shanghai, China. Available online: https://kns.cnki.net/kcms/detail/detail.aspx?FileName=2005086209.nh&DbName=CDFD2005 (accessed on 30 May 2021). (In Chinese).
37. Scavia, D. Documentation of CLEANX a generalized model for simulating the open-water ecosystems of lakes. *Simulation* **1974**, *2*, 51–56. [CrossRef]
38. Christensen, V.; Walters, C.; Ahrens, R. Databasedriven models of the world's Large Marine Ecosystems. *Ecol. Model.* **2009**, *9*, 1984–1996. [CrossRef]
39. Libralato, S.; Christensen, V.; Pauly, D. A method for identifying keystone species in food web models. *Ecol. Model.* **2006**, *195*, 153–171. [CrossRef]
40. Valls, A.; Coll, M.; Christensen, V. Keystone species: Toward an operational concept for marine biodiversity conservation. *Ecol. Monogr.* **2015**, *85*, 29–47. [CrossRef]
41. Power, M.E.; Tilman, D.; Estes, J.A.; Menge, B.A.; Bond, W.J. Challenges in the Quest for Keystones. *BioScience* **1996**, *46*, 609–620. [CrossRef]
42. Jiang, X.H.; Zhang, D.Y. Discussion on the functions of model in ecology. *Resour. Ecol. Environ. Net. Res.* **1992**, *2*, 30–34.
43. Polovina, J.J. Model of a coral reef ecosystem. *Coral Reefs.* **1984**, *3*, 1–11. [CrossRef]
44. Chen, T.; Shen, D.; Jin, Y.; Li, H.; Yu, Z.; Feng, H.; Long, Y.; Yin, J. Comprehensive evaluation of environ-economic benefits of anaerobic digestion technology in an integrated food waste-based methane plant using a fuzzy mathematical model. *Appl. Energy* **2017**, *208*, 666–677. [CrossRef]

Article

Distribution, Formation and Human Health Risk of Fluorine in Groundwater in Songnen Plain, NE China

Jianwei Wang [1], Nengzhan Zheng [1], Hong Liu [1], Xinyi Cao [1], Yanguo Teng [1,2,*] and Yuanzheng Zhai [1,*]

[1] College of Water Sciences, Beijing Normal University, Beijing 100875, China; wang2017bnu@163.com (J.W.); zhengnz07@163.com (N.Z.); liuh6015@163.com (H.L.); 18844120690@163.com (X.C.)

[2] Engineering Research Center of Groundwater Pollution Control and Remediation Ministry of Education, Beijing 100875, China

* Correspondence: teng1974@163.com (Y.T.); diszyz@163.com (Y.Z.)

Abstract: Songnen Plain is one of the three great plains in northeast China with abundant groundwater resources. The continuous population growth and the rapid development of agriculture and economy in China has caused a series of environmental problems in the plain, such as endemic diseases caused by the accumulation of harmful substances in drinking water. This paper conducts a systematic investigation of fluorine in the groundwater of Songnen Plain. The results showed that fluorine was widespread in the groundwater of the plain in the concentration range of BDL–8.54 mg·L^{-1}, at a mean value of 0.63 mg·L^{-1} and detectable at a rate of 85.91%. The highest concentrations of fluorine were found in central and southwest areas of the plain. The concentration exceeded the guideline values for fluorine in drinking water and may have varying degrees of adverse effects on adults, and especially children, in the study area. The fluorine in groundwater mainly came from the dissolution of fluorite and other fluorine-containing minerals, and the concentrations and distribution of fluorine were affected by cation exchange, groundwater flow field and hydrochemical indexes (pH, TDS and HCO$_3^-$). The study provides scientific basis for the investigation, evaluation and prevention of endemic diseases caused by fluorine.

Keywords: Songnen Plain; groundwater; fluorine; distribution; formation; human health risk

Citation: Wang, J.; Zheng, N.; Liu, H.; Cao, X.; Teng, Y.; Zhai, Y. Distribution, Formation and Human Health Risk of Fluorine in Groundwater in Songnen Plain, NE China. *Water* **2021**, *13*, 3236. https://doi.org/10.3390/w13223236

Academic Editor: Fausto Grassa

Received: 16 September 2021
Accepted: 6 November 2021
Published: 15 November 2021

Publisher's Note: MDPI stays neutral with regard to jurisdictional claims in published maps and institutional affiliations.

1. Introduction

Water is an important natural resource to guarantee normal human life and socioeconomic development. As one of the main components of water resources, groundwater is exerting a more and more significant influence on society, and groundwater is also a prerequisite for the development of other resources, especially in arid and semi-arid areas [1]. However, the accumulation of harmful elements caused by geological causes or human activities not only harms the water environment, but also seriously affects the safety of drinking water [2,3]. Fluorine is an essential element of the human body and moderate intake (0.5–1.5 mg·L^{-1}) is beneficial to human health, according to the WHO guidelines [4–6]. However, excessive fluoride, after long-term consumption may lead to fluorosis, and it is also a serious problem for the world's geological environments [2,7]. Studies have shown that over 260 million people are at risk of fluorosis all over the world, reported in locations such as America, Argentina, China, Mexico, and India [8–13]. Therefore, the research on high fluorine groundwater have gradually become a research focus.

Drinking water is the primary route by which fluoride enters the human body [14]. The National Health Commission of the PRC [15] and the WHO [16] guidelines have set values for fluorine in drinking water at 1.0 and 1.5 mg·L^{-1}, respectively. Fluorine is widespread in natural minerals, such as flourite, cryolite, fluorapatite, etc. [8,9,17]. Studies have shown that the main reason for the formation of high-fluorine groundwater in many regions of the world is dissolution of fluorine-bearing minerals [18–20]. However, the distribution, formation and risk of high-fluorine groundwater in Songnen Plain, China have not been systematically studied.

Songnen Plain is one of the largest and most fertile plains in China, and an important agricultural production base [21,22]. The plain has a large groundwater aquifer system with multiple aquifers and abundant groundwater resources [23]. With the continuous growth of population and the rapid development of agriculture and economy, the contradiction between supply and demand of water resources is becoming more and more prominent in the area [24], and has caused a series of environmental problems, such as metal pollution [25], nitrate pollution [26,27], and endemic diseases caused by excessive content of fluorine, iodine and arsenic [28,29].

In this paper, the high content of fluorine in the groundwater of Songnen Plain is investigated systematically. Through a series of processes such as field sampling, index determination, sample preservation, pretreatment, detection and data analysis, the concentration, distribution, formation and human health risk of fluorine in groundwater in Songnen Plain are revealed. This provides scientific basis for the investigation, evaluation, and prevention of endemic diseases caused by fluorine.

2. Materials and Methods

2.1. Study Area

Songnen Plain is in the northeastern part of China and located at longitude 121°21′–128°18′ and latitude 43°36′–49°26′. Songnen Plain is an important grain commodity production base and animal husbandry base of China. It covers an area of 103,200 km^2, and is placed in the Songhua River basin. The plain evolved from the Mesozoic-Cenozoic faulted basin and has accumulated over 8 km of Cretaceous terrestrial clastic deposits. Gravel, sand and loam are the main components of strata in the study area, and cohesive soil interlayers are locally distributed [1,26]. Analysis of the strata minerals revealed that fluorine–bearing minerals are rich in the central and southwest strata of the plain, mainly including flourite, apatite, cryolite, topaz, biotite, hornblende, tourmaline [26]. The regional groundwater resources are abundant and the largest groundwater system of the entire aquifer includes Neogene fissure–pore water, Cretaceous pore–fracture water, Quaternary pore water and Paleogene fissure-pore water. Irrigation and precipitation constitute the main sources of local groundwater recharge [30]. The shallow groundwater (depth is less than 50 m) are greatly affected by anthropogenic activities, complicated and changeable chemical composition. With the rapid development of agriculture and industry in northeastern China, groundwater exploitation in this region is expanding and, coupled with decreasing precipitation, the water table is declining and the groundwater environment in the study area changed greatly.

2.2. Sampling

In this paper, a comprehensive groundwater pollution survey was carried out in Songnen Plain from 2012 to 2014. Sampling time was concentrated in May to October each year, due to the cold winters in northeast China. Groundwater sample collection in the study area relied on local mechanized wells. A total of 2683 groundwater samples were collected; their locations are shown in Figure 1. Prior to groundwater collection, the original well water was pumped more than three times to flush the well's pump [3]. The sampling bottles (500 mL) were made of polyethylene plastic and were soaked in a 10%-sodium hydroxide solution for 3 h [11], then cleaned with deionized water and distilled water in turn, and finally dried at 60 °C for 5 h and stored in ziplock bags. Additionally, 2 mL of concentrated nitric acid (1:1) was added to the sampling bottle for measuring heavy metals [26]. Then, 2 mL of concentrated sulphuric acid (1:1) was added to the sample bottles for the measurement of Fe and Mn [1]. The sample bottles were washed three times with the corresponding water before each sampling. Each water sample collected was refrigerated at −4 °C, and handled within 48 h. Unstable parameters, such as water temperature (T), pH, electrical conductivity (EC) and also water-table depth were measured in situ.

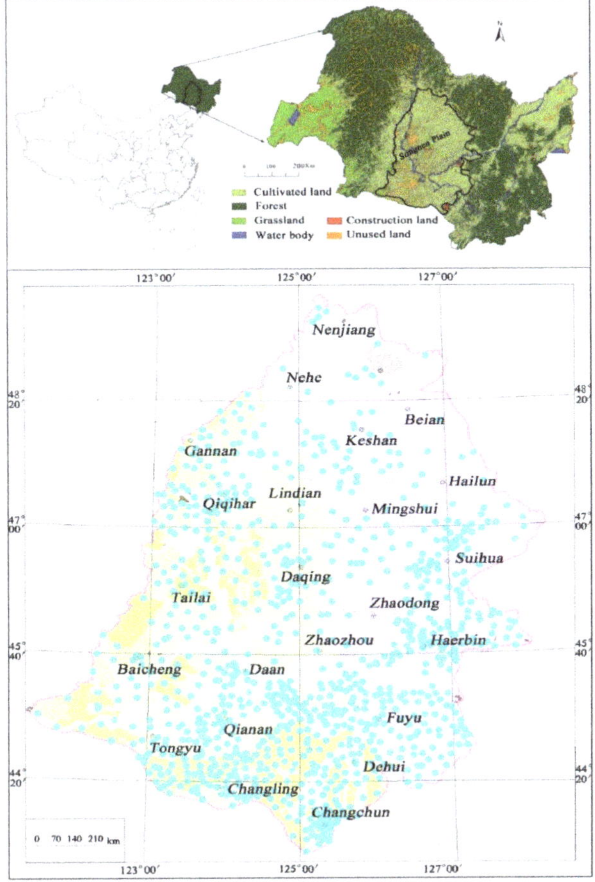

Figure 1. Location of sampling points in the study area.

2.3. Analysis Methods

2.3.1. Data Analysis Method

In this paper, a variety of data analyses are reported; a hydrochemical analysis, simulation methods, spatial analysis and mapping software are used herein toprocess the large volume of groundwater-sample testing data collected. The comprehensive parameters that can reflect groundwater characteristics and kriging method in statistics are selected as spatial interpolation method. Based on variogram theory and structural analysis, the comprehensive parameter zoning maps of groundwater at each layer are drawn with the help of ArcGIS and Surfer professional mapping software. The SPSS software was used to carry out descriptive statistical analysis on the main components of the sampled groundwater. SUPCRTBL and PHREEQC were used to calculate the saturation index (SI) of various rocks in combination with the latest database.

2.3.2. Instrumental Analysis

The instrumental analysis method used for the determinations of the water samples is referenced from a previous study [1,26]: the pH and redox potential were determined by dual-channel multi-parameter water quality analyzer (HQ40D, Field Case, cat. No:58258-00, HACH, Loveland, CO, USA); K, Ca, Na, Mg, Mnand Fe concentrations were measured by ICP-AES (IRIS Intrepid II XSP, Thermo Scientific, Waltham, MA, USA); the concentrations of Cl^- and NO_3^- in the water samples were measured by ion chromatograph (Dionex2500, Dionex, Sunnyvale, CA, USA).

2.3.3. Saturation Index

The saturation index (SI) is, itself, derived from the formula of a theoretical derivation and used to describe the equilibrium state of the water with respect to mineral phases therein and to determine the dissolution or precipitation of minerals in rock–water interactions [31,32]. The index was calculated by Equation (1):

$$SI = \log \left(\frac{IAP}{Ks} \right) \tag{1}$$

where IAP is the ion activity product of the solution and Ks is the solubility product of the mineral. Different SI values indicate different states of ion in solution: $SI > 0$ indicates oversaturation (precipitation), $SI = 0$ indicates equilibrium and $SI < 0$ indicates undersaturation (dissolution).

2.3.4. Human Health Risk Assessment

A human health risk assessment consists in determining potential adverse effects of a target pollutant. The health risk assessment model (RBCA) was created to assess non-carcinogenic risks. We have mainly referred to oral exposures to pollutants in this study because the mouth is considered the primary route thereof. The calculations of non-carcinogenic risk (hazard quotient, HQ) of directly consuming water resources were extrapolated from the oral reference dose (RfD), hazard index (HI) represents the total non-carcinogenic risk to humans when the ADI (average daily intake) was unavailable, as shown in Equation (2). HI > 1 indicates that the exposed individual was adversely affected.

$$HQ = \frac{EDI}{RfD} = \frac{CS \times IR \times EF \times ED}{BW \times AT \times RfD} \ and \ HI = \sum_{i=1}^{n} HQ_i \tag{2}$$

where CS (mg·L^{-1}) is the concentration of OPPs in the water; IR is the average daily water intake (1.5 and 0.7 L·d^{-1} for adults, children, respectively); EF stands for exposure frequency (365 d·y^{-1}). The EDs (exposure durations) for children and adults were 12 and 30 years, respectively; BWs (body weights) for children and adults were 10 and 60 kg, respectively [16]. AT represents average lifetime, and was 4380 and 1095 days for adults and children, respectively. RfD stands for the reference dose of the carcinogen consumed orally. The value of the RfD for fluoride was 0.04 mg·kg^{-1}·day^{-1} [33,34].

2.4. Quality Assurance/Quality Control (QA/QC)

In order to make sure the accuracy of the measurement results, blank samples and standard samples were taken from each batch during sampling. After analysis, the standard deviations ranged from 0.09% to 0.23% and the target pollutant was not detected in the blank samples, which conformed to the stated data processing standards [35,36]. In determining groundwater quality parameters, the recovery indicator was added before the water sample was processed; samples' recovery rates ranged from 84.4 to 95.7%, and the average was 89.4%; moreover, we compiled the standard curve for target objects and the results of the analysis show that the correlation coefficient of the linear equation was over 0.99; all conformed to quality assurance standards for the processing of groundwater [37,38].

3. Results and Discussion

3.1. Hydrochemical Parameters and Types

The concentrations of hydrochemical parameters are shown in Table 1. The main cations in groundwater in the study area were Ca and Na, the concentration of Ca ranged from 1.56 to 567.65 mg·L^{-1}, with an average of 95.04 mg·L^{-1} and the concentration of Na ranged from 4.51 to 1107.36 mg·L^{-1}, with an average of 74.45 mg·L^{-1}. According to the groundwater index detection values, the Piper diagram of groundwater hydrochemistry types in the study area was drawn in Figure 2. Generally speaking, the main groundwater chemical type in the study area was HCO_3–Ca type [1], accounting for 24.83% of the total water samples. Other main groundwater types include HCO_3–Ca·Mg type, HCO_3–Na·Ca type, HCO_3–Na·Mg·Ca type and HCO_3·Cl–Ca type, accounting for 19.43%, 16.78%, 13.92%, 12.11% and 9.56% of the total water samples, respectively.

Table 1. Concentrations of hydrochemical parameters and saturation indexes of minerals.

Parameters	Concentration		
	Minimum	Maximum	Mean
K^+ (mg·L^{-1})	0.85	234.18	8.55
Na^+ (mg·L^{-1})	4.51	1107.36	74.45
Ca^{2+} (mg·L^{-1})	1.56	567.65	95.04
Mg^{2+} (mg·L^{-1})	2.43	589.88	36.77
HCO_3- (mg·L^{-1})	11.61	1838.05	354.69
$SO_4{}^{2-}$ (mg·L^{-1})	0.19	1198.79	86.93
$Cl-$ (mg·L^{-1})	BDL	1831.56	113.45
NO_3- (mg·L^{-1})	BDL	1751.89	100.23
TH (g·L^{-1})	0.15	2.44	0.98
TDS (g·L^{-1})	0.58	6.17	1.46
pH	5.76	9.99	7.37
F^- (mg·L^{-1})	BDL	8.54	0.63
SI (Flourite)	−5.57	−0.48	−1.88
SI (Calcite)	−1.87	−0.10	−0.98
SI (Gypsum)	−6.01	3.11	−0.79
SI (Halite)	−7.16	−0.05	−3.08
SI (Dolomite)	−2.80	−0.35	−1.01

Unit: BDL = below detection limit. TH: total hardness; TDS: total dissolved solids.

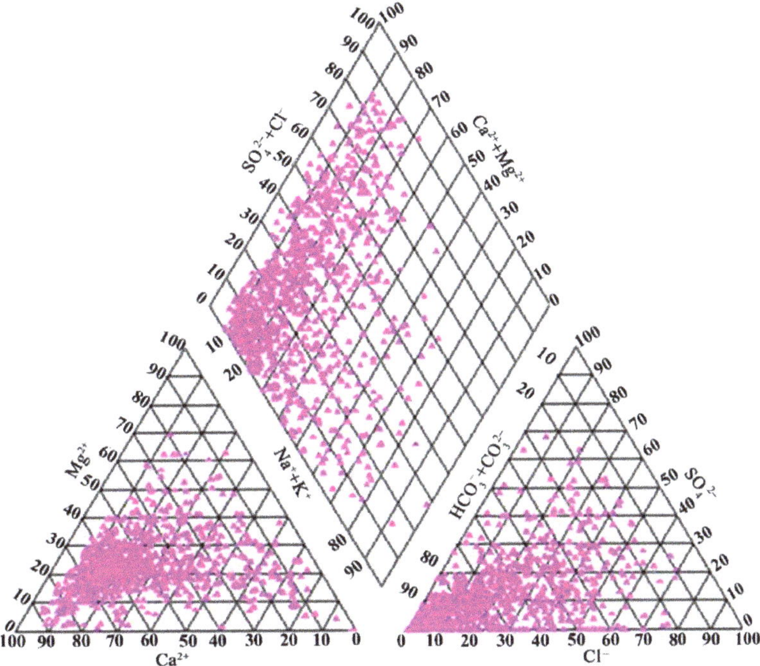

Figure 2. Piper diagram of groundwater samples.

3.2. Distributions of Fluorine

The concentration range of fluorine in groundwater was BDL–8.54 mg·L^{-1}, with a mean value of 0.63 mg·L^{-1}, and its detection rate was 85.91%. The content distributions of fluorine are showed in Figure 3; in the study area's groundwater the highest concentrations of fluorine (over 2 mg·L^{-1}) were found in the central and southwest areas of the Songnen Plain, such as Tongyu, Qianan, Baicheng, Lindian, Daqing, Zhaozhou, Zhaodong—and, the observed concentrations exceeded the maximum fluorine content (1.5 mg·L^{-1}) that is beneficial to human health [6] and within the guideline values set by the National Health Commission of the PRC (1.0 mg·L^{-1}) [15] and the WHO (1.5 mg·L^{-1}) [16].

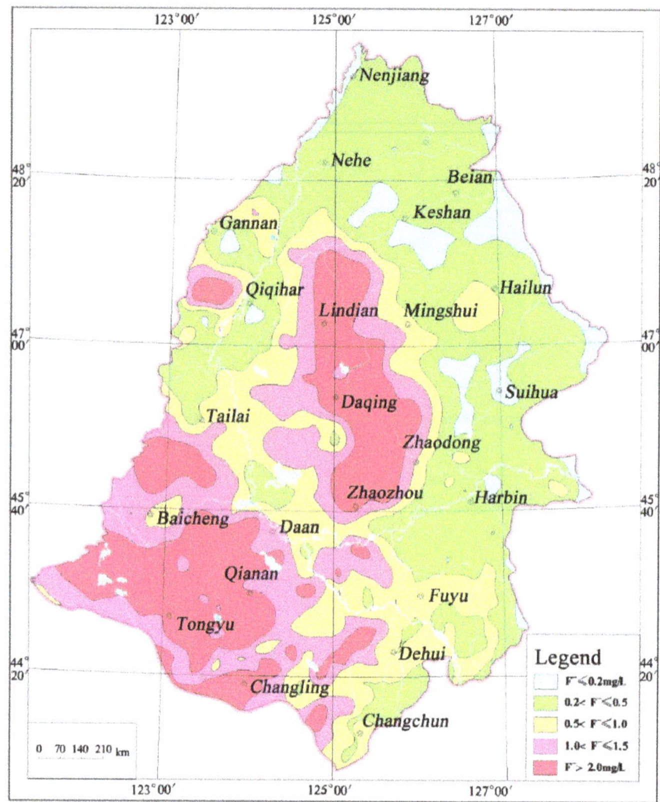

Figure 3. Distribution of fluorine in groundwater.

3.3. Formation and Influencing Factors of Fluorine in Groundwater

3.3.1. Dissolution and Precipitation of Minerals

Research has shown that the dissolution of fluorine-bearing minerals and the precipitation of calcium-bearing minerals are the main influencing factors of F$^-$ enrichment in groundwater [8,20,39,40]. Analysis of the strata minerals in the study area revealed an abundance of fluorine-bearing minerals in the central and southwest strata of the plain, mainly flourite, apatite, cryolite, topaz and hornblende [1,26]. The saturation indices of SI $_{fluorite}$ in almost all groundwater samples in the study area were less than zero, and there was a significant positive correlation between F$^-$ concentration and SI $_{fluorite}$ (Figure 4a), suggesting that the dissolution of fluorite is the main source of F$^-$ in the groundwater of these areas. According to other mineral saturation indices (SI $_{Calcite}$ < 0, SI $_{Halite}$ < 0, SI $_{Dolomite}$ < 0), calcite, halite and dolomite had not reached the saturation state and were easy to dissolve in reaction. SI $_{fluorite}$ had a logarithmic increase, with an increasing concentration of F^- in groundwater (Figure 4a). Where fluorite tended to saturate, the concentration of F^- reached the upper limit, indicating that the concentration of F^- was restricted by the equilibrium constant of fluorite (Ksp = $10^{-10.059}$, 22 °C). By comparing the concentration relationship between Ca^{2+} and F^- (Figure 4b), fluorine in the groundwater samples below the fluorite dissolution curve (dotted line in Figure 4b) mainly came from the dissolution of fluorite, and the fluorine in the groundwater samples above the dissolution curve came not only from the dissolution of fluorite, but also from other sources. The results show that the dissolution of fluorine-bearing minerals is main reason for the deposition of significant fluorine in the groundwater of Songnen Plain—similar to conditions found elsewhere in China [14,41] and the world, such as America [11], Mexico [8] and India [10].

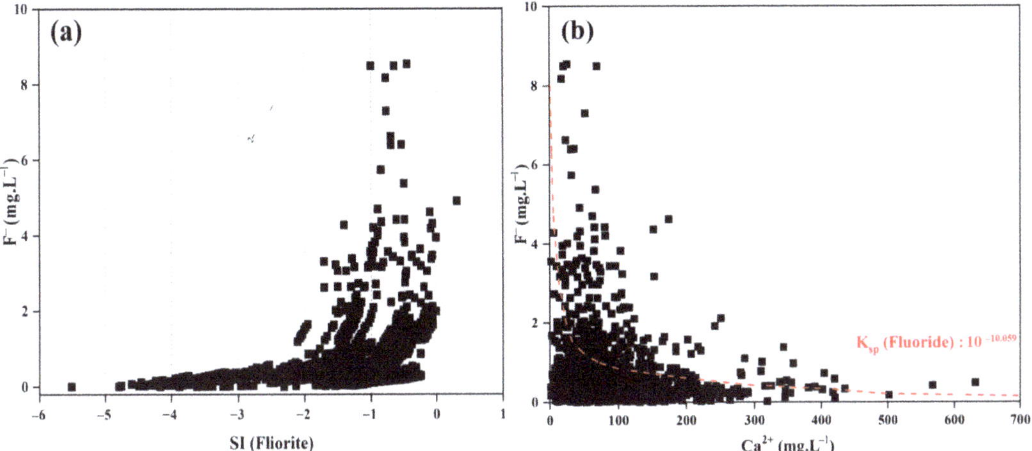

Figure 4. Relationship between F^- concentration and SI(Fluorite) (**a**); Ca^{2+} concentration (**b**).

3.3.2. Cation Exchange

The formation of groundwater hydrochemistry is often closely related to cation exchange [40], and cation exchange is the significant factor affecting the formation of fluorine in this study area. Cation exchange was confirmed with chloro-alkaline CAI 1 and CAI 2, and the indices were calculated by Equations (3) and (4).

$$CAI\ 1 = \frac{Cl^- - (Na^+ + K^+)}{Cl^-} \tag{3}$$

$$CAI\ 2 = \frac{Cl^- - (Na^+ + K^+)}{HCO_3^- + SO_4^{2-} + CO_3^{2-} + NO^{3-}} \tag{4}$$

where, if CAI 1 > 0 and CAI 2 > 0, it is indicated that the dissolved Na^+ and K^+ in the groundwater will exchange cations with the absorbed Mg^{2+} and Ca^{2+}. However, when less than zero, it is indicated that the dissolved Mg^{2+} and Ca^{2+} will exchange with the absorbed Na^+ and K^+. Moreover, the greater the absolute value, the stronger the cation exchange. Figure 5 shows the CAI 1 and CAI 2 of the groundwater samples. All values of CAI 2 do not exceed zero, and most values of CAI 1 were negative. This suggested that the cation exchange process of dissolved Mg^{2+} and Ca^{2+} exchanging cations with the absorbed Na^+ and K^+ was the driving process explaining local mineral concentrations, and it is also responsible for the decreased contents of Mg^{2+} and Ca^{2+} in the groundwater. This process may promote the hydrolysis of fluorite and other fluorine minerals (including apatite, cryolite, topaz, hornblende, tourmaline, etc.), thereby increasing the fluorine in groundwater. This indicates that the cation exchange process can affect the fluorine content in groundwater, as is consistent with previous studies [12,19,40,42].

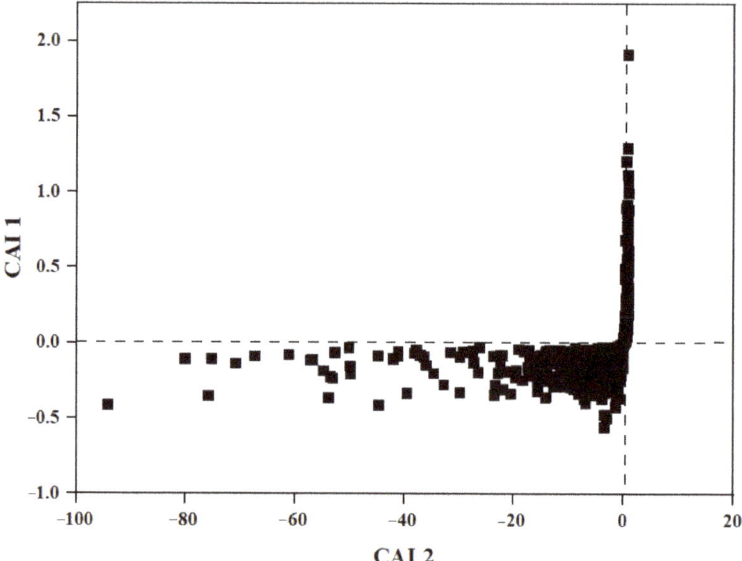

Figure 5. Relationship between CAI 1 and CAI 2.

3.3.3. Hydrochemical and Hydrological Influence Factors

Hydrochemical parameters are also one of the important factors affecting the content and distribution of fluorine [43,44]. It has been found that fluorine accumulates more easily in an alkaline environment [18], therefore, the areas with the highest pH values (pH > 8.5) in the study area (shown in Supplementary materials, Figure S1) also had the highest F^- contents. Correlation analysis between fluorine and other hydrochemical parameters in groundwater showed (Figure 6) that the concentrations of F^- in groundwater were positively correlated with the concentrations of TDS (total dissolved solids) and HCO_3^-. The results indicated that the concentrations and distribution characteristics of F^- in groundwater were closely related to the pH of the groundwater environment and the concentrations of TDS and HCO_3^-, again, as is consistent with previous studies [44]. Correlation analysis of fluorine and other high concentration pollutants (I^-, Mn^{2+}, Fe) in the groundwater showed (Figure S2) that fluorine was only weakly correlated with iodine (having similar properties), indicating that the pollutants in the groundwater had little influence on each other.

Hydrological conditions can partly affect the concentration and distribution of fluorine in groundwater [1,41]. The shallow groundwater system of Songnen Plain belongs to a larger groundwater catchment basin, and groundwater gather in its central low plain [16,29]. Therefore, the groundwater in the surrounding areas, especially the high-fluorine groundwater in the southwest area, will gradually migrate to the central region through the groundwater flow field, further increasing the fluorine concentrations in the already-high-fluorine groundwater in the central plain area.

3.4. Human Health Risk Assessment of Fluorine

Health risk assessments are mainly concerned with oral exposures; to that end, the risk assessment of non-carcinogens performed was based on the concentrations of fluorine in groundwater, and the results are shown in Figure 7. The sampling points (HQ $_{Children}$ > 1) accounted for 92.26% of the total samples, signaling that the groundwater fluorine concentration is high enough to have significant adverse effects on children in the study area. The sampling points (HQ $_{Adults}$ > 1) accounted for 32.18% of the total samples, indicating that the fluorine content was also high enough to adversely affect adults, though much less so than children. In addition, the districts where the fluorine in the groundwater showed the greatest potential influence on children and adults were roughly the same, and were concentrated in the southwest and central Songnen Plain, such as Tongyu Lindian and Daqing.

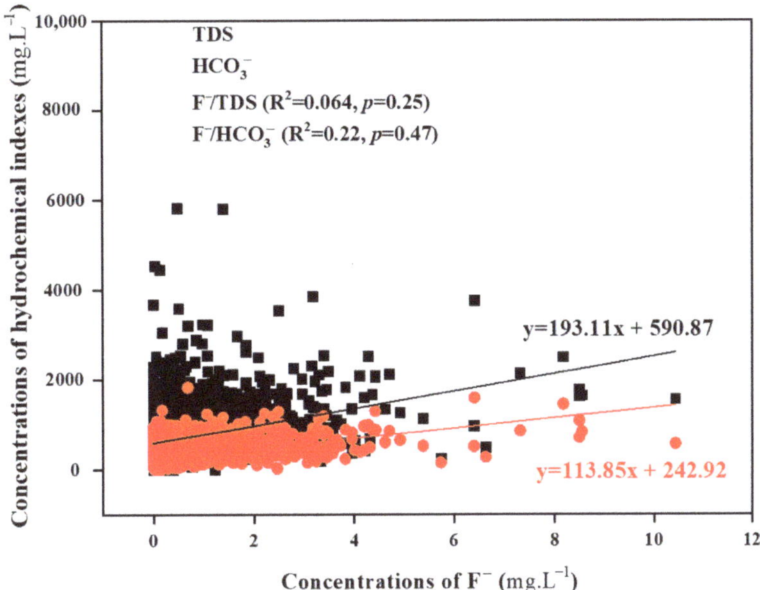

Figure 6. Correlation of F$^-$ and Hydrochemical Indices.

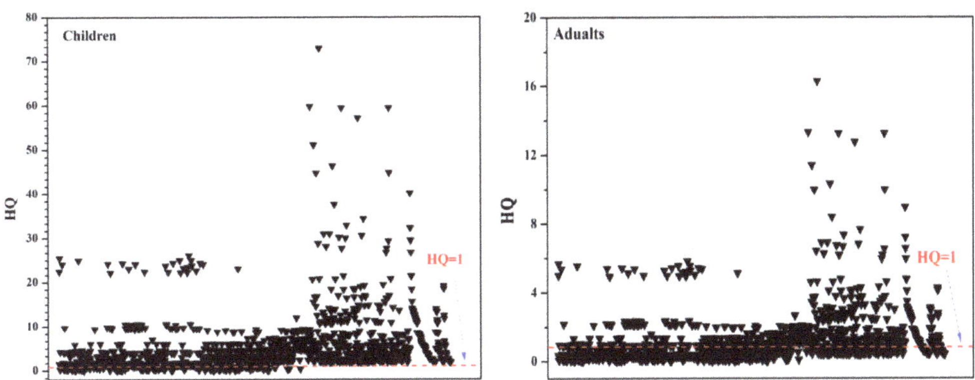

Figure 7. HQs of fluorine to children and Adults.

4. Conclusions

Fluorine is widespread in the groundwater of the Songnen Plain, at a concentration range of BDL–8.54 mg·L^{-1}, with a mean value of 0.63 mg·L^{-1} and detectable at a rate of 85.91%. The highest concentrations of fluorine (over 2 mg·L^{-1}) were found in the central and southwest areas of the plain. The concentrations there exceeded the guideline values for fluorine in drinking water set by both the National Health Commission of the PRC

$(1.0\ \text{mg·L}^{-1})$ and the WHO $(1.5\ \text{mg·L}^{-1})$, and represent varying degrees of adverse effect on adults, and especially children, in the study area. The fluorine in these groundwaters mainly came from the dissolution of fluorite and other fluorine-containing minerals in the study area; additionally, the concentrations and distribution of fluorine were shown to be affected by cation exchange, the groundwater flow field and hydrochemical indexes (pH, TDS and $HCO_3{}^-$). The study provides scientific basis for the investigation, evaluation and prevention of endemic diseases caused by groundwater fluorine.

Supplementary Materials: The following are available online at https://www.mdpi.com/article/10.3390/w13223236/s1, Figure S1: Distribution of pH in groundwater, Figure S2: Correlation of F^- and other pollutants.

Author Contributions: Data curation, X.C.; Formal analysis, H.L.; Funding acquisition, Y.T. and Y.Z.; Investigation, J.W.; Methodology, J.W.; Resources, Y.T.; Software, N.Z.; Writing—original draft, J.W.; Writing—review & editing, Y.T. and Y.Z. All authors have read and agreed to the published version of the manuscript.

Funding: This study was funded by National Natural Science Foundation of China (No. 41877355; 42077170); Major Science and Technology Program for Water Pollution Control and Treatment (2018ZX07101005-04); Beijing Advanced Innovation Program for Land Surface Science of China; the 111 Project of China (B18006).

Institutional Review Board Statement: Not applicable.

Informed Consent Statement: Not applicable.

Data Availability Statement: Not applicable.

Acknowledgments: Authors highly appreciate the detailed and constructive comments from the editors and reviewers to help us improve the quality and English expression of the paper.

Conflicts of Interest: The authors declare no conflict of interest.

References

1. Zuo, R.; Liu, X.; Yang, J.; Zhang, H.K.; Li, J.; Teng, Y.G.; Yue, W.F.; Wang, J.S. Distribution, origin and key influencing factors of fluoride groundwater in the coastal area, NE China. *Hum. Ecol. Risk Assess.* **2019**, *25*, 104–119. [CrossRef]
2. Rajveer, S.D.; Manan, S. A holistic study on fluoride-contaminated groundwater models and its widespread effects in healthcare and irrigation. *Environ. Sci. Pollut. Res.* **2021**, *28*, 60329–60345. [CrossRef]
3. Wang, J.W.; Zhang, C.X.; Liao, X.P.; Teng, Y.G.; Zhai, Y.Z.; Yue, W.F. Influence of surface-water irrigation on the distribution of organophosphorus pesticides in soil-water systems, Jianghan Plain, central China. *J. Environ. Manag.* **2021**, *281*, 111874. [CrossRef]
4. Dar, M.A.; Sankar, K.; Dar, I.A. Fluorine contamination in groundwater: A major challenge. *Environ. Monit. Assess.* **2011**, *173*, 955–968. [CrossRef]
5. Alekseyev, V.A.; Kochnova, L.N.; Cherkasova, E.V.; Tyutyunnik, O.A. Possible reasons for elevated fluorine concentrations in groundwaters of carbonate rocks. *Geochem. Int.* **2010**, *48*, 68–82. [CrossRef]
6. Reimann, C.; Banks, D. Setting Action Levels for Drinking Water: Are We Protecting Our Health or Our Economy (or Our Backs!)? *Sci. Total Environ.* **2004**, *332*, 13–21. [CrossRef] [PubMed]
7. Finch, E.G.; Tomkins, A.G. Fluorine and chlorine behaviour during progressive dehydration melting: Consequences for granite geochemistry and metallogeny. *J. Metamorph. Geol.* **2017**, *40*, 739–757. [CrossRef]
8. Sarah, E.; Smith, S.; Appold, M.S. Determination of fluorine concentrations in mineralizing fluids of the Hansonburg, New Mexico Ba-F-Pb district via SEM-EDS analysis of fluid inclusion decrepitates. *J. Geochem. Explor.* **2021**, *230*, 106861.
9. Li, X.; Zhang, W.; Qin, Y.; Ma, T.; Zhou, J.; Du, S. Fe–colloid cotransport through saturated porous media under different hydrochemical and hydrodynamic conditions. *Sci. Total Environ.* **2019**, *647*, 494–506. [CrossRef] [PubMed]
10. Deepali, M.; Malpe, D.B.; Subba, R.N. Geochemical Assessment of Fluoride Enriched groundwater and health implications from a part of Yavtmal District, India. *Hum. Eco. Risk Assess.* **2020**, *26*, 673–694.
11. Withanachchi, S.S.; Ghambashidze, G.; Kunchulia, I.; Urushadze, T.; Ploeger, A. Water Quality in Surface Water: A Preliminary Assessment of Heavy Metal Contamination of the Mashavera River, Georgia. *Int. J. Environ. Res. Public Health* **2018**, *15*, 621. [CrossRef]
12. Zabala, M.E.; Manzano, M.; Vives, L. Assessment of processes controlling the regional distribution of fluoride and arsenic in groundwater of the Pampeano Aquifer in the Del Azul Creek basin (Argentina). *J. Hydrol.* **2016**, *541*, 1067–1087. [CrossRef]
13. Amini, M.; Mueller, K.; Abbaspour, K.C. Statistical modeling of global geogenic fluoride contamination in groundwaters. *Environ. Sci. Technol.* **2008**, *42*, 3662–3668. [CrossRef]

14. Jia, H.; Qian, H.; Qu, W.G.; Zheng, L.; Feng, W.W.; Ren, W.H. Fluoride Occurrence and Human Health Risk in Drinking Water Wells from Southern Edge of Chinese Loess Plateau. *Int. J. Environ. Res. Public Health* **2019**, *16*, 1683. [CrossRef] [PubMed]
15. *Standards for Drinking Water Quality (GB5749-2006)*; National Health Commission of the People's Republic of China: Beijing, China, 2006.
16. World Health Organisation (WHO). *Guidelines for Drinking-Water Quality*, 4th ed.; WHO: Geneva, Switzerland, 2011.
17. Raju, N.J. Prevalence of fluorosis in the fluoride enriched groundwater in semi-arid parts of eastern India: Geochemistry and health implications. *Quat. Int.* **2016**, *443*, 265–278. [CrossRef]
18. Zhou, L.; Zheng, S.S.; Chen, D.; Yuan, X.K.; Lu, M.S.; Feng, Q.Y. Hydrogeochemistry of fluoride in shallow groundwater of the abandoned Yellow River delta, China. *Hydrol. Res.* **2021**, *52*, 572–584. [CrossRef]
19. Liu, J.H.; Wang, X.F.; Weidong Xu, W.D.; Mian Song, M. Hydrogeochemistry of Fluorine in Groundwater in Humid Mountainous Areas: A Case Study at Xingguo County, Southern China. *J. Chem.* **2021**, *2021*, 5567353. [CrossRef]
20. Saxena, V.; Ahmed, S. Dissolution of fluoride in groundwater: A water-rock interaction study. *Environ. Geol.* **2001**, *40*, 1084–1087.
21. Zhai, Y.; Zheng, F.; Zhao, X.; Xia, X.; Teng, Y. Identification of hydrochemical genesis and screening of typical groundwater pollutants impacting human health: A case study in Northeast China. *Environ. Pollut.* **2019**, *252*, 1202–1215. [CrossRef]
22. Su, X.; Wang, H.; Zhang, Y. Health risk assessment of nitrate contamination in groundwater: A case study of an agricultural area in Northeast China. *Water Resour. Manag.* **2013**, *27*, 3025–3034. [CrossRef]
23. Yin, W.; Teng, Y.; Zhai, Y.; Hu, L.; Zhao, X.; Zhang, M. Suitability for developing riverside groundwater sources along Songhua River, Northeast China. *Hum. Ecol. Risk Assess.* **2018**, *24*, 2088–2100. [CrossRef]
24. Jiang, L.; Guo, S.; Wang, G.; Kan, S.; Jiang, H. Changes in agricultural land requiremen ts for food provision in China 2003–2011: A comparison between urban and rural residents. *Sci. Total Environ.* **2020**, *725*, 138293. [CrossRef] [PubMed]
25. Zhai, Y.; Ma, T.; Zhou, J.; Li, X.; Liu, D.; Wang, Z.; Qin, Y.; Du, Q. Impacts of leachate of landfill on the groundwater hydrochemistry and size distributions and heavy metal components of colloids: A case study in NE China. *Environ. Sci. Pollut. Res.* **2019**, *26*, 5713–5723. [CrossRef]
26. Zhai, Y.; Xia, X.; Yang, G.; Lu, H.; Ma, G.; Wang, G.; Teng, Y.; Yuan, W.; Shrestha, S. Trend, seasonality and relationships of aquatic environmental quality indicators and implications: An experience from Songhua River, NE China. *Ecol. Eng.* **2020**, *145*, 105706. [CrossRef]
27. Ji, X.; Xie, R.; Hao, Y.; Lu, J. Quantitative identification of nitrate pollution sources and uncertainty analysis based on dual isotope approach in an agricultural watershed. *Environ. Pollut.* **2017**, *229*, 586–594. [CrossRef] [PubMed]
28. Zhang, B.; Song, X.; Zhang, Y.; Han, D.; Tang, C.; Yu, Y.; Ma, Y. Hydrochemical characteristics and water quality assessment of surface water and groundwater in Songnen plain, Northeast China. *Water Res.* **2012**, *46*, 2737–2748. [CrossRef]
29. Zhu, W. *Research on the Development Trend of Shallow Groundwater Quality in Songnen Plain, NE China*; Jilin University: Changchun, China, 2011.
30. Zhang, G.; Deng, W.; He, Y.; Ramsis, S. Hydrochemical characteristics and evolution laws of groundwater in Songnen Plain, Northeast China. *Adv. Water Sci.* **2006**, *17*, 20–28.
31. Qian, C.; Wu, X.; Mu, W.P.; Fu, R.Z.; Zhu, G.; Wang, Z.R.; Wang, D.D. Hydrogeochemical characterization and suitability assessment of groundwater in an agro-pastoral area, Ordos Basin, NW China. *Environ. Earth Sci.* **2016**, *75*, 1356. [CrossRef]
32. Jabal, M.S.A.; Abustan, I.; Rozaimy, M.R.; Al-Najar, H. Fluoride enrichment in groundwater of semi-arid urban area: Khan Younis City, southern Gaza Strip (Palestine). *J. Afr. Earth Sci.* **2014**, *100*, 259–266. [CrossRef]
33. Li, P.; He, X.; Li, Y.; Xiang, G. Occurrence and Health Implication of Fluoride in Groundwater of Loess Aquifer in the Chinese Loess Plateau: A Case Study of Tongchuan, Northwest China. *Expo. Health* **2019**, *11*, 95–107. [CrossRef]
34. Qasemi, M.; Afsharnia, M.; Zarei, A.; Farhang, M.; Allahdadi, M. Non-Carcinogenic Risk Assessment to Human Health Due to Intake of Fluoride in the Groundwater in Rural Areas of Gonabad and Bajestan, Iran: A Case Study. *Hum. Ecol. Risk Assess* **2019**, *25*, 1222–1233. [CrossRef]
35. Katarzyna, O.; Iwona, S.; Joanna, S. Impact of Catchment Area Activities on Water Quality in Small Retention Reservoirs. In *E3S Web of Conferences*; EDP Sciences: Paris, France, 2018; Volume 30, p. 01013.
36. Xi, B.D.; Huo, S.L.; Chen, Q.; Chen, Y.Q.; Zan, F.Y.; Xia, X.F. U.S Water Quality Standard System and Its Revelation for China. *Enuivon. Sci. Technol.* **2011**, *34*, 100–103, 120.
37. Marek, G.; Krystyna, Z.; Honorata, D.; Anna, S. Weed Infestation and Yielding of Potato Under Conditions of Varied Use of Herbicides and Bio-Stimulants. *J. Ecol. Eng.* **2018**, *19*, 191–196.
38. Kutseva, N.K.; Kartashova, A.V.; Tchamaev, A.W. Standard and Methodological Provision of the Quality Control of Water. *J. Anal. Chem.* **2005**, *60*, 788–795. [CrossRef]
39. Wu, J.; Li, J.; Teng, Y.; Chen, H.; Wang, Y. A partition computing-based positive matrix factorization (PC-PMF) approach for the source apportionment of agricultural soil heavy metal contents and associated health risks. *J. Hazard. Mater.* **2019**, *388*, 121766. [CrossRef]
40. Wu, C.; Wu, X.; Qian, C.; Zhu, G. Hydrogeochemistry and groundwater quality assessment of high fluoride levels in the Yanchi endorheic region, northwest China. *Appl. Geochem.* **2018**, *98*, 404–417. [CrossRef]
41. Li, Q.; Tao, H.F.; Aihemaitia, M.; Jiang, Y.W.; Su, Y.P.; Yang, W.X. Spatial distribution characteristics and enrichment factors of high-fluorine groundwater in the Kuitun River basin of Xinjiang Uygur Autonomous Region in China. *Desalination Water Treat.* **2021**, *223*, 208–217. [CrossRef]

42. Rao, N.S.; Rao, P.S.; Dinakar, A.; Rao, P.V.N. Fluoride occurrence in the groundwater in a coastal region of Andhra Pradesh, India. *Appl. Water Sci.* **2017**, *7*, 1467–1478. [CrossRef]
43. Wang, Z.; Guo, H.M.; Xing, S.P.; Liu, H.Y. Hydrogeochemical and geothermal controls on the formation of high fluoride groundwater. *J. Hydrol.* **2021**, *598*, 126372. [CrossRef]
44. Saxena, V.K.; Ahmed, S. Inferring the chemical parameters for the dissolution of fluoride in groundwater. *Environ. Geol.* **2003**, *43*, 731–736. [CrossRef]

water

MDPI

Article

Evaluation of Groundwater Suitability for Irrigation and Drinking Purposes in an Agricultural Region of the North China Plain

Haipeng Guo [1,2], Muzi Li [1,2,*], Lu Wang [1,2], Yunlong Wang [1,2], Xisheng Zang [1,2], Xiaobing Zhao [1,2], Haigang Wang [1,2] and Juyan Zhu [1,2]

[1] Hebei Cangzhou Groundwater and Land Subsidence National Observation and Research Station, Cangzhou 061000, China; pengfei7971@sohu.com (H.G.); esperanzall@163.com (L.W.); wangyunlong@mail.cgs.gov.cn (Y.W.); zangxisheng@mail.cgs.gov.cn (X.Z.); zhaoxiaobing@mail.cgs.gov.cn (X.Z.); wanghaigang@mail.cgs.gov.cn (H.W.); zhujuyan@mail.cgs.gov.cn (J.Z.)
[2] China Institute of Geo-Environment Monitoring, Beijing 100081, China
* Correspondence: limzsky@163.com; Tel.: +86-010-15210930156

Abstract: Groundwater is an irreplaceable resource for irrigation and drinking in the North China Plain, and the quality of groundwater is of great importance to human health and social development. In this study, using the information from 59 groups of groundwater samples, groundwater quality conditions for irrigation and drinking purposes in an agricultural region of the North China Plain were analyzed. The groundwater belongs to a Quaternary loose rock pore water aquifer. The depths of shallow groundwater wells are 20–150 m below the surface, while the depths of deep groundwater wells are 150–650 m. The sodium adsorption ratio (SAR), sodium percentage (%Na), residual sodium carbonate (RSC), magnesium hazard (MH), permotic index (PI) and electrical conductivity (EC) were selected as indexes to evaluate the shallow groundwater suitability for irrigation. What's more, the deep groundwater suitability for drinking was assessed and the human health risk of excessive chemicals in groundwater was studied. Results revealed that SAR, Na% and RSC indexes indicated the applicability of shallow groundwater for agricultural irrigation in the study area. We found 57.1% of the shallow groundwater samples were located in high salinity with a low sodium hazard zone. The concentrations of fluorine (F^-) in 79.0% of the deep groundwater samples and iodine (I^-) in 21.1% of the deep groundwater samples exceeded the permissible limits, respectively. The total hazard quotient (HQ) values of fluorine in over half of the deep groundwater samples exceeded the safety limits, and the health risk degree was ranked from high to low as children, adult females and adult males. In addition to natural factors, the soil layer compression caused by groundwater over-exploitation increased the fluorine concentration in groundwater. Effective measures are needed to reduce the fluorine content of the groundwater of the study area.

Keywords: groundwater quality; groundwater hydrochemistry; drinking suitability; irrigation suitability; health risk assessment

Citation: Guo, H.; Li, M.; Wang, L.; Wang, Y.; Zang, X.; Zhao, X.; Wang, H.; Zhu, J. Evaluation of Groundwater Suitability for Irrigation and Drinking Purposes in an Agricultural Region of the North China Plain. *Water* **2021**, *13*, 3426. https://doi.org/10.3390/w13233426

Academic Editors: Yuanzheng Zhai, Jin Wu and Huaqing Wang

Received: 2 November 2021
Accepted: 1 December 2021
Published: 3 December 2021

Publisher's Note: MDPI stays neutral with regard to jurisdictional claims in published maps and institutional affiliations.

1. Introduction

With the development of society, groundwater plays an increasingly important role in agricultural irrigation and domestic drinking [1,2]. Access to high-quality groundwater is indispensable to human health, agricultural irrigation and sustainable social development [3,4]. The shallow groundwater is often developed for irrigation in agricultural areas with the advantages of a low well-forming cost and being easy to obtain. In recent decades, groundwater quality and hydrochemical characteristics have been under the increased influence of pollution in an area of intense agricultural activities [5]. The pollutants pass through the soil and unsaturated zones and penetrate the aquifer, causing groundwater quality deterioration [6]. On the other side, the quality of groundwater directly affects the soil permeability, soil fertility and crop production when used for irrigation because of the ions exchange reaction between groundwater and soil [7–10]. Sodium concentration in groundwater is important because sodium reacts with soil to reduce its permeability and then the hydraulic conductivity declines, causing soil with poor internal drainage [11,12]. In general, the type of sodium-enriched soil will not support plant growth. The evaluation of groundwater suitability for irrigation has been studied by many researchers, which is mainly based on the important indicator values of total salinity, sodium and other related ions in irrigation water assessment [13,14]. According to the previous study, different indicators reflect different water quality results for irrigation. Although irrigation water quality is confirmed to be good by some indicators, the evaluation results by other indicators may be inappropriate [7,15–17].

In addition to irrigation, another important function of groundwater is domestic drinking, with over a third of the world population using groundwater as a drinking water resource [18,19]. Compared with shallow groundwater, deep groundwater is less contaminated by human activities and can better play a drinking function. The quality and safety conditions of drinking water have been a public concern all over the world especially in developing countries, where many kinds of diseases are directly associated with unsanitary conditions in drinking water [3,20–22]. Therefore, the evaluation of groundwater quality for drinking is significant for health [23,24].

Contaminants in groundwater can directly enter the human body by oral, inhalation, skin contact or indirectly accumulate in the body through the food chain [2,25]. The problem of contaminants intake from groundwater are more serious in agricultural irrigation areas and rural areas, as residents tend to drink groundwater directly through simple filtration measures. Given the adverse effects of such groundwater contamination on human health, closer links between pollutant concentrations and related health effects are needed to implement more effective risk assessment and mitigation measures [26,27]. Since the 1990s, health risk assessment (HRA) has become a hot topic with a crucial guiding function in determining whether contaminants pose adverse risks to health and whether groundwater can be drunk directly [28,29]. In recent years, human health risk assessment has been developed and widely used in many countries to determine the adverse effects of chemicals taken from groundwater in different populations [30–34].

Fluoride contamination in groundwater has been recognized as a widespread international problem, which influences millions of people in many regions [35–38]. Drinking groundwater with a high concentration of fluoride may endanger human health and cause fluorosis or other diseases. On the other hand, the use of groundwater with a high concentration of fluoride for irrigation makes the aeration zone and shallow groundwater contaminated. Food, vegetables and fruits irrigated by high fluoride groundwater enter the human body through the food chain, which may cause food-type poisoning [39]. In recent years, many scholars have assessed the health risk of fluoride in groundwater exposure to people of different age groups [35,36,40,41].

The North China Plain is a significant agricultural area in China, and also a typical water shortage area. Groundwater is used as the main resource for irrigation and drinking in this area. Meanwhile, groundwater with high fluoride concentrations exists in some areas in central and eastern of the North China Plain, which threatens the health of residents. At present, the contradiction between supply and demand of groundwater resources is prominent in this area, and the shortage and pollution problems of groundwater are increasingly serious [39,42,43]. Thus, a comprehensive and detailed evaluation of groundwater suitability for agricultural irrigation and drinking is important for groundwater scientific management and sustainable utilization, which is urgently needed in the North China Plain [44].

This paper selected a typical agricultural area in the central region of the North China Plain, where groundwater played an increasingly important role in recent years. In the study area, groundwater accounts for 65% of the total available water resources, so the quality of groundwater has a significant effect on irrigation and domestic drinking [45,46].

The investigation findings showed that the groundwater was over-exploited in the study area, and some kinds of chemicals in groundwater exceeded the standard levels, especially fluorine. However, the groundwater suitability for drinking and agriculture in this region has not been well studied and there are fewer relevant references can be found. Thus, more detailed research work on this topic is necessary. In this study, the hydrochemical characteristics of the groundwater were analyzed based on 59 groundwater samples collected in the field. According to the different practical uses of shallow and deep groundwater, the irrigation suitability of shallow groundwater and drinking suitability of deep groundwater were evaluated separately and the health risk of major excessive ions to different human groups was studied. Multiple evaluation methods (irrigation water suitability indexes, groundwater quality index, groundwater quality standards and health risk model) were selected to reflect the groundwater suitability situation more comprehensively and objectively, and to find typical indicators affecting the groundwater suitability. The findings can be used in the formulation of new policies and strategies for groundwater quality management in the North China Plain.

2. Materials and Methods

2.1. Study Area

The water samples used in this paper were taken from Bazhou irrigation district in the central North China Plain. The study region extends between E 116°15′–116°55′ longitudes and N 38°59′–39°13′ latitudes, covering a total area of about 780 km^2 (Figure 1). The study region is located in the alluvial plain of lower reaches of the Haihe River basin, with four seasonal rivers including Hongjiang River, Xiongguba New River, Mangniu River and Zhongting River, as well as more than 50 large drainage and irrigation canals. The terrain slopes from northwest to southeast in this area, and the ground elevation slowly drops from 11.1 m to 2.1 m. The climate is suitable for the growth of a variety of crops, with an annual average temperature of 11.5 °C. The annual average precipitation is 543.6 mm and the mean annual potential evaporation is 1060 mm [46,47].

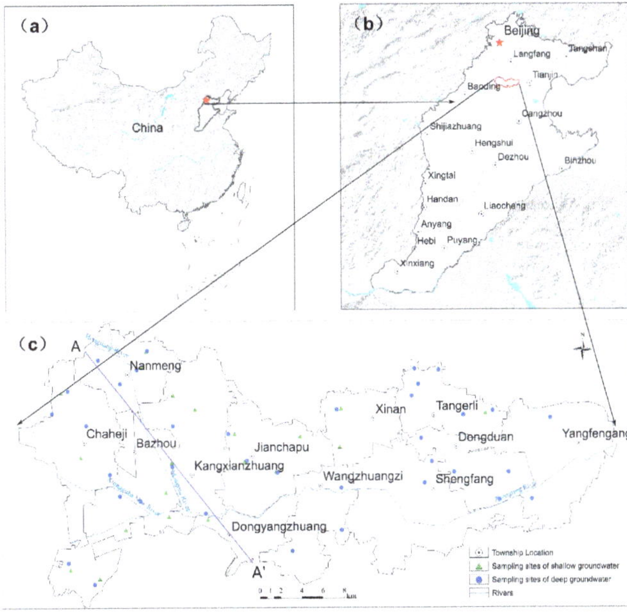

Figure 1. Location of the study area and the sampling sites ((**a**) China, (**b**) the North China Plain, (**c**) the study area and the sampling sites).

The study area possesses typicality in the aspect of hydrogeological conditions and groundwater utilization for the research of groundwater quality, function or suitability. Field investigation shows that groundwater in this region is mainly deposited in the loose sand layer pore of the Quaternary system. The aquifer exhibits spatial gradient characteristics, changing from a single structure composed of freshwater to a multilayer structure composed of freshwater and saltwater from northwest to southeast. The flow direction of both shallow and deep groundwater is generally from northwest to southeast, and the groundwater flow field is changed in the local groundwater funnel area under the influence of over-exploitation.

Groundwater in the study region can be divided into shallow groundwater and deep groundwater according to the regional hydrogeological conditions and groundwater exploitation. The depths of shallow groundwater wells are 20–150 m below the surface, while the depths of deep groundwater wells are 150–650 m. In general, the groundwater level is greatly affected by exploitation and precipitation. The shallow groundwater levels are 2–30 m below the surface, while the deep groundwater levels are within the depths of 30–90 m.

The aquifer system can be classified into four different aquifers by its lithological properties and geological age, named I, II, III and IV from top to bottom vertically [48] (Figure 2). Aquifer I is unconfined, composed of sand gravel, medium sand, fine sand and silty-fine sand, and the depths of the aquifer bottom are 30–50 m. Aquifer II, 140–160 m deep, is a semi-confined aquifer. Aquifer III consists of sandy gravel and medium to fine sand with depths of 360–380 m. Aquifer IV is made up of cemented sand gravel and medium to fine sand, and the aquifer bottom is below 380 m. Both third and fourth aquifers are confined aquifers. Groundwater in Aquifer I and II is classified as shallow groundwater, while groundwater in Aquifer III and IV is named deep groundwater based on aquifer distribution and groundwater exploitation depth [49].

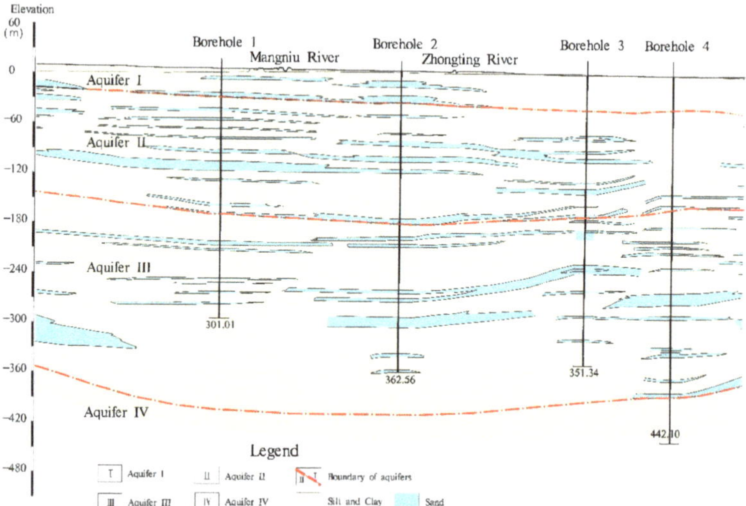

Figure 2. Hydrogeological cross-section along A–A'.

The study region has gradually changed from a single all-freshwater aquifer structure to a multi-layer structure composed of freshwater on the top, saltwater in the middle and freshwater on the bottom from northwest to southeast. Saltwater is mainly deposited in the lower part of Aquifer I and upper part of Aquifer II, and all deep groundwater is freshwater situated below the saltwater bottom boundary.

According to the field investigation, the source of irrigation water is mainly from shallow groundwater, and the source of drinking water is mainly from deep groundwater. Groundwater provides about 65% of the total water supply, and the proportions of groundwater used for agricultural irrigation and drinking are about 74% and 14%, respectively, in the study area [45,46]. Long-term groundwater over-exploitation causes a general decline of groundwater level, forming shallow groundwater funnel of agricultural exploitation type and deep groundwater funnel of domestic drinking exploitation type. Due to the geological conditions and anthropogenic activities in the study region, the fluorine content in groundwater exceeds the standard limits, having a serious impact on groundwater function.

2.2. Methods of Suitability Evaluation for Irrigation Purpose

In this study, six indicators were selected for comprehensive analysis to evaluate whether groundwater is suitable for irrigation, containing sodium adsorption ratio (SAR), sodium percentage (%Na), residual sodium carbonate (RSC), magnesium hazard (MH), permeability index (PI) and electrical conductivity (EC). According to the suitability classification of groundwater agricultural irrigation (Table 1), the proportion of different categories of each index was statistically analyzed. The spatial distribution figures of different categories of each index were drawn by the Kriging interpolation method in ArcGIS software, and this method was often used when analyzing and interpreting groundwater quality spatial variations [6,50].

Table 1. Classification of groundwater suitability for agricultural irrigation.

Index Range Classification	Index Range Classification	Index Range Classification	Standard Formulas	References
SAR (meq/L)	<10 10–18 18–26 >26	Excellent Good Doubtful Unsuitable	$SAR = \dfrac{Na^+}{\sqrt{\frac{Ca^{2+}+Mg^{2+}}{2}}}$	Richards (1954)
%Na (%)	<20 20–40 40–60 60–80 >80	Excellent Good Permissible Doubtful Unsuitable	$\%Na = \dfrac{(Na^++K^+)}{Ca^{2+}+Mg^{2+}+Na^++K^+} \times 100\%$	Wilcox (1955)
RSC (meq/L)	<1.25 1.25–2.5 >2.5	Good Doubtful Unsuitable	$RSC = \left(HCO_3^-+CO_3^{2-}\right)-\left(Ca^{2+}+Mg^{2+}\right)$	Richards (1954)
MH (%)	<50 >50	Desirable Undesirable	$MH = \dfrac{Mg^{2+}}{Ca^{2+}+Mg^{2+}} \times 100\%$	Szaboles and Darab (1964)
PI (%)	<25 25–75 >75	Unsuitable Moderately suitable Suitable	$PI = \dfrac{Na^++\sqrt{HCO_3^-}}{Ca^{2+}+Mg^{2+}+Na^+} \times 100\%$	Doneen (1964)
EC (us/cm)	<250 250–750 750–2250 >2250	Excellent Good Doubtful Unsuitable	Measured by instruments	Wilcon (1955)

The concentrations of Na^+, K^+, Ca^{2+}, Mg^{2+}, HCO_3^- and CO_3^{2-} are expressed in meq/L; SAR, %Na, RSC, MH and PI indicate the sodium adsorption ratio, sodium percentage, residual sodium carbonate, magnesium hazard and permeability index, respectively.

SAR is introduced, from the U.S. Department of Agriculture, which can reflect the relative activity of the alternate adsorption effect between Na^+ and soil components in groundwater. SAR predicts the Na^+ hazard of high carbonate waters, especially if they contain no residual alkali [1], and can act as a good indicator of the alkalization ability of groundwater. The higher the SAR value is, the stronger the alkalization ability of the groundwater.

Sodium concentration is usually expressed in the form of %Na [51], which affects the permeability and structure of the soil. Sodium filling in the soil would limit water and airflow in the soil, alter the permeability structure of the soil and inhibit crop growth.

RSC is an indicator that determines the harmful effect of carbonate and bicarbonate on groundwater quality for irrigation [7,51]. In general, high concentrations of carbonate and bicarbonate in groundwater along with calcium and magnesium can affect the suitability of groundwater for crop rising [2]; If the concentration of carbonate in groundwater is too high, excess carbonate may combine with sodium ion to form bicarbonate which would affect the permeability structure of the soil.

The magnesium hazard (MH) was suggested by Szaboles and Darab, which was also used to assess the water suitability for agricultural use [7]. MH values in groundwater render the soil to become alkaline, thus resulting in low crop production. If the Mg^{2+} concentration in irrigation water reaches a certain level, MH may affect the soil structure and produce bad effects on crops.

The existence of sodium, calcium, magnesium and bicarbonate in irrigation water may influence the soil permeability. If the soil accumulates large amounts of salts, the soil structure will be destroyed and crop growth will be affected [52]. Permeability index (PI) can be used to determine the water movement capability in soil based on the concentration of Ca^{2+}, Mg^{2+}, Na^+ and HCO_3^- [16]. PI is also a criterion for water quality suitability for agricultural irrigation, which is used to assess the permeability and drainage capacity of the soil [51].

In addition, EC is usually used as the indicator of salinity hazard to reflect water quality for irrigation.

In addition, an irrigation water classification diagram (USSL diagram and Wilcox diagram) is used to classify the groundwater suitability for irrigation according to the irrigation water quality classification standard from the U.S. Department of Agriculture [53,54]. The relationship between EC and SAR is indicated by the USSL diagram, and the relationship between EC and %Na is indicated by the Wilcox diagram.

2.3. Methods of Suitability Evaluation for Drinking

The suitability of groundwater for domestic drinking evaluated by comparing the values of different water quality parameters with the Class III water limits (suitable for drinking directly) stipulated by the Standard for Groundwater Quality of China (SGQC, GB/T 14848-2017) and permissible values for drinking water recommended by the World Health Organization (WHO, 2017) guidelines [55] presented in Section 3.1.

In this study, groundwater quality index (GQI) values are calculated using the World Health Organization standard (WHO, 2011) [50,56] and the suitability for drinking purposes is investigated. GQI variation graphs, computed by using the WHO standard and Kriging interpolation method, were provided by ArcGIS software. According to the WHO (2011) standard, GQI is expressed as follows:

$$GQI = \sum_{i=1}^{n} w_i \left(\frac{C_i}{S_i} \right) \times 100 \tag{1}$$

where C is the observed groundwater quality parameters (GQP), S is the standard value of GQP based on the WHO (2011) standard and w indicates the weight of each GQP based on the WHO (2011) standard. The WHO (2011) standards for each GQP and the weight values were presented in Table 2. The various groundwater classification grading for drinking consumptions based on computed GQI values were presented in Table 3.

Table 2. The WHO (2011) standards for groundwater quality parameters and the weight values.

Parameters	WHO Standards (S)	Weight (w_i)	Relative Weight (W_i)
Cl^-	250 (mg/L)	3	0.083
SO_4^{2-}	200 (mg/L)	4	0.112
HCO_3^-	150 (mg/L)	3	0.083
pH	6.5–9.2 (mg/L)	4	0.112
TDS	500 (mg/L)	5	0.139
EC	500 (us/cm)	5	0.139
Na^+	200 (mg/L)	2	0.055
K^+	10 (mg/L)	2	0.055
Mg^{2+}	30 (mg/L)	2	0.055
Ca^{2+}	75 (mg/L)	2	0.055
TH	100 (mg/L)	4	0.112
Total weight		36	1

Table 3. Groundwater quality classifications for drinking based on GQI values.

GQI	Type of Water
<50	Excellent
50–100	Good
100–200	Poor
200–300	Very poor
>300	Water unsuitable for drinking

According to the results of drinking water quality evaluation, typical pollutants are selected for health risk assessment using the mode recommended by the U.S. EPA [57]. Based on the actual situation of water utilization in the study area, the exposure pathways of drinking water intake and skin contact are considered in health risk assessment, while the respiratory exposure pathway is negligible due to the low risk for human health [24,34]. Human health risk through drinking intake and skin exposure pathways was calculated using the following formulas [57,58]:

$$CDI = \frac{C_w \times IR \times EF \times ED}{BW \times AT} \tag{2}$$

$$DAD = \frac{C_w \times K_i \times ET \times SA \times EF \times ED \times EV \times CF}{BW \times AT} \tag{3}$$

$$HQ_c = \frac{CDI}{RfD_c} \tag{4}$$

$$HQ_d = \frac{DAD}{RfD_d} \tag{5}$$

$$HQ = \sum\left(HQ_c + HQ_d\right) \tag{6}$$

where C_w (mg/L) indicates the concentration of the typical pollutant in groundwater. IR (L/day) is the ingestion rate for drinking water. EF (day/year) is the exposure frequency for ingestion and dermal pathways. ED (year) is the average exposure duration. BW (kg) is the human average body weight. AT (AT = 365 × ED, day) indicates average exposure time for ingestion and dermal pathways. CDI (mg/(kg·day)) and DAD (mg/(kg·day)) indicate daily average exposure dosage through drinking water and dermal contact, respectively. K_i (cm/hour) indicates the dermal permeability coefficient. ET (hour/day) means the exposure time during the shower. SA (cm^2) specifies exposed skin surface area during bathing. EV (times/day) is the bathing frequency. CF (L/cm^3) is the conversion factor. RfD_c (mg/(kg·day)) and RfD_d (mg/(kg·day)) are the reference dose absorbed by drinking water and skin contact, respectively. HQ_c and HQ_d are the non-carcinogens hazard quotient through ingestion and dermal absorption of water, respectively. HQ is the total hazard quotient including exposure routes of drinking water and skin contact. HQ <1 suggests an acceptable non-carcinogenic risk, while the value above 1 indicates a higher probability of adverse health effects.

2.4. Water Sampling and Measurement

In August 2020, 59 groups of groundwater samples were collected in the study area including 21 groups of shallow groundwater samples and 38 groups of deep groundwater samples. The shallow groundwater samples were selected within the depth of 20–130 m, while the deep groundwater samples were selected within the depth of 170–650 m. The spatial distribution of sampling points were shown in Figure 1. The shallow groundwater samples were collected from local irrigation wells and the deep groundwater samples were collected from drinking water wells. Sampling procedures, samples preservation and treatment methods were conducted in accordance with groundwater sampling technical standards.

To make sure the collected samples reflect the actual situation of the chemicals in groundwater, groundwater in wells would be pumped out more than three minutes before sampling. The 2.5 L plastic sampling bottles were used as containers and were washed three times with deionized water to keep clean. Haver rapid water quality detector was used to measure water temperature, pH and redox potential in the field. All groundwater samples were sealed with sealing membranes and kept in a cryogenic incubator.

All samples were delivered to the laboratory (Tianjin geological and mineral testing center) to test within 48 h. The test indicators mainly include potassium (K^+), sodium (Na^+), calcium (Ca^{2+}), magnesium (Mg^{2+}), chloride (Cl^-), sulfate (SO_4^{2-}), bicarbonate (HCO_3^-), carbonate (CO_3^{2-}), orthophosphate (PO_4^{3-}), fluorine (F^-), iodine (I^-), ammonia-nitrogen (NH_4^+-N), nitrate-nitrogen (NO_3^--N), nitrite-nitrogen (NO_2^--N), total alkalinity, total hardness (TH), iron (Fe), manganese (Mn) and chemical oxygen demand (COD_{Mn}). The analysis technology and equipment referred to the detection indexes and methods recommended by the Standard for Groundwater Quality of China (SGQC) [44].

When analyzing groundwater samples, quality control was performed with less than 5% error for all duplicate samples. Quality assurance is achieved by implementing laboratory standard procedures and applying quality control methods. In the test of indicators, average values were obtained from multiple test records. The verification of the analysis method was based on subsequent criteria for detection quality control, including external calibration, precision, percent accuracy, linearity, detection limit (DL), quantitative limit (QL) and blank reagents.

3. Results and Discussion

3.1. Hydrochemical Characteristics of Main Ions in Groundwater

The statistical results of the main anions, cations and hydrochemical indicators in groundwater are shown in Table 4. The measured pH values ranged from 7.31 to 8.50 with a mean value of 7.85 in the shallow groundwater and ranged from 8.20 to 9.15 with an average value of 8.60 in the deep groundwater, indicating that groundwater in the study area is generally in a partial alkaline environment. The measured EC values ranged from 3.56 to 3717.10 μs/cm with an average value of 1709.77 μs/cm in the shallow groundwater and changed from 1.59 to 1556.00 μs/cm with a mean of 698.95 μs/cm in the deep groundwater, showing that shallow groundwater is more susceptible to fertilization and irrigation. The total hardness (TH, by $CaCO_3$) values of shallow groundwater ranged from 18.73 to 1482.85 mg/L with an average value of 673.82 mg/L and ranged from 10.12 to 82.96 mg/L with an average value of 24.81 mg/L in the deep groundwater. The TH value in shallow groundwater is over 27 times the value in deep groundwater, probably due to the salinity and mineral dissolution in aquifers under the influence of climate, precipitation, evaporation, topography and human activity [7]. The measured COD_{Mn} values ranged from 0.66 to 5.58 mg/L with an average value of 1.74 mg/L in the shallow groundwater and ranged from 0.37 to 1.66 mg/L with an average value of 0.67 mg/L in the deep groundwater, indicating that shallow groundwater is more polluted by organic matter than deep groundwater.

Table 4. Statistical summary of hydrochemical characteristics of groundwater.

Parameter	Shallow Groundwater					Deep Groundwater					Standard Values	
	Min	Max	Mean	SD	CV	Min	Max	Mean	SD	CV	GB	WG
pH	7.31	8.50	7.85	0.34	0.04	8.20	9.15	8.60	0.17	0.02	6.5–8.5	6.5–8.5
EC (µs/cm)	3.56	3717.10	1709.77	730.56	0.43	1.59	1556.0	698.95	269.50	0.39		1500
Total hardness (mg/L)	18.73	1482.85	673.82	351.95	0.52	10.12	82.96	24.81	14.67	0.59	450	500
Total alkalinity (mg/L)	268.76	712.1	494.2	120.83	0.24	185.19	418.35	286.67	51.08	0.18		
COD_{Mn} (mg/L)	0.66	5.58	1.74	1.09	0.63	0.37	1.66	0.67	0.24	0.36	3	
Soluble silica (mg/L)	3.60	21.20	13.24	4.69	0.35	9.80	15.90	12.67	1.13	0.09		
Carbon dioxide (mg/L)	0.00	85.40	38.64	27.52	0.71	0.00	2.20	0.06	0.35	6.08		
K^+ (mg/L)	0.30	31.40	2.49	6.51	2.61	0.20	29.10	1.22	4.59	3.76		12
Na^+ (mg/L)	44.80	569.70	240.88	134.48	0.56	40.58	314.60	160.48	49.62	0.31	200	200
Ca^{2+} (mg/L)	4.70	331.20	122.91	81.78	0.67	2.90	327.10	14.44	51.49	3.57		200
Mg^{2+} (mg/L)	1.70	175.60	89.14	40.69	0.46	0.70	9.60	2.62	2.07	0.79		150
Cl^- (mg/L)	55.30	550.50	220.30	132.42	0.60	13.80	231.10	53.83	48.62	0.90	250	250
SO_4^{2-} (mg/L)	16.00	1037.36	344.71	307.02	0.89	0.35	50.07	26.43	12.09	0.46	250	250
CO_3^{2-} (mg/L)	0.00	12.00	1.57	3.16	2.01	0.00	31.20	7.59	5.48	0.72		
HCO_3^- (mg/L)	315.50	868.30	599.39	147.50	0.25	213.60	491.80	334.09	62.24	0.19		500
F^- (mg/L)	0.35	2.15	1.02	0.50	0.49	0.29	4.21	2.23	1.20	0.54	1	1.5
I^- (mg/L)	0.00	0.25	0.05	0.08	1.48	0.00	0.36	0.05	0.08	1.52	0.08	
NH_4^+ (mg/L)	0.04	0.40	0.12	0.07	0.60	0.00	0.16	0.08	0.05	0.56	0.5	
NO_3^- (mg/L)	0.00	126.80	18.16	31.83	1.75	0.00	7.63	0.62	1.67	2.69	20	50
NO_2^- (mg/L)	0.00	0.36	0.05	0.08	1.68	0.00	0.10	0.01	0.02	2.64	1	3
PO_4^{3-} (mg/L)	0.02	0.57	0.08	0.12	1.50	0.02	0.46	0.20	0.10	0.51		30
Fe (mg/L)	0.01	0.07	0.02	0.02	0.91	0.01	0.10	0.02	0.02	0.87	0.3	
Mn (mg/L)	0.00	0.60	0.08	0.14	1.87	0.00	0.04	0.01	0.01	1.39	0.1	0.4

SD: standard deviation, CV: variation coefficient, GB: class III water limits (suitable for drinking directly) stipulated by the Standard for Groundwater Quality of China (GB/T 14848-2017), WG: WHO guideline (2017).

The field investigation findings show that there is continuous aquiclude (confining stratum) between shallow and deep groundwater, obstructing the chemical exchange between shallow and deep groundwater. Obvious differences in water levels can be seen between shallow groundwater and deep groundwater, which illustrate the aquiclude's impact. The deep groundwater is in a more closed environment than shallow groundwater. As a result, the difference in hydrochemistry exists between shallow and deep groundwater.

The average concentrations of cations (expressed as meq/L) in shallow and deep groundwater in the study area were in the following order of $Na^+ > Mg^{2+} > Ca^{2+} > K^+$ and $Na^+ > Ca^{2+} > Mg^{2+} > K$, respectively. The high content of Na^+ in groundwater is attributed to the dissolution of sodium-containing minerals, cation exchange among minerals and high weathering processes of rocks [13,52,59]. Ca^{2+} is primarily derived from calcium-rich minerals (including pyroxene, feldspar and amphibole), and Mg^{2+} is mainly derived from ions exchange between groundwater and minerals in rocks and soil [13]. Furthermore, the high concentration values of Na^+, Ca^{2+} and Mg^{2+} in groundwater may be due to the effects of domestic wastewater and irrigation water [14].

The average concentrations of anions (expressed as meq/L) in shallow and deep groundwater in the study area were in the following order of $HCO_3^- > SO_4^{2-} > Cl^- > CO_3^{2-}$ and $HCO_3^- > Cl^- > SO_4^{2-} > CO_3^{2-}$, respectively. Strong weathering and dissolution of carbonate, and the reaction of soil CO_2 with the dissolution of silicate minerals are all responsible for the high concentration of bicarbonate in groundwater [60,61]. The strong evaporation effect and dissolution of gypsum, as well as human activity (including the utilization of agricultural fertilizers and wastewater discharge) in semi-arid areas, may increase the SO_4^{2-} concentration in groundwater [59].

3.2. Hydrochemical Types of Groundwater

The hydrochemical data are shown on the Piper trilinear diagram to determine the groundwater hydrochemical facies in the study area. Piper trilinear diagram contains two triangles, one for plotting cations, and the other for plotting anions, where the hydrochemical facies can be identified within the diamond-shaped field [26,62]. The groundwater samples collected from shallow and deep aquifers were plotted for comparison (Figure 3). The results showed that most samples of the shallow and deep groundwater were distributed in the lower right of the cation triangle, indicating that the groundwater samples are mainly concentrated in the Na^+ and Ca^{2+} cation facies. In the anion triangle, most of the samples were distributed on the left, indicating that the samples were mainly concentrated in the high equivalent percentage region of HCO_3^- and SO_4^{2-}. The weathering of carbonate minerals, dissolution of gypsum and evaporation are important factors in controlling groundwater chemistry characteristics [10]. Based on the analysis results, the hydrochemical facies of groundwater in the study area could be classified into HCO_3-Na, HCO_3-Ca•Mg, HCO_3•SO_4-Na•Mg and SO_4•Cl-Ca•Mg types in the shallow groundwater as well as HCO_3-Na, HCO_3•Cl-Na and HCO_3-Ca types in the deep groundwater according to the naming rules of the Schukalev classification.

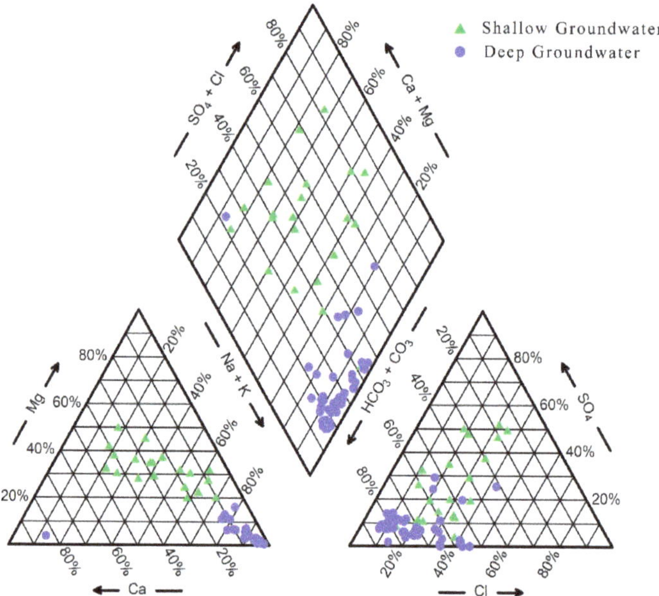

Figure 3. Piper diagram showing the groundwater hydrochemical types.

The continuous aquiclude between shallow and deep groundwater has obstructing impacts on the ions exchange. In addition, there are significant differences in the depths of the shallow and deep groundwater samples. As a result, sharp variation in hydrochemistry can be seen between shallow and deep groundwater samples.

3.3. Suitability Evaluation for Agricultural Irrigation

The shallow groundwater suitability evaluation for agricultural irrigation was conducted based on Table 1 using six evaluation indexes: SAR, %Na, RSC, MH, PI and EC. Irrigation suitability proportions based on different indexes are shown in Figure 4.

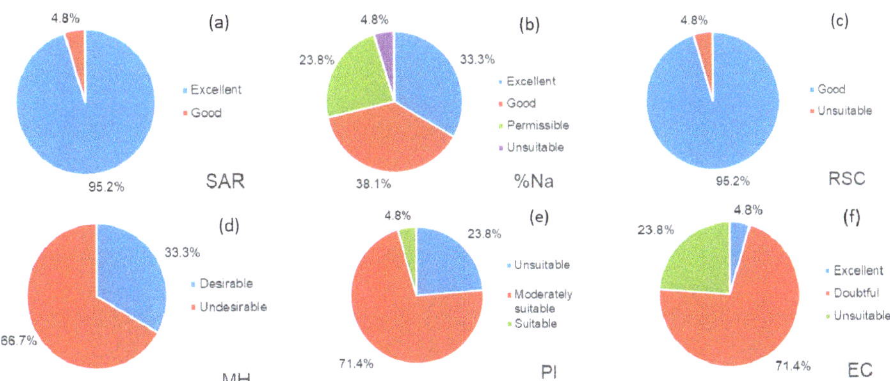

Figure 4. Proportion of groundwater suitable for irrigation based on different indexes: (**a**) sodium adsorption ratio (SAR), (**b**) sodium percentage (%Na), (**c**) residual sodium carbonate (RSC), (**d**) magnesium hazard (MH), (**e**) permotic index (PI) and (**f**) electrical conductivity (EC).

SAR can be used to test the suitability of groundwater for irrigation, and a higher SAR value indicates a stronger alkalinization capacity. The SAR values of groundwater in the study area ranged from 0.64 to 14.28 meq/L, showing that the groundwater in the study area has a low sodium hazard and is suitable for irrigation.

%Na is another manifestation of sodium hazard, where high concentrations of sodium can lead to magnesium and calcium deficiency in plants. To obtain a higher crop yield, generally, the %Na value of irrigation water should not exceed 60%. The %Na values of 95.2% of the shallow groundwater samples in the study area were less than 60%, indicating their suitability for irrigation. Moreover, the %Na values of 71.4% of the samples were less than 40%, showing the groundwater quality was excellent or good for irrigation.

RSC is used to describe the amount of carbonate and bicarbonate present in groundwater, which would reduce soil permeability when the concentration is too high. According to Table 2, RSC values in 95.2% of the shallow groundwater samples were less than 1.25 meq/L, indicating that the groundwater quality in this region is very suitable for irrigation. The RSC values in the remaining samples were greater than 2.5 meq/L, indicating that they were not suitable for irrigation.

MH is one of the important parameters used to estimate the groundwater suitability for irrigation [7]. When Mg^{2+} content in irrigated water reaches a certain level, magnesium alkalinization may occur in the soil affecting the soil structure. Only 33.3% of the groundwater samples in the study area had MH values of less than 50%, which were suitable for irrigation. The remaining region, with MH values greater than 50%, may lead to soil magnesinization during long-term irrigation using groundwater.

PI is also an important parameter for measuring the groundwater suitability for irrigation. Based on computational analysis, PI values in 4.8% of the shallow groundwater samples were greater than 75%, indicating that the groundwater in these areas is suitable for irrigation. In addition, PI values in 23.8% of the shallow groundwater samples were less than 25%, indicating that the groundwater in these areas was not suitable for long-term irrigation.

Only 4.8% of the shallow groundwater samples were in the good category according to EC values. We found 71.4% of the samples had EC values between 750 μs/cm and 2250 μs/cm, suggesting that the shallow groundwater at these locations may not be suitable for drainage restricted soil. More seriously, 23.8% of the shallow groundwater samples were unsuitable for irrigation based on EC analysis.

An irrigation water classification diagram was drawn according to the U.S. irrigation water quality classification criteria to evaluate the feasibility of groundwater for irrigation. A USSL diagram (Figure 5a), where the SAR values were plotted against the EC values in irrigation groundwater, was used to comprehensively reflect sodium and salinity hazards [7]. The results showed that 57.1% of the shallow groundwater samples fell into C_3-S_1 zone (high salinity with low sodium hazard) where using groundwater for irrigation would not bring sodium harm. However, it was necessary to select crops with good salt tolerance for planting. In total, 14.3%, 9.5% and 9.5% of the shallow groundwater samples fell into the C_4-S_1 zone (very high salinity with low sodium hazard), C_3-S_2 zone (high salinity with medium sodium hazard) and C_4-S_2 zone (very high salinity with medium sodium hazard), respectively, where it was also necessary to select plants with good salt tolerance and take drainage measures. Overall, using shallow groundwater for agricultural irrigation would give rise to a serious salinity hazard, but the degree of sodium hazard would be low. The process of salt deposition and soil salinization is mainly caused by the salinity in irrigation water, which may reduce the effective absorption of water and nutrients by plants [20]. Reasonable drainage mode and good soil permeability are needed for better agricultural irrigation and lower salinity hazards [18].

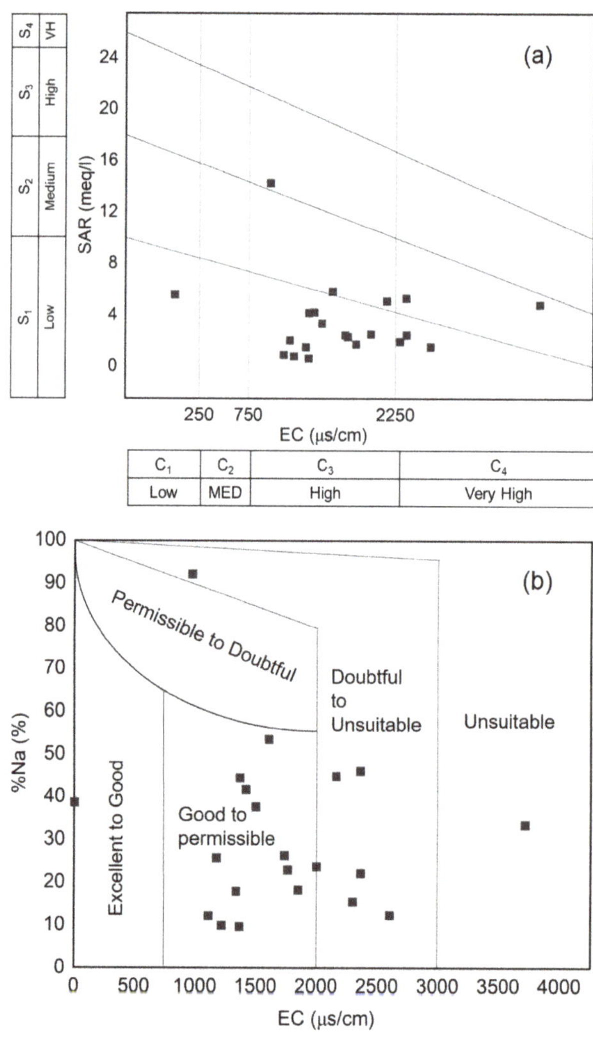

Figure 5. (**a**) USSL diagram and (**b**) Wilcox diagram showing the irrigation water quality classification.

The relationship between the electrical conductivity (EC) and the sodium percentage (%Na) was given by the Wilcox diagram for classifying irrigation water quality (Figure 5b). We found 57.1% and 33.3% of shallow groundwater samples were within good to permissible zone and doubtful to unsuitable zone, respectively. In general, most of the shallow groundwater in the study area was appropriate for irrigation based on the contrasting relationship between EC and %Na.

The classification zoning map about groundwater suitability for irrigation based on six irrigation indexes was shown in Figure 6. As can be seen from Figure 6a, almost all SAR values are less than 10 meq/L (Excellent level), indicating that groundwater in the study region has a low sodium hazard for irrigation. Based on Figure 6b, almost all groundwater is suitable for irrigation (%Na values less than 60%, Excellent, Good and Suitable levels), except for groundwater in the northeast part of the study (%Na values exceed 80%, Unsuitable level). According to the on-site investigation, there are factories in the northeast outside the study area, which may have effects on groundwater quality. Also, almost all RSC values are less than 1.25 meq/L (Good level) (Figure 6c), showing groundwater quality has a low harmful effect of carbonate on irrigation. MH values larger than 50% (Undesirable level) distribute in many parts of the study area (Figure 6d), which means the Mg^{2+} concentration in irrigation water reaches a certain level in many regions and may affect the soil structure. PI values in most areas range from 25% to 75% (Figure 6e), indicating that groundwater is moderately suitable for irrigation. However, PI values less than 25% (Unsuitable level) exist in some parts of the northwest and south of the study area, thus the permeability and drainage capacity of the soil in these areas need more attention. Almost all EC values range from 750 to 2250 (Doubtful level) (Figure 6f), meaning the groundwater suitability for irrigation is doubtful when assessed by the EC index.

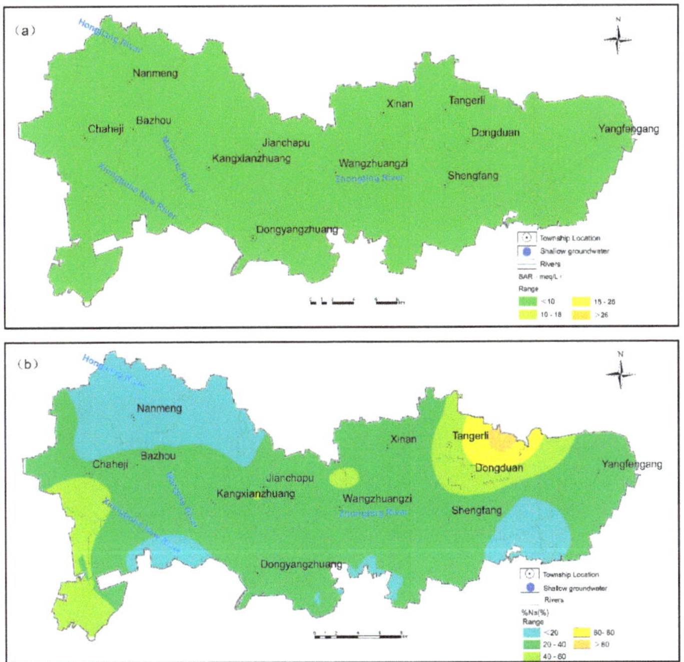

Figure 6. *Cont.*

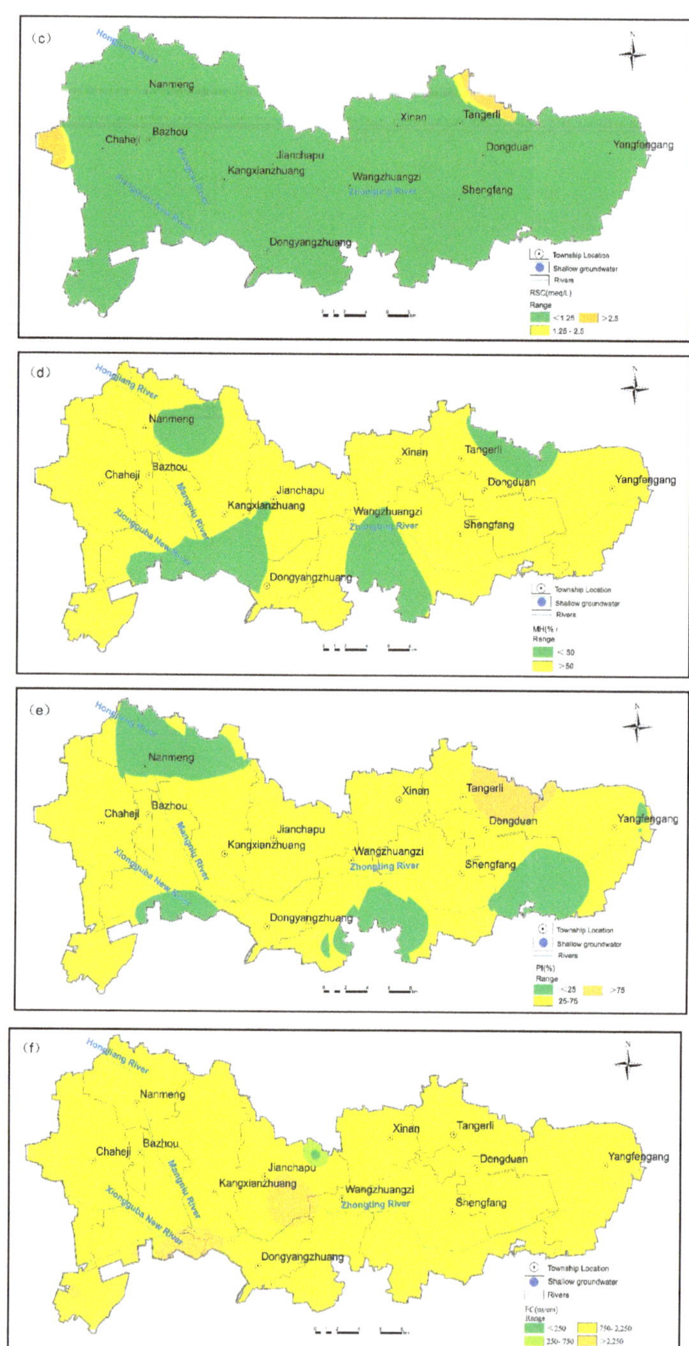

Figure 6. Classification of groundwater suitability for irrigation based on different indexes of (**a**) sodium adsorption ratio (SAR), (**b**) sodium percentage (%Na), (**c**) residual sodium carbonate (RSC), (**d**) magnesium hazard (MH), (**e**) permotic index (PI), and (**f**) electrical conductivity (EC).

In conclusion, the utilization of shallow groundwater for agricultural irrigation in the study area will bring a low degree of sodium hazard but a high degree of salinity hazard, thus salt-resistant planting mode and sufficient drainage measures are required to solve the irrigation problems from shallow groundwater. Mg^{2+} concentration in irrigation water and the permeability and drainage capacity of soil should cause more attention. Combined with the actual situation of crop irrigation and utilization of groundwater in the study area, some effective measures are suggested to take such as making appropriate irrigation management policies and cultivating salinity-friendly crops. What's more, it is significant to apply reasonable fertilizer according to the needs of crops and soil characteristics, to improve the crop yield and promote the long-term development of agriculture.

3.4. Suitability Evaluation for Drinking and Health Risk Assessment

3.4.1. Suitability Evaluation for Drinking

Statistical analyses of the hydrochemical characteristics of groundwater and permissible limits of water quality are presented in Table 2. According to the SGQC and WHO, the permissible limit of pH is from 6.5 to 8.5. In the study area, 71.4% of the deep groundwater samples presented a high pH value exceeding 8.5. According to the suitability of groundwater based on hardness classification [13,30], the total hardness (TH) as $CaCO_3$ (mg/L) could be divided into four types: soft (<75 mg/L), moderately hard (75–150 mg/L), hard (150–300 mg/L) and very hard (>300 mg/L). In this study area, 97.4% of the deep groundwater samples fell in the soft water category; 0.6% and 2% of the deep groundwater samples belonged to moderately hard and hard water, respectively. The main cations (Na^+, K^+, Ca^{2+} and Mg^{2+}) and anions (Cl^-, SO_4^{2-} and HCO_3^-) in the deep groundwater samples were within the permissible concentration range in drinking water recommended by the WHO. The contents of COD_{Mn}, iron and manganese in the samples were lower than the Class III water limits of SGQC. NH_4^+, NO_3^-, NO_2^- and PO_4^{3-} in the samples were all below the permissible limits of drinking water recommended by the SGQC and WHO, reflecting little effect of agricultural activity on deep groundwater quality.

The physico-chemical basic parameters include pH, Cl^-, SO_4^{2-}, HCO_3^-, Ca^{2+}, Mg^{2+}, Na^+, K^+, TH, EC and TDS were calculated to get GQI values according to the WHO (2011). GQI values indicated the groundwater's suitability for drinking purposes. The spatially distributed GQI values were interpolated using the Kriging interpolation method (Figure 7). It is shown that the GQI values of most deep groundwater samples range from 50 to 100, meaning the groundwater quality is good for drinking purposes and meets the World Health Organization (WHO) standard. The GQI values of groundwater in the east part of the study area range from 100 to 200, showing the poor quality of groundwater for drinking. The main reason is the higher content of Na^+, Cl^- and TDS in groundwater of the east region.

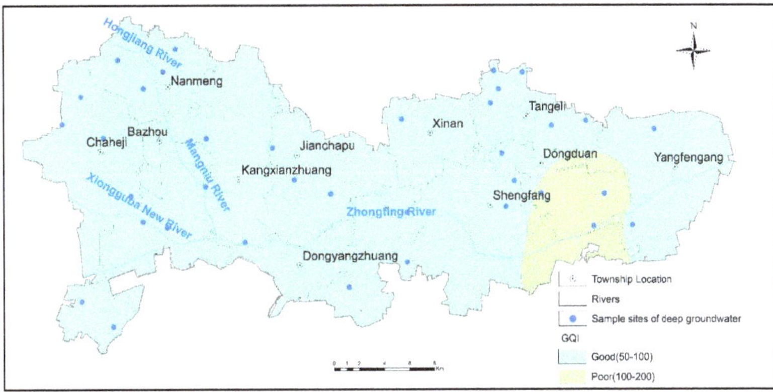

Figure 7. Spatially distributed groundwater quality index (GQI) values in the study area.

The fluoride concentration of deep groundwater in the study area ranged from 0.29 to 4.21 mg/L and the mean value was 2.23 mg/L. We found 79.0% of the groundwater samples with a fluoride concentration higher than the Class III water limit of SGQC (1 mg/L) and 65.8% of the samples with a fluoride concentration exceeding the permissible limit-concentration specified by the WHO (1.5 mg/L), indicating that there is a risk of excessive fluoride in groundwater. Physical or chemical methods need to be used to reduce the fluorine concentration in the groundwater of the study area. Iodine test results showed that 21.1% of deep groundwater samples exceeded the Class III standard (0.08 mg/L) according to SGQC. Excessive iodine intake can cause diseases such as thyroid function disease [63]. Monitoring work for iodine concentration in food should be further strengthened, and it is significant to take iodine reduction measures such as stopping iodine salt and water source modification in high iodine areas.

Based on a comprehensive water quality evaluation method recommended by SGQC, 9 of the 38 groundwater sampling points in the study area were classified as the Class III category, which were mainly distributed in the west of the study area. Twenty of the deep groundwater samples were classified as Class IV, and nine were classified as Class V, which were mainly distributed in the east of the study area. According to the SGQC groundwater quality classification standard and the current situation of deep groundwater quality, the study area could be divided into three parts: Class III zone, where groundwater is suitable for direct drinking water supply; Class IV groundwater area, where groundwater is suitable for drinking water supply after proper treatment; Class V zone, where groundwater is unsuitable for drinking water supply. The spatial distribution of drinking water quality suitability was interpolated using the Kriging interpolation method and visualized using ArcGIS software (Figure 8). It can be seen that the deep groundwater quality in the western study area is suitable for direct drinking. The deep groundwater in the central parts of the study area including Kangxinzhuang Town, Jianchapu Town, Dongyangzhuang Town and Tangerli Town could be used as a drinking water supply source after proper treatment. The deep groundwater in the central and eastern parts of the study area including Dongduan Town, Yangfengang Town, Wangzhuangzi Town and Shengfang Town may be inappropriate for a drinking water supply. Fluorine is the main factor excessing the standard values in groundwater of these locations., The fluorine concentration in deep groundwater for long-term drinking water supply deserved close attention due to the risk of skeletal fluorosis.

Figure 8. Division diagram of deep groundwater suitable for drinking.

3.4.2. Drinking Water Health Risk Assessment

This study focused on exposure assessment of drinking water intake and skin contact, in which drinking water intake is considered as the major pathway of chemicals entering the human body. Health risk assessment was conducted for different age groups (children, adult men and adult women) based on the results of deep groundwater quality analysis in the study area. The corresponding potential non-carcinogenic health risk values for F^- and I^- through ingestion of water and dermal absorption were evaluated according to the International Center for the Study of Cancer (IARC), as shown in Table 5.

Table 5. Personal health risk values of typical pollutants in the deep groundwater.

Statistics	F^-			I^-		
	HQ_c	HQ_d	HQ	HQ_c	HQ_d	HQ
			Children			
Min	0.22	0.00037	0.22	0.000031	0.018	0.018
Max	3.1	0.0054	3.1	0.0056	3.3	3.3
Mean	1.7	0.0029	1.7	0.00084	0.49	0.49
			Male			
Min	0.15	0.00028	0.15	0.000023	0.013	0.013
Max	2.2	0.004	2.2	0.0042	2.3	2.3
Mean	1.2	0.0021	1.2	0.00062	0.35	0.35
			Female			
Min	0.17	0.00028	0.17	0.000023	0.014	0.014
Max	2.4	0.0041	2.4	0.0042	2.5	2.5
Mean	1.3	0.0022	1.3	0.00063	0.37	0.37

HQ_c and HQ_d are non-carcinogens hazard quotient through ingestion of water and dermal absorption of water, respectively. HQ is the total hazard quotient including exposure routes of drinking water and dermal absorption of water. $HQ = HQ_c + HQ_d$.

According to Table 3, the HQ values of F^- and I^- in deep groundwater varied greatly in populations with different age groups (children, adult males and adult females). The health risk values of F^- ranged from 0.22 to 3.10 with an average value of 1.70 for children. Meanwhile, the HQ values of 71.1% of the deep groundwater sampling sites outstripped the acceptable safety limit. If groundwater is used directly for drinking purposes, significant health risks will be brought to children. For adult males, the HQ values of F^- ranged between 0.15 and 2.20, with the mean value 1.2, and the health risk of 65.8% of the deep groundwater sampling sites was unacceptable. For adult women, the HQ levels of F^- varied from 0.17 to 2.40 with the mean level of 1.3, and the HQ values of 65.8% of the deep groundwater sampling sites exceeded the acceptable safety limits. Chronic use of deep groundwater would pose potential harm to female adults. The HQ values of I^- ranged from 0.018 to 3.30 with an average value of 0.49 for children, and the HQ values of 18.4% of the deep groundwater sampling sites exceeded the safety limits. Long-term exposure to groundwater with a high iodine concentration can have potentially harmful and adverse effects on children. The HQ values of I^- for adult males ranged from 0.013 to 2.30 with an average value of 0.35, and the HQ values of 10.5% of the deep groundwater sampling sites were unacceptable. For adult women, the HQ levels of I^- varied from 0.014 to 2.50 with the mean level of 0.37, and the HQ values of 10.5% of the deep groundwater sampling sites exceeded the permissible level.

The above results indicated that children have a higher health risk from excessive intake of high fluoride and iodine in groundwater than in adults, in descending order of children, adult females and adult males. This phenomenon is mainly attributed to children's bodies being more sensitive, lower weight and less immune than adults, so they are more susceptible to health risks by ingesting water containing high concentrations of fluoride and iodine. This finding is consistent with many previous studies in other regions such as China, India, Iran and so on [64–66]. Long-term drinking of high fluoride groundwater will endanger human health, so it is suggested that before groundwater is used as drinking water, adsorption method and chemical sedimentation method can be applied to reduce the concentration of fluorine and protect the safety of the drinking water environment.

3.5. Cause of High Fluoride Groundwater Formation

In terms of geological background, high fluorinated groundwater is mainly caused by weathering, dissolution and groundwater-rock interactions containing fluorine minerals (apatite, fluorite, mica, etc.) in Quaternary deposits [13,38,61]. The groundwater moves slowly in the study area because of the poor dynamic conditions, and deep groundwater in a closed environment moves more slowly, which is conducive to the enrichment of fluorine elements.

In terms of chemical type and chemical composition of groundwater, alkaline groundwater is conducive to the dissolution of fluorine-containing minerals. There are more OH^-, which are easy to replace F^- in fluorine-containing minerals. The alkalinity environment with high HCO_3^- and Na^+ may accelerate the solubility of fluorite in groundwater in the study area, and competitive adsorption between fluoride and bicarbonate promoted the release of fluoride in the sediment, leading to an increased concentration of fluoride in groundwater [39,67]. The pH value of groundwater in the study area is about eight, which indicates the alkaline characteristics. Ca^{2+} and CO_3^{2-} produce calcium carbonate precipitation, improving the molar concentration of Na^+. The characteristics of high HCO_3^- and Na^+ concentrations in the study area are conducive to the formation and stability of fluoride in groundwater.

The effect of human activity on the fluoride content in groundwater cannot be negligible. Fluorinated solid waste and wastewater emissions from industrial activity can lead to fluorine concentration increase in soil and groundwater. In addition, groundwater over-exploitation leads to water level decline and soil layer compression, which is another important factor for fluorine concentration increasing in groundwater [68,69]. With long-term exploitation of deep groundwater, the water level of the sand aquifer in the main exploitation layer is declining. Due to the difference in water level, water is released from the clay soil layer and the fluorine ions adsorbed by the clay soil are also released, increasing the fluorine content in the groundwater. It can be learned from this study that regions with higher F^- concentrations were consistent with regions of lower deep groundwater levels and larger ground subsidence based on spatial distribution analysis. Good correspondence was found between fluorine concentration changes with groundwater level and soil compression in deep groundwater in the Hebei Plain [70,71]. The layout and exploitation quantity of deep groundwater exploitation wells should be scientifically standardized because the soil layer compression caused by groundwater over-exploitation may increase the fluoride concentration in deep groundwater.

4. Conclusions

In this study, the suitability of shallow groundwater for irrigation and deep groundwater for drinking in a typical agricultural area of North China Plain were analyzed. What's more, the human health risks associated with over-standard chemicals in groundwater were evaluated. The groundwater belongs to a Quaternary loose rock pore water aquifer. The depths of shallow groundwater wells are 20–150 m below the surface, while the depths of deep groundwater wells are 150–650 m. The main conclusions are as follows:

Hydrochemical analysis revealed that groundwater in the study area was generally in an alkaline environment. According to the analysis of SAR, Na% and RSC indexes, the shallow groundwater was suitable for irrigation in the study area. According to the irrigation water quality classification, 57.1% of the shallow groundwater samples fell into high salinity with a low sodium hazard zone, and 14.3% of the samples fell into very high salinity with a low sodium hazard zone. Crops with good salt tolerance and drainage measures were necessary for sustainable agricultural development.

The evaluation of drinking water quality suitability showed that F^- concentrations in 79.0% of the deep groundwater samples exceeded the Class III water limits of SGQC (1 mg/L) and 65.8% of the deep groundwater samples contained F^- levels exceeding the permissible value of 1.5 mg/L recommended by the WHO for drinking. Groundwater with a high concentration of fluoride was mainly distributed in the east of the study area.

The total hazard quotient HQ values of F^- exceeded the safety limits (HQ >1) in over half of the deep groundwater samples, and the degree of risk varied greatly in populations of different age groups, in descending order of children, adult females and adult males. Except for natural factors, the soil layer compression caused by groundwater over-exploitation is an important reason for high fluoride concentration in deep groundwater.

The rational exploitation of limited groundwater resources is a significant challenge. Effective measures about groundwater management should be strengthened, such as carrying out long-term groundwater quality investigation and monitoring, establishing specialized research projects about fluoride in groundwater and controlling the amount of groundwater exploitation for irrigation. Physical or chemical methods should be better used to reduce the fluorine concentration in groundwater and improve the quality of drinking water. The study on the relationship between fluorine concentration change and soil layer compression needs more attention. Although this study answers important questions about the suitability of groundwater quality for irrigation and drinking, its temporal trends in this region have remained unsolved. More research work about this subject is suggested to be conducted in the future.

Author Contributions: Conceptualization, H.G. and M.L.; data curation, H.G., L.W. and X.Z. (Xisheng Zang); formal analysis, H.G. and M.L.; funding acquisition, H.G., M.L. and Y.W.; investigation, Y.W., X.Z. (Xiaobing Zhao), H.G., M.L., H.W. and J.Z.; methodology, H.G. and M.L.; supervision, L.W.; writing—original draft, M.L. and H.G.; writing—review and editing, H.G., M.L. and L.W. All authors have read and agreed to the published version of the manuscript.

Funding: This work was supported by the National Natural Science Foundation of China (Grant No. 41877294) and the China Geological Survey (Grant Nos. DD20190679 and DD20160235).

Institutional Review Board Statement: Not applicable.

Informed Consent Statement: Not applicable.

Data Availability Statement: Not applicable.

Acknowledgments: The authors are appreciative of the valuable comments and advice given by editors and anonymous reviewers. The authors also thank Xin Yan for revising the English text of this manuscript.

Conflicts of Interest: The authors declare no conflict of interest.

References

1. Beyene, G.; Aberra, D.; Fufa, F. Evaluation of the suitability of groundwater for drinking and irrigation purposes in Jimma Zone of Oromia, Ethiopia. *Groundw. Sustain. Dev.* **2019**, *9*, 1–11. [CrossRef]
2. Karunanidhi, D.; Aravinthasamy, P.; Kumar, D.; Subramani, T.; Roy, P. Sobol sensitivity approach for the appraisal of geomedical health risks associated with oral intake and dermal pathways of groundwater fluoride in a semi-arid region of south India. *Ecotoxicol. Environ. Saf.* **2020**, *194*, 1–12. [CrossRef]
3. Maghrebi, M.; Noori, R.; Partani, S.; Araghi, A.; Haghighi, A. Iran's Groundwater Hydrochemistry. *Earth Space Sci.* **2021**, *8*, 1–17. [CrossRef]
4. Noori, R.; Maghrebi, M.; Mirchi, A.; Tang, Q.; Bhattarai, R.; Sadegh, M.; Noury, M.; Haghighi, A.; Kløve, B.; Madani, K. Anthropogenic depletion of Iran's aquifers. *Proc. Natl. Acad. Sci. USA* **2021**, *118*, 1–7. [CrossRef] [PubMed]
5. Medici, G.; Baják, P.; West, L.; Chapman, P.; Banwart, S. DOC and nitrate fluxes from farmland; impact on a dolostone aquifer KCZ. *J. Hydrol.* **2021**, *595*, 125658. [CrossRef]
6. Ghahremanzadeh, H.; Noori, R.; Baghvand, A.; Nasrabadi, T. Evaluating the main sources of groundwater pollution in the southern Tehran aquifer using principal component factor analysis. *Environ. Geochem. Health* **2018**, *40*, 1317–1328. [CrossRef] [PubMed]
7. Bian, J.; Nie, S.; Wang, R.; Wan, H.; Liu, C. Hydrochemical characteristics and quality assessment of groundwater for irrigation use in central and eastern Songnen Plain, Northeast China. *Environ. Monit. Assess.* **2018**, *190*, 1–6. [CrossRef] [PubMed]

8.   Dev, R.; Bali, M. Evaluation of groundwater quality and its suitability for drinking and agricultural use in district Kangra of Himachal Pradesh, India. *J. Saudi. Soc. Agric. Sci.* **2019**, *18*, 462–468. [CrossRef]

9.   Jasmin, I.; Mallikarjuna, P. Evaluation of groundwater suitability for irrigation in the Araniar River Basin, South India—A case study using Gis approach. *Irrig Drain.* **2015**, *64*, 600–608. [CrossRef]

10.   Li, P.; Wu, J.; Qian, H. Assessment of groundwater quality for irrigation purposes and identification of hydrogeochemical evolution mechanisms in Pengyang County, China. *Environ. Earth Sci.* **2013**, *69*, 2211–2225. [CrossRef]

11.   Gowd, S. Assessment of groundwater quality for drinking and irrigation purposes: A case study of Peddavanka watershed, Anantapur District, Andhra Pradesh, India. *Environ. Geol.* **2005**, *48*, 702–712. [CrossRef]

12.   Srinivasamoorthy, K.; Vasanthavigar, M.; Vijayaraghavan, K.; Sarathidasan, R.; Gopinath, S. Hydrochemistry of groundwater in a coastal region of Cuddalore district, Tamilnadu, India: Implication for quality assessment. *Arab. J. Geosci.* **2013**, *6*, 441–454. [CrossRef]

13.   Aghazadeh, N.; Mogaddam, A. Assessment of groundwater quality and its suitability for drinking and agricultural uses in the Oshnavieh Area, Northwest of Iran. *J. Environ. Prot.* **2010**, *1*, 30–40. [CrossRef]

14.   Gu, X.; Xiao, Y.; Yin, S.; Hao, Q.; Liu, H.; Hao, Z.; Meng, G.; Pei, Q.; Yan, H. Hydrogeochemical characterization and quality assessment of groundwater in a long-term reclaimed water irrigation area, North China Plain. *Water* **2018**, *10*, 1209. [CrossRef]

15.   Anim-Gyampo, M.; Anornu, G.; Appiah-Adjei, E.; Agodzo, S. Quality and health risk assessment of shallow groundwater aquifers within the Atankwidi basin of Ghana. *Groundw. Sustain.* **2019**, *9*, 100217. [CrossRef]

16.   Chidambaram, S.; Prasanna, M.; Venkatramanan, S.; Nepolian, M.; Pradeep, K. Groundwater quality assessment for irrigation by adopting new suitability plot and spatial analysis based on fuzzy logic technique. *Environ. Res.* **2022**, *204*, 1–15.

17.   Subramani, T.; Elango, L.; Damodarasamy, S. Groundwater quality and its suitability for drinking and agricultural use in Chithar River Basin, Tamil Nadu, India. *Environ. Geol.* **2005**, *47*, 1099–1110. [CrossRef]

18.   Adimalla, N.; Li, P.; Venkatayogi, S. Hydrogeochemical evaluation of groundwater quality for drinking and irrigation purposes and integrated interpretation with water quality index studies. *Environ. Process.* **2018**, *5*, 363–383. [CrossRef]

19.   Adimalla, N.; Vasa, S.; Li, P. Evaluation of groundwater quality, Peddavagu in Central Telangana (PCT), South India: An insight of controlling factors of fluoride enrichment. *Model. Earth Syst. Environ.* **2018**, *4*, 841–852. [CrossRef]

20.   Abd El-Aziz, S. Evaluation of groundwater quality for drinking and irrigation purposes in the north-western area of Libya (Aligeelat). *Environ. Earth Sci.* **2017**, *76*, 1–7. [CrossRef]

21.   Ehya, F.; Marbouti, Z. Hydrochemistry and contamination of groundwater resources in the Behbahan plain, SW Iran. *Environ. Earth Sci.* **2016**, *75*, 1–13. [CrossRef]

22.   Zhang, L.; Huang, D.; Yang, J.; Wei, X.; Qin, J.; Ou, S.; Zhang, Z.; Zou, Y. Probabilistic risk assessment of Chinese residents' exposure to fluoride in improved drinking water in endemic fluorosis areas. *Environ. Pollut.* **2017**, *222*, 118–125. [CrossRef]

23.   Aradpour, S.; Noori, R.; Tang, Q.; Bhattarai, R.; Hooshyaripor, F.; Hosseinzadeh, M.; Haghighi, A.; Klöve, B. Metal contamination assessment in water column and surface sediments of a warm monomictic man-made lake: Sabalan Dam Reservoir, Iran. *Hydrol. Res.* **2020**, *51*, 799–814. [CrossRef]

24.   Li, P.; He, X.; Li, Y.; Xiang, G. Occurrence and health implication of fluoride in groundwater of loess aquifer in the Chinese loess plateau: A case study of Tongchuan, Northwest China. *Expo. Health* **2019**, *11*, 95–107. [CrossRef]

25.   Yu, Y.; Yang, J. Health risk assessment of fluorine in fertilizers from a fluorine contaminated region based on the oral bioaccessibility determined by Biomimetic Whole Digestion-Plasma in-vitro Method (BWDPM). *J. Hazard Mater.* **2020**, *383*, 121124. [CrossRef] [PubMed]

26.   Emenike, C.; Tenebe, I.; Jarvis, P. Fluoride contamination in groundwater sources in Southwestern Nigeria: Assessment using multivariate statistical approach and human health risk. *Ecotoxicol. Environ. Saf.* **2018**, *156*, 391–402. [CrossRef] [PubMed]

27.   Zhai, Y.; Lei, Y.; Wu, J.; Teng, Y.; Wang, J.; Zhao, X.; Pan, X. Does the groundwater nitrate pollution in China pose a risk to human health? A critical review of published data. *Environ. Sci. Pollut. Res.* **2016**, *24*, 3640–3653. [CrossRef]

28.   Jianmin, B.; Yu, W.; Juan, Z. Arsenic and fluorine in groundwater in western Jilin Province, China: Occurrence and health risk assessment. *Nat. Hazards* **2015**, *77*, 1903–1914. [CrossRef]

29.   Qin, L.; Pang, X.; Zeng, H.; Liang, Y.; Mo, L.; Wang, D.; Dai, J. Ecological and human health risk of sulfonamides in surface water and groundwater of Huixian karst wetland in Guilin, China. *Sci. Total Environ.* **2020**, *708*, 134552. [CrossRef]

30.   Adimalla, N.; Qian, H. Groundwater quality evaluation using water quality index (WQI) for drinking purposes and human health risk (HHR) assessment in an agricultural region of Nanganur, south India. *Ecotoxicol. Environ. Saf.* **2019**, *176*, 153–161. [CrossRef]

31.   Aradpour, S.; Noori, R.; Naseh, M.; Hosseinzadeh, M.; Safavi, S.; Ghahraman-Rozegar, F.; Maghrebi, M. Alarming carcinogenic and non-carcinogenic risk of heavy metals in Sabalan dam reservoir, Northwest of Iran. *Environ. Pollut. Bioavailab.* **2021**, *33*, 278–291. [CrossRef]

32.   Fang, H.; Lin, Z.; Fu, X. Spatial variation, water quality, and health risk assessment of trace elements in groundwater in Beijing and Shijiazhuang, North China Plain. *Environ. Sci. Pollut. Res.* **2021**, *28*, 57046–57059. [CrossRef]

33.   Liu, J.; Gao, Z.; Zhang, Y.; Sun, Z.; Sun, T. Hydrochemical evaluation of groundwater quality and human health risk assessment of nitrate in the largest peninsula of China based on high-density sampling: A case study of Weifang. *J. Clean. Prod.* **2021**, *322*, 1–12. [CrossRef]

34. Zhai, Y.; Zhao, X.; Teng, Y.; Li, X.; Zhang, J.; Wu, J.; Zuo, R. Groundwater nitrate pollution and human health risk assessment by using HHRA model in an agricultural area, NE China. *Ecotoxicol. Environ. Saf.* **2017**, *137*, 130–142. [CrossRef]
35. Adimalla, N.; Li, P. Occurrence, health risks, and geochemical mechanisms of fluoride and nitrate in groundwater of the rockdominant semi-arid region, Telangana State, India. *Hum. Ecol. Risk Assess.* **2019**, *25*, 81–103. [CrossRef]
36. Aghapour, S.; Bina, B.; Tarrahi, M.; Amiri, F.; Ebrahimi, A. Distribution and health risk assessment of natural fluoride of drinking groundwater resources of Isfahan, Iran, using GIS. *Environ. Monit. Assess.* **2018**, *190*, 1–3. [CrossRef] [PubMed]
37. Li, P.; Qian, H.; Wu, J.; Chen, J.; Zhang, Y.; Zhang, H. Occurrence and hydrogeochemistry of fluoride in alluvial aquifer of Weihe River, China. *Env. Earth Sci.* **2014**, *71*, 3133–3145. [CrossRef]
38. Narsimha, A.; Sudarshan, V. Assessment of fluoride contamination in groundwater from Basara, Adilabad District, Telangana State, India. *Appl. Water. Sci.* **2017**, *7*, 2717–2725. [CrossRef]
39. Li, J.; Zhou, H.; Qian, K.; Xie, X.; Xue, X.; Yang, Y.; Wang, Y. Fluoride and iodine enrichment in groundwater of North China Plain: Evidences from speciation analysis and geochemical modeling. *Sci. Total Environ.* **2017**, *598*, 239–248. [CrossRef]
40. Fallahzadeh, R.; Miri, M.; Taghavi, M.; Gholizadeh, A.; Anbarani, R. Spatial variation and probabilistic risk assessment of exposure to fluoride in drinking water. *Food Chem. Toxicol.* **2018**, *113*, 314–321. [CrossRef] [PubMed]
41. Yadav, K.; Kumar, V.; Gupta, N.; Kumar, S.; Rezania, S.; Singh, N. Human health risk assessment: Study of a population exposed to fluoride through groundwater of Agra city, India. *Regul. Toxicol. Pharmacol.* **2019**, *106*, 68–80. [CrossRef]
42. Gu, X.; Xiao, Y.; Yin, S.; Pan, X.; Niu, Y.; Shao, J.; Cui, Y.; Zhang, Q.; Hao, Q. Natural and anthropogenic factors affecting the shallow groundwater quality in a typical irrigation area with reclaimed water, North China Plain. *Environ. Monit. Assess.* **2017**, *189*, 1–12. [CrossRef]
43. Liu, H.; Guo, H.; Yang, L.; Wu, L.; Li, F.; Li, S.; Ni, P.; Liang, X. Occurrence and formation of high fluoride groundwater in the Hengshui area of the North China Plain. *Environ. Earth Sci.* **2015**, *74*, 2329–2340. [CrossRef]
44. Zhao, X.; Guo, H.; Wang, Y.; Wang, G.; Wang, H.; Zang, X.; Zhu, J. Groundwater hydrogeochemical characteristics and quality suitability assessment for irrigation and drinking purposes in an agricultural region of the North China plain. *Environ. Earth Sci.* **2021**, *80*, 162. [CrossRef]
45. Shi, S.; Zhang, J. The current situation and countermeasures about the utilization of groundwater resources in Bazhou. *Hebei Water Res.* **2010**, *32*, 1–2. (In Chinese)
46. Zhang, G.; Yan, X.; Tian, Y.; Yan, M.; Wang, W. Response characteristic and mechanism of shallow groundwater level on the successive years with less rainfall in the agricultural areas of the central Hebei piedmont plain. *J. Earth Sci. Environ.* **2015**, *37*, 68–74, (In Chinese with English abstract).
47. Yan, L.; Qin, Y.; Yang, H.; Zhang, J.; Le, Z. Characteristics of Bazhou ground temperature change from 1964–2015 and its relationship with temperature and precipitation. *Mod. Agricult. Sci. Technol.* **2018**, *3*, 223–229. (In Chinese)
48. Zhang, Z.; Shi, D.; Fuhong, R.; Yin, Z.; Sun, J.; Zhang, C. Evolution of quaternary groundwater system in North China Plain. *Sci. China* **1997**, *40*, 276–283. [CrossRef]
49. Xing, L.; Guo, H.; Zhan, Y. Groundwater hydrochemical characteristics and processes along flow paths in the North China Plain. *J. Asian Earth Sci.* **2013**, *70–71*, 250–264. [CrossRef]
50. Najafzadeh, M.; Homaei, F.; Mohamadi, S. Reliability evaluation of groundwater quality index using data-driven models. *Environ. Sci. Pollut. Res.* **2021**. [CrossRef]
51. Raju, N. Hydrogeochemical parameters for assessment of groundwater quality in the upper Gunjanaeru River basin, Cuddapah District, Andhra Pradesh, South India. *Environ. Geol.* **2007**, *52*, 1067–1074. [CrossRef]
52. Abboud, I. Geochemistry and quality of groundwater of the Yarmouk basin aquifer, north Jordan. *Environ. Geochem. Health* **2018**, *40*, 1405–1435. [CrossRef]
53. USSL Salinity Laboratory. *Diagnosis and Improvement of Saline and Alkaline Soils*; US Department of Agriculture Handbook; No. 60; US Department of Agriculture: Washington, DC, USA, 1954.
54. Wilcox, L. *Classification and Use of Irrigation Water*, 4th ed.; USDA, Circular: Washington, DC, USA, 1955; p. 969.
55. *WHO Guidelines for Drinking Water Quality: Fourth Edition Incorporating the First Addendum*; World Health Organization: Geneva, Switzerland, 2017.
56. *WHO Guidelines for Drinking-Water Quality*; World Health Organization: Geneva, Switzerland, 2011.
57. U.S. EPA. *Risk Assessment Guidance for Superfund Volume I—Human Health Evaluation Manual*; Office of Emergency and Remedial Response: Washington, DC, USA, 1989.
58. U.S. EPA. *Risk Assessment Guidance for Superfund Volume I: Human Health Evaluation Manual*; Office of Superfund Remediation and Technology Innovation, U.S. Environmental Protection Agency: Washington, DC, USA, 2004.
59. Li, C.; Gao, X.; Wang, Y. Hydrogeochemistry of high-fluoride groundwater at Yuncheng Basin, northern China. *Sci. Total Environ.* **2015**, *508*, 155–165. [CrossRef]
60. Alaya, M.; Saidi, S.; Zemni, T.; Zargouni, F. Suitability assessment of deep groundwater for drinking and irrigation use in the Djeffara aquifers (Northern Gabes, south-eastern Tunisia). *Environ. Earth Sci.* **2014**, *71*, 3387–3421. [CrossRef]
61. Narsimha, A.; Sudarshan, V. Contamination of fluoride in groundwater and its effect on human health: A case study in hard rock aquifers of Siddipet, Telangana State, India. *Appl. Water Sci.* **2017**, *7*, 2501–2512. [CrossRef]
62. Wu, J.; Li, P.; Qian, H. Hydrochemical characterization of drinking groundwater with special reference to fluoride in an arid area of China and the control of aquifer leakage on its concentrations. *Environ. Earth Sci.* **2015**, *73*, 8575–8588. [CrossRef]

63. Li, J.; Wang, Y.; Xie, X.; Depaolo, D. Effects of water-sediment interaction and irrigation practices on iodine enrichment in shallow groundwater. *J. Hydrol.* **2016**, *543*, 293–304. [CrossRef]
64. Mahmood, Y.; Asghari, F.B.; Zuccarello, P.; Conti, G.O.; Ejlali, A.; Mohammadi, A.A.; Ferrante, M. Spatial Distribution Variation and Probabilistic Risk Assessment of Exposure to Fluoride in Ground Water Supplies: A Case Study in an Endemic Fluorosis Region of Northwest Iran. *Int. J. Environ. Res. Public Health* **2019**, *16*, 564.
65. Rao, N.; Sunitha, B.; Sun, L. Mechanisms controlling groundwater chemistry and assessment of potential health risk: A case study from South India. *Chemie. Der Erde–Geochem.* **2019**, *80*, 125568.
66. Gao, Z.; Han, C.; Xu, Y.; Zhao, Z.; Luo, Z.; Liu, J. Assessment of the water quality of groundwater in Bohai Rim and the controlling factors—a case study of northern Shandong Peninsula, north China. *Environ. Pollut.* **2021**, *285*, 117482. [CrossRef] [PubMed]
67. Narsimha, A.; Rajitha, S. Spatial distribution and seasonal variation in fluoride enrichment in groundwater and its associated human health risk assessment in Telangana State, South India. *Hum. Ecol. Risk Assess. Int. J.* **2018**, *24*, 2119–2132. [CrossRef]
68. Guo, H.; Zhang, Z.; Cheng, G.; Li, W.; Li, T.; Jiao, J. Groundwater-derived land subsidence in the North China Plain. *Environ Earth Sci.* **2015**, *74*, 1415–1427. [CrossRef]
69. Han, Y.; Zhai, Y.; Guo, M.; Cao, X.; Lu, H. Hydrochemical and Isotopic Characterization of the impact of water diversion on water in drainage channels, groundwater, and Lake Ulansuhai in China. *Water* **2021**, *13*, 3033. [CrossRef]
70. Wang, Y.; Chen, Z.; Fei, Y.; Liu, J.; Wei, W. Estimation of water released from aquitard compaction indicated by fluorine in Cangzhou. *J. Jilin Univ.* **2011**, *41*, 298–302. (In Chinese with English Abstract)
71. Zhang, R. Study of fluorine increase mechanism in deep groundwater in the Eastern Hebei Plain. *Hydrogeol. Engin. Geol.* **1991**, *18*, 56–58. (In Chinese with English Abstract)

Article

A SEEC Model Based on the DPSIR Framework Approach for Watershed Ecological Security Risk Assessment: A Case Study in Northwest China

Bin Wang [1,2], Fang Yu [2,*], Yanguo Teng [1,*], Guozhi Cao [2], Dan Zhao [2] and Mingyan Zhao [3]

[1] Engineering Research Center of Groundwater Pollution Control and Remediation of Ministry of Education, College of Water Sciences, Beijing Normal University, Beijing 100875, China; wangbin@caep.org.cn
[2] State Environmental Protection Key Laboratory of Eco-Environmental Damage Identification and Restoration, Chinese Academy of Environmental Planning, Beijing 100012, China; Caogz@caep.org.cn (G.C.); zhaodan@caep.org.cn (D.Z.)
[3] Hebei Institute of Water Science, Shijiazhuang 050051, China; mingyanzhao99@126.com
* Correspondence: yufang@caep.org.cn (F.Y.); teng1974@163.com (Y.T.)

Abstract: The DPSIR model is a conceptual model established by the European Environment Agency to solve environmental problems. It provides an overall framework for analysis of environmental problems from five aspects: driving force (D), pressure (P), state (S), impact (I), and response (R). Through use of the DPSIR model framework, this paper presents the SEEC model approach for evaluating watershed ecological security. The SEEC model considers four aspects: socioeconomic impact (S), ecological health (E), ecosystem services function (E), and control management (C). Through screening, 38 evaluation indicators of the SEEC model were determined. The evaluation results showed that the ecological security index of the study area was >80, indicating a generally safe level. The lowest score was mainly attributable to the low rate of treatment of rural domestic sewage. The water quality status was used to evaluate the applicability of the SEEC model, and the calculation results indicated that the higher the score of the ecological security evaluation results, the better the water quality status. The findings show that the SEEC model demonstrates satisfactory applicability to evaluation of watershed ecological security.

Keywords: watershed ecological security assessment; DPSIR model framework; environmental management

Citation: Wang, B.; Yu, F.; Teng, Y.; Cao, G.; Zhao, D.; Zhao, M. A SEEC Model Based on the DPSIR Framework Approach for Watershed Ecological Security Risk Assessment: A Case Study in Northwest China. *Water* 2022, 14, 106. https://doi.org/10.3390/w14010106

Academic Editor: Tammo Steenhuis

Received: 30 November 2021
Accepted: 31 December 2021
Published: 4 January 2022

Publisher's Note: MDPI stays neutral with regard to jurisdictional claims in published maps and institutional affiliations.

1. Introduction

The footprints of human activities have covered the world [1]. In the process of rapid development of both industry and agriculture, the ecological environment has suffered unprecedented damage [2,3]. Globally, the soil [4,5], water [6,7], air [8,9], and other environmental media in areas with frequent human activity are in a state of continuous deterioration [10,11]. Ecosystem degradation and environmental pollution are gradually threatening and destroying human socioeconomic progress, survival, and development [12,13]. In recent years, researchers have attempted to evaluate the consequences and degree of risk associated with current changes of the ecological environment but without reaching consensus [14,15].

China remains in the process of rapid economic growth and urbanization. However, various ecological and environmental problems continue to emerge, threatening to destroy China's sustainable development and affecting the living conditions of the population [16]. China has experienced ecological and environmental crises in the past and it now faces many new challenges regarding environmental protection. Therefore, President Xi proposed the idea of an ecological civilization, which means that China's model of development has changed from that of "grow first, clean up later" to one of sustainable development [17,18]. At the policy level and in everyday life, the expectation is for a safer and cleaner living environment. In the past, researchers often used the concept of environmental risk to assess whether the environment of a region might pose a threat to human health. Specifically, such assessments can be used to evaluate whether the current ecological situation of an area in which humans survive continues to be safe and whether it can ensure the environmental needs of human life and development. Therefore, evaluation of ecological safety is vital.

The main objective of ecological security assessment is to determine the ecological status and ecological pressure faced by a region under normal human activities [19]. It was formally proposed by the International Institute for Applied Systems Analysis in 1989 [20]. Evaluation results can be expressed using the ecological security index (ESI). A high ESI value indicates that the ecological state of the evaluation receptor is able to not only ensure the needs of human survival and development but also resist the pressures brought by human development. Therefore, through scientific evaluation of the factors on which the ESI is based, policies can be formulated to improve the situation. This approach also makes the work of ecological protection more refined and targeted, which is of great importance considering China's current state of development and environmental protection [21,22].

In accordance with different evaluation objects, ecological security can be divided into water ecological security [23], land ecological security [24], coastal ecological security [25], and urban ecological security [26], and all these aspects of ecological security assessment have been widely studied and applied. Broadly, ecological security includes natural ecological security, economic ecological security, and social ecological security, which mainly refers to a state in which human life, health, and resources are not threatened in terms of the above aspects. In this study, we were interested in the ecological security of a watershed, which refers to the ecological state of the lakes, rivers, and other areas within a catchment, and its ability to resist the ecological pressure brought by human activities.

Currently, model evaluation methods are used in ecological security assessment, and the most commonly used evaluation models include the PSR model, DSR model, and DPSIR model [27]. In 1979, Rapport and Friend proposed a model framework for analyzing and describing the interaction between socioeconomic development and the ecological environment, which was further improved by the Organization for Economic Co-operation and Development and United Nations Environment Programme, forming the PSR model framework [28]. The basic connotation of the PSR model is that human activities exert pressure on the environment and its natural resources (P-pressure), which changes the state of both the environment and the quality of the natural resources (S-state), forcing human society to respond to these state changes through adoption of policies, decisions, or management measures that affect the environment, economy, and land (R-response) [29]. The PSR model is suitable for ecological security evaluation on a small spatial scale and with few influencing factors. However, because it simplifies the causal relationship between indicators, it ignores the complexity of the system, especially the driving force factors of ecological security [30]. To overcome this weakness, the United Nations Conference on Sustainable Development established the DSR framework in 1996, in which the driving force factors (D) refer to the regional socioeconomic objectives which represent the fundamental environmental pressure. The DSR model can better characterize the impact of the driving force factors on ecosystem evolution, but the definitions of the driving force factors and the response factors in the model were vague. Therefore, to improve the applicability of the DSR model, in 1999, the European Environment Agency officially adopted the DPSIR model (driving force–pressure–state–influence–response), in which influence refers to the impact of changes in environmental status on environmental receptors [31]. The model combines the characteristics of the PSR and DSR frameworks. It has the advantages of comprehensiveness, systematicness, and flexibility, and covers five assessment factors and constructs a causal network between them to reflect the impact of socioeconomic development and human activities on the system state and the human response to adverse impacts [32].

In the DPSIR model framework suitable for watershed ecological security assessment, factor D generally includes population, socioeconomic, and other indicators; factor P generally includes pollutant discharge and other indicators; factor S generally includes water quality status, sediment status, and other indicators; factor I generally includes water service function and other indicators; and factor R generally includes river protection policy, ecological restoration, and other indicators. Generally, the DPSIR model framework is a circular system, i.e., the driving force leads to pressure, then the pressure changes the state, and the change of state has a consequential impact, which promotes a response that leads to adjustment of the driving force [33]. In recent years, the DPSIR model has been widely accepted and used in the process of ecological security research because it can reveal causal relationships between ecology and human activities [33–40]. It provides a conceptual model for a research scheme for evaluation of human activities, resources, the ecology, and sustainable development [41,42] and is also applied to interdisciplinary research [36,43]. Through application of the DPSIR model framework, many studies have performed ecological security evaluation of lakes, rivers, land, and oceans, thereby providing a scientific basis for further expansion of the connotation and application of the DPSIR model framework [44,45].

Although the DPSIR model has been used widely in many fields, applicability of the evaluation method has been limited owing to inconsistent selection of indicators, poor analysis of the reasons for the selection of indicators, and unclear determination of the process of index weighting [27,46–48]. Additionally, in previous watershed ecological security assessments, factors D and P were usually evaluated using socioeconomic and other related indicators, and ecological indicators were ignored, which resulted in overestimation of the impact of policy and economic development and underestimation of the ability of the ecosystem to deal with the pressure (factor P). Furthermore, the existing evaluation method lacks verification of the evaluation results, thereby diminishing the reliability and guidance of the evaluation results [49,50]. To resolve the problems of poor applicability of the DPSIR model to watershed ecological security evaluation and lack of a verification method, this study adopted the following research methods. By identifying the key factors affecting the ecological security of a watershed, and through analyzing the DPSIR model framework, the SEEC model including the process of indicator selection and the determination of weights was established. A study area was selected for application of the new model, and a method of water quality evaluation and analysis was innovatively used to evaluate the applicability of the SEEC model. The study area, located in an arid area in Northwest China, comprised a watershed that represents an important water supply source for a large city. However, owing to the specific geographical location and the harsh natural environment of the watershed, research data were scant, and therefore a watershed ecological security assessment was not undertaken. The ESI of the study area was obtained, and the reasons for low ESI values were analyzed, on the basis of which, suggestions for improvement of the ESI of the watershed were proposed. The results could serve both as a reference for subsequent environmental planning and management and as a scientific basis for comprehensive pollution control and ecological environmental protection of the study area.

2. Materials and Methods

2.1. Construction of the Evaluation Model

2.1.1. Identification of the Model

Currently, methods used for watershed ecological security assessment are not unified. By identifying the connotation of the model, this article established a method suitable for watershed ecological security assessment under the framework of the DPSIR model. However, the DPSIR model only provides an evaluation framework, and it does not offer methods for selecting and evaluating the applicability of indicators. Therefore, using the DPSIR model framework, this study identified the primary indicators of each factor and, in combination with consideration of the key issues of watershed ecological security assessment, the SEEC model was constructed. The essence of the SEEC model is that it is a representation of the DPSIR model framework specifically suited to watershed ecological security assessment. Therefore, in building the SEEC model, the connotation of the DPSIR model should be identified first, and an index system suitable for watershed ecological security assessment should also be established.

The essence of the DPSIR model is to identify the main factors affecting ecological security under the influence of human activities. It needs to determine the ecological state under the action of these factors, identify the impact, select relevant indicators of the response, evaluate the state of ecological security in terms of the five aspects, and obtain the ESI. Therefore, the key to using the DPSIR model to evaluate watershed ecological security is to accurately identify representative indicators that can characterize watershed ecological security.

The driving force factors (D) in the DPSIR model mainly represent social development and economic growth, reflecting the trends of population change, socioeconomic activity, and industrial economic development. These factors represent potential causes of environmental change, and they are also the most primitive and important indicators of change of the water environment security system.

Pressure factors (P) refer to the pressure applied directly to the water ecosystem through the driving force (D). Similar to D, P is an external force that affects the development and change of water ecosystem security. In previous research on urban ecological security, D and P were regarded as two separate factors, because the driving force and pressure can directly affect urban ecological security [51]. However, it is difficult to observe and calculate the impact of driving forces on watershed ecological security, mainly because most areas of many watersheds are not located in urban built-up areas and there are few human activities around. In this case, it needs to redefine the meaning of P used for watershed ecological security assessment under the DPSIR framework. Therefore, P mainly reflects the pollution load in this article.

State factors (S) refer to the state of the water ecosystem under the influence of both D and P. Thus, S can be illustrated directly through the characteristics of environmental media such as water quality and sediment, which are indicators of ecological health.

Impact factors (I) refer to the impact of the state of the water environment ecosystem on the economy and the livelihood and health of the population, which is the inevitable result of the interaction of the first three factors (D, P, and S). The ecological state changes caused by the above factors are mainly reflected in changes of the watershed ecological service functions. Therefore, the impact factors mainly include watershed ecological service functions such as water resources supply and the cultural landscape.

Response factors (R) refer to the countermeasures taken by humans to improve or adapt to the ecological state, which reflect the process of human regulation and management. Therefore, R mainly includes supervision of the ecological environment, pollution control, and industrial adjustment.

Because the DPSIR model overestimates nonecological factors such as the economy and population change, it is considered that the nature of the ecosystem itself is a more direct factor of ecological security, and thus it should receive greater attention. Moreover, with improvement of the level of ecological management, government departments also consider ecological factors when determining economic and population objectives, resulting in gradual closing of the relationship between the driving force (D) and pressure (P) factors. Considering D and P as a comprehensive factor (i.e., S—socioeconomic impact) allows more detailed analysis of the impact of human social activities on ecology. At the same time, by choosing ecological health (E) as the characterizing factor of S, ecosystem services function (E) as the characterizing factor of I, and control management (C) as the characterizing factor of R, the SEEC model can be established. The framework of the SEEC model is shown in Figure 1.

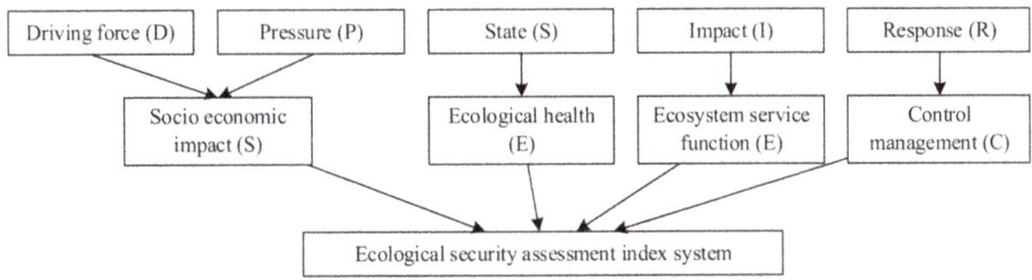

Figure 1. Framework of the SEEC model.

2.1.2. Construction of the System of Indicators

After determining the four factors of the SEEC model, the representative indicators must be screened to obtain the evaluation results of the SEEC model.

(1) Socioeconomic impact

As mentioned in the introduction, the factors D and P in the original DPSIR model were generally related to the socioeconomic factors in the conventional sense [52,53]. However, when assessing the ecological security of the watershed, the particularity of this assessment objective needs to be considered, that is, for the ecological security of the watershed, the driving force and pressure are not very intuitive in the general sense. This paper needs to consider the driving force and pressure indicators that have a more significant impact on the ecological security of the watershed. Moreover, in the assessment of watershed ecological security, the applicability of each assessment factor should be considered in a balanced manner, and the original DPSIR model cannot be relied on completely, which will over amplify the impact of D and P. Therefore, socioeconomic impact factors mainly include the socioeconomic development and pollution load attributable to human activities in this article. Socioeconomic development mainly includes population, economic, and social indicators. Population indicators include the quantity, density, and natural growth rate of the population. Economic indicators are mainly used to determine the level of economic development and the intensity of economic activities. Therefore, economic indicators that can fully represent the structure and scale of the economy should be selected, e.g., GDP, per capita GDP, and total industrial and agricultural output value. Social indicators are relatively comprehensive and can be expressed by the urbanization rate.

Watershed pollution load is the primary way in which human activities affect water ecology. Generally, point source or nonpoint source discharge of pollutants can have serious impact on a watershed.

(2) Ecological health

Ecological health is reflected by the water quality and water ecology. In the process of watershed monitoring, water quality is commonly assessed using dissolved oxygen, total nitrogen, total phosphorus, the permanganate index, ammonia nitrogen, transparency, suspended solids, chlorophyll a, and heavy metal indicators.

In addition to the above indexes, water ecological indicators also include zooplankton, phytoplankton, and the benthic biomass. However, because it is very difficult to monitor the above indicators within many watersheds, the quality of sediment is often used to characterize the ecological status of water.

(3) Ecosystem services

The ecological services function of a watershed is mainly reflected in purifying the water quality, providing aquatic products, and providing cultural tourism services. Considering that aquaculture and fishing are no longer allowed in many watersheds, inclusion of this indicator was dependent on the specific situation of the watershed. The water purification function of a watershed is generally realized through natural shoreline filtration on both banks of the river. Cultural tourism services are also related to the geographical location of the watershed.

(4) Control management

Control management is mainly reflected in policies related to the economy, ecology, and environment, including ecological protection capital investment, industrial structure adjustment, and ecological supervision capacity building.

Through analysis of the connotation, scientificity, representativeness, and applicability of each indicator and following literature-based research, 34 evaluation indicators were determined. See Table A1 for the definition, reasons for selection, and calculation or acquisition method of each indicator.

2.1.3. Data Processing

(1) Data standardization

All indicators must be standardized to facilitate ecological security assessment. The concept of a reference standard should be introduced, meaning the value of each evaluation index in the ideal state (conducive to ecological security) or at the average level in a large-scale region [54].

The process of standardization of the indicators is conducted as follows:

$$\text{Positive indicator: } R_{ij} = X_{ij}/S_{ij}, \tag{1}$$

$$\text{Negative indicator: } R_{ij} = S_{ij}/X_{ij}, \tag{2}$$

where positive indicator means that the larger the value of the indicator, the more favorable it is to ecological security; negative indicator means that the larger the value of the indicator, the more unfavorable it is to ecological security; X_{ij} is the measured value of indicator i at sampling point j; S_{ij} is the reference standard of indicator i; and R_{ij} is the dimensionless value of the evaluation indicator, where $0 < R_{ij} < 1$; when $R_{ij} > 1$, we take $R_{ij} = 1$.

(2) Weight calculation

The main methods used to determine the weights are the subjective weight method and the objective weight method. The most common subjective weighting method is the Delphi method, also known as the expert scoring method [55]. Its advantages are its clear concept, simplicity, and ease of operation, which can grasp the main factors of ecological security assessment, but it needs a certain number of experts with experience to produce the scores. The objective weighting method determines the index weights using a judgment matrix composed of evaluation index values. The most commonly used objective weighting method is the entropy method, which uses the utility values of the index information in the calculation; the higher the utility value, the more important it is to the evaluation. The SEEC model involves four factors, each of which contains information with differing complexity. Thus, the objective weighting method will lead to some factors with less impact on ecological security obtaining higher weighting, making it difficult to truly reflect the importance of the factors. Therefore, the weights of the four factors were determined using the expert scoring method. The full score was set at 10 and the higher the score, the more important the factor. Using a judgment matrix after sorting the consultation results, the weighting coefficient of the index was calculated, and the entropy method was used to determine the weights of the 34 indicators.

The process of application of the entropy method is as follows [56]:

(a)   Construct the judgment matrix Z for n samples and m evaluation indicators:

$$Z = \begin{bmatrix} X_{11} & X_{12} & \cdots & X_{1m} \\ X_{21} & X_{22} & \cdots & X_{2m} \\ \cdots & \cdots & \cdots & \cdots \\ X_{n1} & X_{n2} & \cdots & X_{nm} \end{bmatrix} \tag{3}$$

(b) The dimensionless data are used to obtain a new judgment matrix, in which the expression of the element is:

$$R = (r_{ij}\, n \times m) \tag{4}$$

(c) According to the definition of entropy, for n samples and m evaluation indicators, the entropy of the evaluation indicators can be determined as follows:

$$H_i = -\frac{1}{\ln(n)}\left[\sum_{i=1}^{n} f_{ij}\, \ln f_{ij}\right] \tag{5}$$

$$f_{ij} = \frac{r_{ij}}{\sum\limits_{i=1}^{n} r_{ij}} \tag{6}$$

where $0 \leq H_i \leq 1$.

To make $\ln f_{ij}$ meaningful, it is assumed that $f_{ij} = 0$, $f_{ij}\ln f_{ij} = 0$, $i = 1,2, \ldots ,m$, and $j = 1,2, \ldots ,n$.

(d) Calculate the entropy weight (W_i) of the evaluation indicators:

$$W_i = \frac{1 - H_i}{m - \Sigma H_i} \tag{7}$$

where W_i is the weighting coefficients of the evaluation indicators that meet the following requirement:

$$\Sigma W_i = 1 \tag{8}$$

(3) Expression of the evaluation results

The evaluation results are expressed using the ESI; the higher the ESI value, the safer the ecological status. On the basis of the general expression of river and lake ecosystem assessment results in China, the ESI is divided into five levels from 0–100, as shown in Table 1.

Table 1. Classification standard of the ESI.

Classification	Ecological Security Index (ESI)	Safety Status
I	$80 \leq \text{ESI} \leq 100$	Safe
II	$60 \leq \text{ESI} < 80$	Relatively safe
III	$40 \leq \text{ESI} < 60$	Generally safe
IV	$20 \leq \text{ESI} < 40$	Relatively unsafe
V	$\text{ESI} < 20$	Unsafe

2.2. Study Area

The study area is located in Northwest China. The watershed is 214 km long and 25–50 km wide from east to west. The drainage area is 4684 km^2 and the annual runoff is 237 million m^3. The average elevation is 2545.63 m (Figure 2).

Figure 2. Overview of the study area.

The study area is affected mainly by the mid-latitude near-surface atmospheric circulation. The annual average temperature of the watershed is 3.5 °C. The temporal distribution of precipitation is uneven, falling mainly in June–August. The annual average precipitation of the watershed is approximately 420 mm, and its spatial distribution is very uneven. Solid precipitation accounts for approximately 54.2% of the annual total precipitation. According to the results of the "Water Resources Bulletin," the average annual precipitation in the study area was 1.28 billion m^3 during 2006–2013, with no obvious trend of increase or decrease. In 2015, the amount of surface water resources within the study area was 1.160 billion m^3, and the amount of groundwater resources was 574 million m^3. The total amount of the water supply was maintained at more than 1.1 billion m^3, and more than half of the water supply was derived from surface water.

The soil distribution in the upper reaches of the study area has vertical zonation. In addition to ice and snow cover, at elevations above 3600 m, exposed rocks and stone mounds are widely distributed, although some areas have poorly developed soil. The soil in the elevation range of 3100–3400 m is mainly alpine meadow soil; 2000–3100 m is mainly subalpine meadow soil; 1700–2000 m is mountain chernozem; 1200–1700 m is mountain chestnut soil; and 800–1200 m is brown calcareous soil. The soil at 1700–2900 m has mosaic distribution characteristics. Taupe forest soil is mainly found on shady slopes at elevation of 1700–2900 m, and it is distributed in a compound area with chernozem and subalpine meadow soil. The climate and terrain in the watershed vary markedly, and the corresponding vegetation types are relatively complete with obvious regularity in terms of geographical distribution. The main vegetation types are coniferous forest, broadleaved forest, shrub, grassland, meadow, and desert grassland. We used ArcGIS 10.3 to interpret the land use of the study area in 2014, and the results showed that grassland was the main vegetation cover type, followed by woodland and cultivated land (Figure 3).

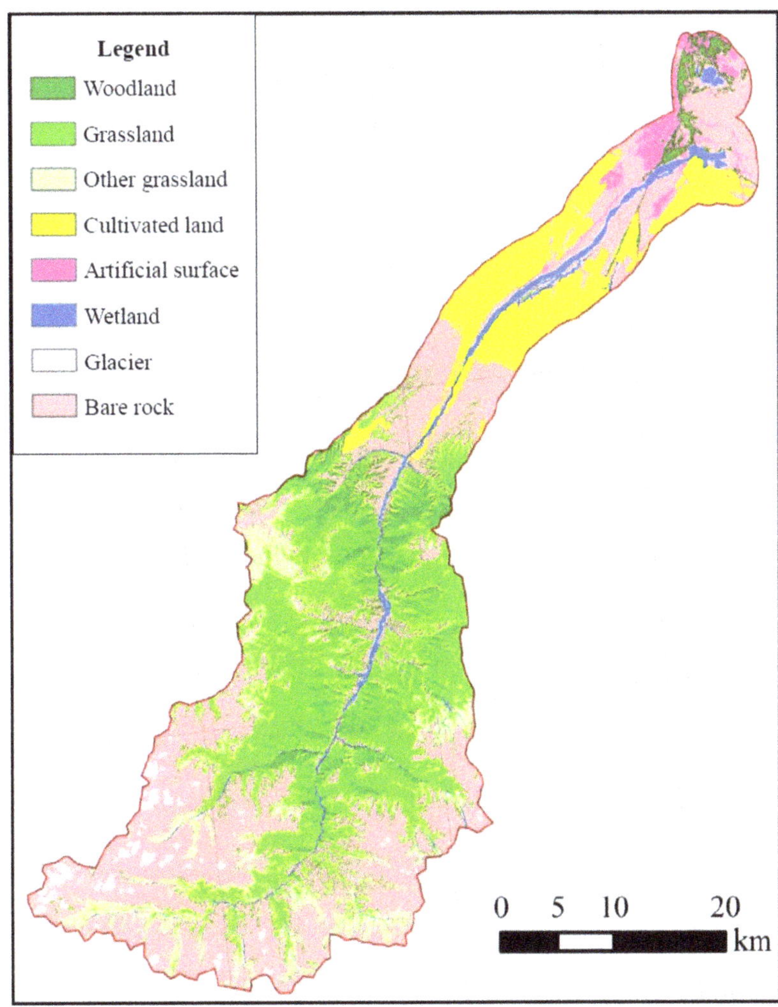

Figure 3. Land use types within the study area (2014).

In addition to the main stream, the study area includes one main glacier and three reservoirs (Figure 2). The main glacier, which is the source of the river, presently covers an area of 1.62 km^2. Taking this glacier as the center, 109 large and small modern glaciers are developed within an area of 300 km^2, comprising a total glacier area of 38.3 km^2. Among them, 22 glaciers are distributed within the study area, covering a total area of 9.7 km^2. The annual runoff of glacier meltwater into the trunk stream is 17.69 million m^3. Therefore, glaciers are not only an important water source in the study area but they also represent a "solid reservoir" regulating runoff. Reservoir 1 (Figure 2) is located in the upper reaches of the study area. It is a water conservancy project that has flood control and irrigation as its primary and secondary purposes. Its operation is mainly divided into reservoir closure, dam flood control, and sluice water storage. The dam crest elevation is 2189.2 m, and the total storage capacity is 69.9 million m^3. The highest dam height is 98 m. Reservoir 1 can exploit its flood-retention capability to cut the peak flood and regulate and store the flood during the flood season. Reservoir 2 (Figure 2) is located in a mountain depression of the middle reaches of the study area. It is a series regulating reservoir upstream of the urban area. It undertakes the three tasks of flood control, drought relief, and urban water supply for the city. The catchment area above the section of Reservoir 1 is 2596 km^2, including 1070 km^2 in the mountainous area of the main stream, 950 km^2 in the piedmont plain area, and 576 km^2 in the mountainous area of tributary gullies. Reservoir 3 (Figure 2) is located 6 km downstream of Reservoir 2. It is a through-injection reservoir built using natural depressions. The current water surface area is 4.5 km^2 and the maximum storage capacity is 53 million m^3. It is a large reservoir for comprehensive flood discharge irrigation, power generation, fish aquaculture, and urban water supply.

To obtain accurate data, a number of investigations were conducted in the study watershed during 2016–2017. These investigations included discussion with local research departments, data acquisition through visits to local government departments, collection of water and sediment samples, and investigation of pollution sources. Thus, data regarding the 34 evaluation indicators were obtained. Details were given in Table A2, which described the source of original data. Tables A3 and A4, respectively, introduced the weights and reference standards of each indicator.

3. Results and Discussion

3.1. Results of the Ecological Security Assessment

Through evaluation of the SEEC model, it was determined that the ESI value of the study area is 80.9–94.2. In accordance with the geographical characteristics, the watershed was divided into three sections:

- Upstream area: the section from Glacier No. 1 to the region downstream of Reservoir 1 (hereafter, the upstream area);
- Midstream area: downstream of Reservoir 1 to the region upstream of Reservoir 2 (hereafter, the midstream area);
- Downstream area: the section from the region upstream of Reservoir 2 to Reservoir 3 (hereafter, the downstream area).

The assessment results indicate that the status of the entire watershed was in the "safe" state, as defined in Table 1, indicating that the ecological security level was high. The ESI values of the entire study area are shown in Figure 4. It can be seen that the highest ESI values in the entire watershed are in the upstream area, and that the lowest ESI values are near Reservoir 2 in the downstream area. The ESI values of the midstream area are at an intermediate level. To identify the causes of the low ESI scores, we examined the scores for the four evaluation factors in the SEEC model using radar charts, as shown in Figure 5. It can be seen that the scores for the four evaluation factors are uneven. The lowest score is for C, indicating deficiencies in watershed regulation and management, and the second lowest value is for E (ecological health), indicating the need for attention to improve the state of ecological health. The scores for S and E (ecological services function) are high, indicating that the current socioeconomic factors have not impacted negatively on the ecological security of the watershed. Although the score for ecological health is low, it might not have affected the ecological services function of the watershed. Nevertheless, to improve the ESI score of the watershed, factors C and E (ecological health) need to be the focus of attention.

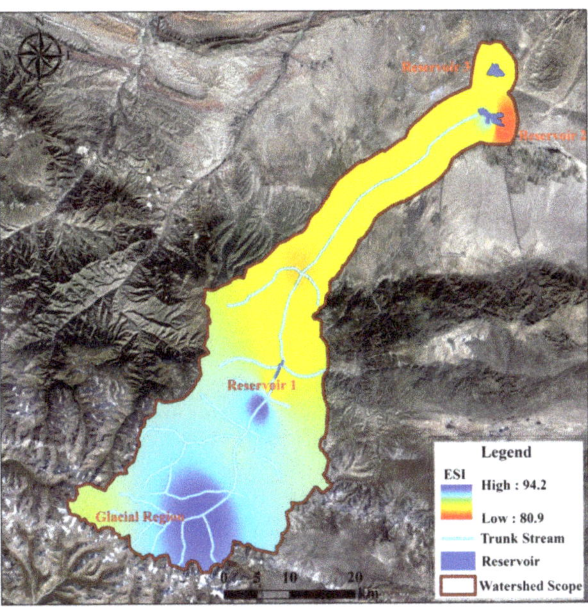

Figure 4. Ecological security index (ESI) values throughout the entire study area.

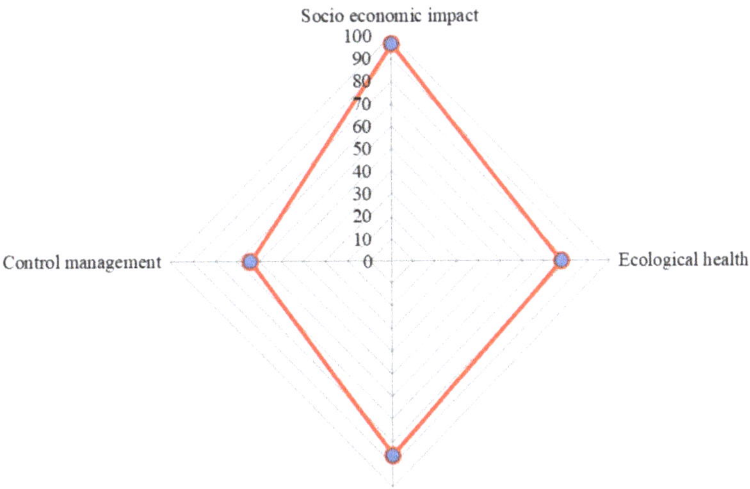

Figure 5. Radar diagram for the four factors in the SEEC model.

3.2. Score for Each Evaluation Factor

(a)   Assessment Results of Socioeconomic Impact (S)

The score for S is approximately 96, indicating the positive role it plays in ensuring ecological security. The score distribution of the 10 indicators is shown in Figure 6a. Among the 10 indicators, those with relatively low scores are population growth rate and per capita GDP. The lowest values of both are in the upstream and midstream areas, which are regions with poor natural conditions, sparse population, and low per capita GDP. In the past, it was often believed that if the population growth rate and per capita GDP were low, the pressure on ecology would be small and ecological security would not be threatened. The assessment reveals that levels of population growth rate and per capita GDP that are too low are not conducive to ecological security and thus they should be maintained at reasonable levels.

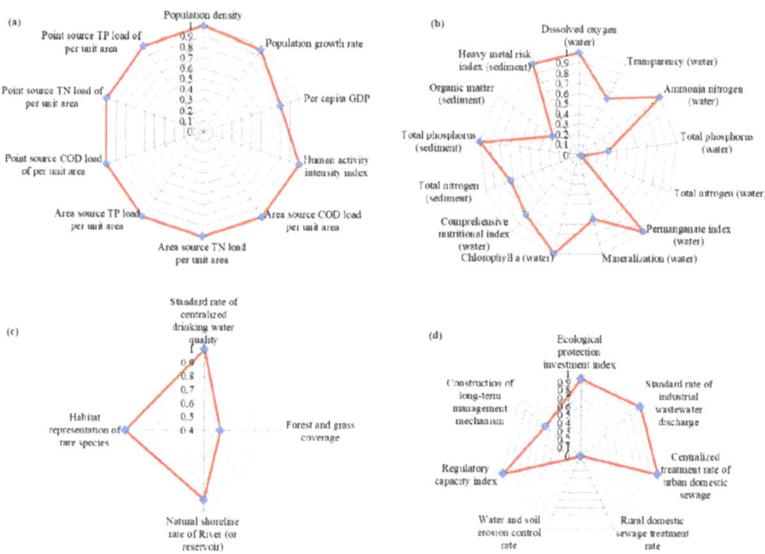

Figure 6. Radar diagrams for the four factors in the SEEC model: (**a**) socioeconomic impact assessment results, (**b**) ecological health assessment results, (**c**) ecological services function assessment results, and (**d**) control management assessment results.

(b)   Assessment Results of Ecological Health (E)

The score for E (ecological health) is about 78, indicating the negative role it has in ecological security. The score distribution of the 13 indicators is shown in Figure 6b. Among the 13 indicators, the value for the total nitrogen in the water is too high, followed by the comprehensive nutritional index, organic matter, and heavy metal risk index. In terms of the spatial distribution, the comprehensive nutritional index in the upstream area is relatively low, but it increases gradually from the midstream area to the downstream area, reaching its highest value near Reservoir 3, which mainly reflects the intensity of human activity and tourism development [57]. The heavy metal risk index in the upstream and midstream areas is low, while the highest value is near Reservoir 2. The evaluation results of the comprehensive nutritional index and heavy metal risk index are shown in Figure 7.

(c)   Assessment Results of Ecosystem Services Function (E)

The calculation result of E (ecosystem services function) is approximately 86, indicating its negative role in ecological security. The score distribution for the four indicators is shown in Figure 6c. The main reason for the low score for E is the low coverage of forest and grass, especially in the upstream and midstream areas, which mainly reflects the relatively intense water and soil erosion in these areas in recent years. Moreover, human activities such as livestock grazing and free-range poultry breeding have aggravated grassland degradation [58]. Additionally, the natural shoreline rate of the river is also low, which is mainly attributable to the construction of embankments, diversion channels, and other human projects in the downstream area [59].

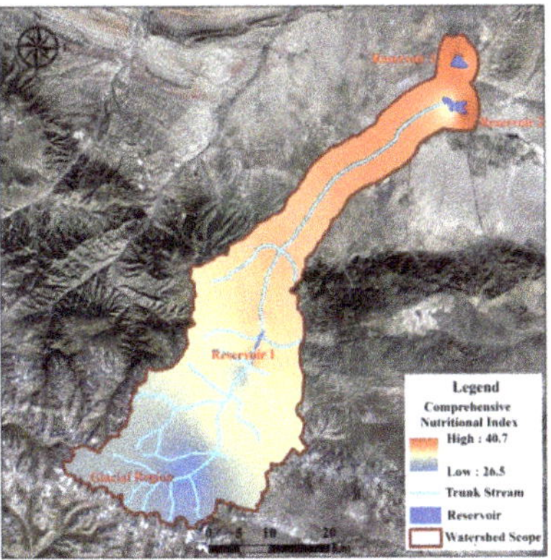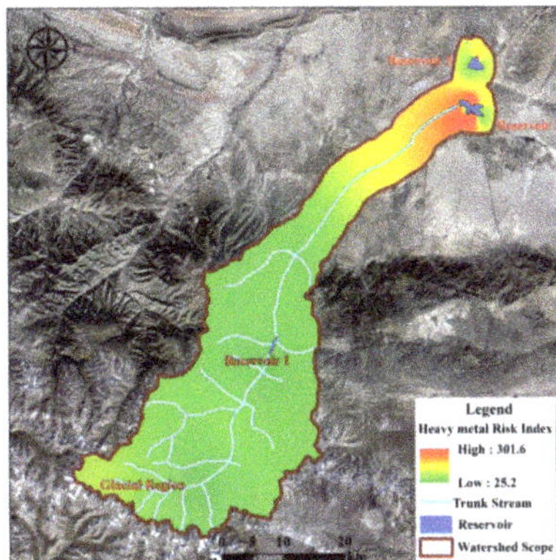

Figure 7. Scoring results for the comprehensive nutritional index (**left**) and potential heavy metals ecological risk index (**right**).

(d) Assessment Results of Control Management (C)

The score for C is approximately 64, indicating its negative role in ecological security. The score distribution for the seven indicators is shown in Figure 6d. The low rate of treatment of rural domestic sewage is the main problem in relation to the low C score. The population of the watershed is sparse and scattered, meaning that the high cost of construction of environmental infrastructure make it difficult to collect and treat domestic sewage. The score for the soil and water erosion control rate is the second biggest problem in relation to the low C score. Water and soil erosion in the upstream and midstream areas is serious, and the grassland is deteriorating steadily. The effect of implemented mitigation measures has been limited because the dry climate and steep terrain in the upstream and midstream areas are not conducive to the control of water and soil erosion.

3.3. Applicability Evaluation of the SEEC Model

To evaluate the applicability of the SEEC model, factors that could characterize the level of watershed ecological security were selected. If the assessment results of the SEEC model indicated that these factors were relevant, then the evaluation method could be considered to have satisfactory applicability. In assessment of watershed ecology, water quality status is often used as an important comprehensive assessment factor with which to characterize whether the current ecology is threatened, i.e., whether the ecological status is safe. Therefore, the water quality of the study watershed was selected as the assessment factor for evaluation of the applicability of the SEEC method. To keep the assessment independent, we gave priority to the correlation analysis between the selected water quality indicators for evaluation (hereafter, the water evaluation indicators) and the water quality indicators participating in the evaluation of the SEEC model (hereafter, the water model indicators). The water evaluation indicators comprised pH, conductivity, biochemical oxygen demand, petroleum, volatile phenol, mercury, lead, copper, zinc, fluoride, selenium, arsenic, cadmium, hexavalent chromium, cyanide, anionic surfactant, sulfide, fecal Escherichia coli, sulfate, chloride, nitrate, and suspended solids. The water model indicators comprised dissolved oxygen, transparency, ammonia nitrogen, total phosphorus, total nitrogen, the permanganate index, mineralization of water, chlorophyll a, and the comprehensive nutritional index. The water quality data of the evaluation indicators were obtained from the local environmental monitoring department.

First, the monitoring data at the same point and for the same period were selected, and IBM SPSS 20.0 software was used to calculate the Pearson correlation coefficient of the two datasets (i.e., water evaluation indicators and water model indicators). Evaluation indicators that showed obvious correlation with the water model indicators were eliminated.

Additionally, because the SEEC model included the heavy metal risk index of sediment, the heavy metal indicators were also removed from the water evaluation indicators to avoid affecting the independence of the index. Through the screening process, the evaluation indicators were determined as follows: pH, petroleum, volatile phenol, fluoride, cyanide, anionic surfactant, sulfide, fecal Escherichia coli, sulfate, chloride, and suspended solids. There was no obvious correlation between the final water evaluation indicators and the water model indicators, indicating that the final water evaluation indicators were independent and could be used to evaluate the applicability of the SEEC model.

Second, using the water evaluation indicators, the Nemero index method [60] was used to evaluate water quality, and the evaluation results were expressed in terms of China's surface water environmental quality standard. The evaluation results were characterized as the higher the score, the worse the water quality. Obviously, when the water quality of the watershed is poor, its ecological security level is low. In other words, under natural conditions, the water quality score results should be negatively correlated with the watershed ecological security index. If the evaluation result of this paper also conforms to the above rules, it shows that the evaluation result of this paper is relatively accurate, and the evaluation method has good applicability.

Then, ArcGIS 10.3 was used for spatial interpolation to obtain the spatial distribution characteristics of water quality, as shown in Figure 8. Finally, the Pearson correlation coefficient between the data illustrated in Figure 8 and the SEEC evaluation results (Figure 4) was calculated using the ArcGIS 10.3 spatial analysis module. Through calculation, the Pearson correlation coefficient was determined to be approximately −0.4, which was consistent with the rules described above, indicating that the comprehensive evaluation results of water quality were better in areas with high ecological security scores. The evaluation results showed that the SEEC model has satisfactory performance when applied to evaluation of watershed ecological security.

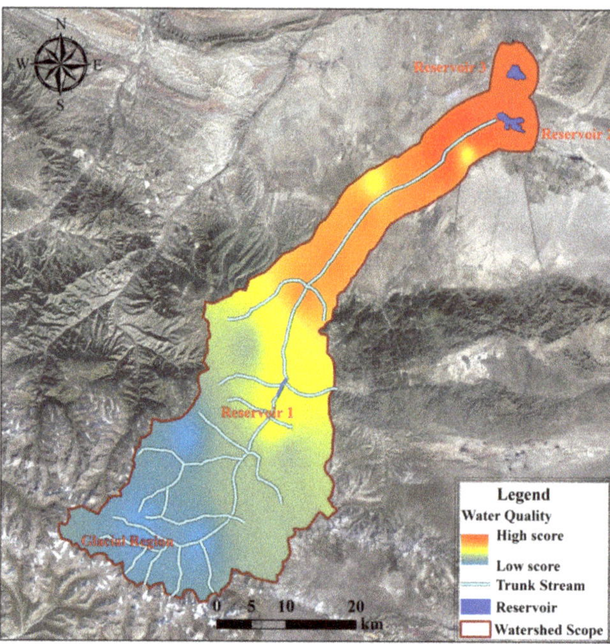

Figure 8. Comprehensive evaluation results of water quality in the study area in 2014 (the lower the score, the better the water quality).

4. Conclusions

Using the framework of the DPSIR model, and considering the connotation of watershed ecological security assessment, this article established the SEEC model that incorporates four factors: socioeconomic impact (S), ecological health (E), ecosystem services function (E), and control management (C). The SEEC model contains 34 evaluation indicators. We selected a watershed in the arid region of Northwest China for application of the research method. Through evaluation, the ESI value of the study area was approximately 81–94. Through analysis of the evaluation results, it was elucidated that the factors leading to the low score were mainly the low rate of treatment of rural domestic sewage and the low rate of mitigation of soil and water erosion. The results of the evaluation of the applicability of the SEEC model showed that the SEEC model has satisfactory performance in evaluation of watershed ecological security.

This article is successful from these aspects: it puts forward the method approach for watershed ecological security assessment and gives the specific evaluation index, index weight, and other key information; through application, it identifies the key factors affecting the ecological security of the study area, which has good guiding significance for local environmental management and can also provide reference for similar studies. However, owing to limited research funds and other reasons, this article fails to verify the SEEC model, such as through comparison of analysis of the ecological security status in different years. Therefore, more in-depth research is needed regarding the applicability and verification of the SEEC model.

Author Contributions: Conceptualization, F.Y., B.W. and G.C.; methodology, B.W. and F.Y.; software, B.W.; validation, B.W.; formal analysis, B.W.; investigation, B.W.; data curation, Y.T.; writing—original draft preparation, B.W.; writing—review and editing, Y.T. and M.Z.; visualization, B.W. and D.Z.; supervision, Y.T.; project administration, G.C. and D.Z.; funding acquisition, M.Z. All authors have read and agreed to the published version of the manuscript.

Funding: This research was funded by the National Key Research and Development Program of China (2019YFC1804404), Beijing Advanced Innovation Program for Land Surface Science of China and Hebei Province Water Conservancy Science and Technology Project (2018-09).

Institutional Review Board Statement: Not applicable.

Informed Consent Statement: Not applicable.

Data Availability Statement: Not applicable.

Acknowledgments: During the investigation, the local Environmental Protection Bureau, local environmental monitoring station, local Agriculture and Animal Husbandry Bureau, local Water Affairs Bureau, local Health Bureau, local water industry group, local Institute of Ecology and Geography, and the Chinese Academy of Sciences provided help regarding the data used in the study. In the process of establishing research methods and analyzing data, experts in environmental science, hydrology, water ecology, biology, and statistics from the local Institute of Ecology and Geography, Chinese Academy of Sciences, Ecological Center of the Chinese Academy of Sciences, Chinese Academy of Environmental Sciences, and other institutions provided assistance. We thank James Buxton, from Liwen Bianji (Edanz) (accessed on 29 November 2021), for editing the English text of a draft of this manuscript.

Conflicts of Interest: The authors declare no conflict of interest.

Appendix A

Table A1. Definition, reasons for selection, and calculation or acquisition method of each indicator.

Factors	Involved Aspects	Indicators	Definition and Reasons for Selection	Method of Calculation (within the Study Area)
Socioeconomic impact	Population	Population density	Population of per unit area. It is an important factor in the impact of social economy on the environment. It affects the allocation of resources and the surplus of environmental capacity.	Total population/area
		Population growth rate	The ratio of population growth to the total population in a given period of time (usually one year). It is an important indicator of population growth.	(Population at the end of the year - population at the beginning of the year)/annual average population $\times$ 1000‰
	Economy	Per capita GDP	Regional GDP per capita in the study area. It is the most common indicator to measure the level and pressure of social and economic development. It can not only reflect the development of social economy but also indirectly reflect the pressure of social and economic activities on the environment to a certain extent.	Total GDP/total population
	Social	Human activity intensity index	Proportion of the sum of construction land area and agricultural land area in the total land area, which are the main land types reflecting the intensity of human activities, which can reflect the pressure of social and economic activities on the environment at present and in the next few years.	(Construction land area + agricultural land area)/total area
	Watershed pollution load	Area source chemical oxygen demand (COD) load per unit area	COD load per unit land area, which mainly includes COD emissions from area sources such as livestock and poultry free range breeding, planting, rural residents' life, and so on. It is one of the most important evaluation indexes of environmental pollution. Considering the horizontal comparison between different watersheds and different statistical units, the COD load per unit area is used as the evaluation index.	(COD emission of livestock and poultry free range breeding + COD emission of planting + COD emission of rural residents' life)/total area
		Area source TN load per unit area	TN load per unit land area, which mainly includes TN emissions from area sources such as livestock and poultry free range breeding, planting, rural residents' life, and so on. It is one of the main factors leading to river eutrophication.	(TN emission of livestock and poultry free range breeding + TN emission of planting + TN emission of rural residents' life)/total area
		Area source TP load per unit area	TP load per unit land area, which mainly includes TP emissions from area sources such as livestock and poultry free range breeding, planting, rural residents' life, and so on. It is one of the main factors leading to river eutrophication.	(TP emission of livestock and poultry free range breeding + TP emission of planting + TP emission of rural residents' life)/total area
		Point source COD load of per unit area	Point source COD load of per unit area, which mainly includes urban industrial COD emission and urban living COD emission. The reason for selection is similar to area source but point source and area source characterize different processes of pollution.	(Urban industrial COD emission and urban living COD emission)/total area

Table A1. *Cont.*

Factors	Involved Aspects	Indicators	Definition and Reasons for Selection	Method of Calculation (within the Study Area)
Ecological health		Point source TN load of per unit area	Point source TN load of per unit area, which mainly includes urban industrial TN emission and urban living TN emission. The reason for selection is same as above.	(Urban industrial COD emission and urban living COD emission)/total area
		Point source TP load of per unit area	Point source TP load of per unit area, which mainly includes urban industrial TP emission and urban living TP emission. The reason for selection is same as above.	(Urban industrial COD emission and urban living COD emission)/total area
	Water quality	Dissolved oxygen	Molecular oxygen dissolved in water is an important index to judge water quality and an important item of water quality monitoring. The growth and reproduction of phytoplankton in water and pollutants in water will affect dissolved oxygen. It is an important indicator of water quality.	Field measurement
		Transparency	It reflects the clarification degree of the water body, which is related to the content of suspended solids and colloids in the water. It is an important indicator for evaluating water eutrophication.	Field measurement
		Ammonia nitrogen	Refers to nitrogen in the form of free ammonia (NH_3) and ammonium ion (NH_4^+) in water. It is an important indicator for evaluating water quality.	Testing
		Total phosphorus	The total amount of various organic and inorganic phosphorus in water, which is generally expressed by the determination results after various forms of phosphorus are transformed into orthophosphate after digestion. It is the key indicator to evaluate the degree of water eutrophication and water quality.	Testing
		Total nitrogen	The total amount of various forms of inorganic and organic nitrogen in water. The reason for selection is same as above.	Testing
		Permanganate index	It refers to the amount of oxidant consumed when treating water samples with potassium permanganate ($KMnO_4$) as oxidant under certain conditions. It is an important indicator to evaluate water quality.	Testing
		Mineralization of water	It refers to the amount of carbonate, bicarbonate, chloride, sulfate, nitrate, and various sodium salts containing calcium, magnesium, aluminum, manganese, and other metals in water. The reason for selection is same as above.	Testing
	Water eutrophication	Chlorophyll a	It is an important photosynthetic pigment in plant photosynthesis. By measuring chlorophyll a of phytoplankton, it can master the primary productivity of water. At the same time, chlorophyll a content is also one of the indicators of water eutrophication. It is an important indicator to reflect eutrophication and algal biomass of water.	Testing

Table A1. *Cont.*

Factors	Involved Aspects	Indicators	Definition and Reasons for Selection	Method of Calculation (within the Study Area)
		Comprehensive nutritional index	It is a comprehensive indicator to reflect water eutrophication.	Taking the state index TLI (Chla) of chlorophyll a as the benchmark, the nutritional state indexes of TP, TN, COD, SD, and other parameters close to the benchmark parameters (with small absolute deviation) are selected for weighted synthesis with TLI (Chla). The comprehensive weighted index model is: $$TLI(\Sigma) = \sum_{j=1}^{M} W_j \cdot TLI(j)$$ where: TLI (å) is the comprehensive weighted nutritional status index; TLI (j) is the nutritional status index of the j-th parameter; W_j is weight of nutritional status index of the j-th parameter. $$W_j = \frac{R_{ij}^2}{\sum_{j=1}^{M} R_{ij}^2}$$ where: R_{ij} means correlation coefficient between the jth parameter and the benchmark parameter, M is number of main parameters close to the benchmark parameter.
		Total nitrogen	Nitrogen content in sediments. It is an important indicator for evaluating sediment quality.	Testing
		Total phosphorus	Phosphorus content in sediments. The reason for selection is same as above.	Testing
	Sediment	Organic matter	It generally refers to substances derived from life in sediments, including sediment. Microorganisms, benthos, and their secretions, as well as plant residues and plant secretions in soil. The reason for selection is same as above.	Testing
		Heavy metal risk index	It is a relatively fast, simple, and standard method to classify the degree of sediment pollution and its potential ecological risk. The reason for selection is same as above.	The Hakandson risk index method [61] is used for calculation. For detailed method introduction, please refer to the references cited here.
	Drinking water service function	Standard rate of centralized drinking water quality	It refers to the inspection frequency of the water quality monitoring of all centralized drinking water sources in the watershed that meets or exceeds the class II water quality standard of the "environmental quality standard for surface water" (GB 3838-2002), accounting for the proportion of the total inspection frequency in the whole year. It is important data of drinking water service function survey.	(Sum of compliance frequency of all monitoring sections/total monitoring frequency of all sections throughout the year) $\times$ 100%
Ecosystem services function	Water conservation function	Forest and grass coverage	It refers to the proportion of the sum of forest and grass vegetation areas such as arbor forest, shrub forest, and grassland in the regional land area. It is an important indicator to reflect the function of water conservation.	(Forest land area + grassland area)/total land area $\times$ 100%

Table A1. *Cont.*

Factors	Involved Aspects	Indicators	Definition and Reasons for Selection	Method of Calculation (within the Study Area)
Control management	Interception and purification function	Natural shoreline rate of River (reservoir)	The riverbank zone is divided into natural zone (undeveloped or natural shoreline length) and artificial zone. Here, it refers to the proportion of the length of natural riverbank zone in the total length of riverbank shoreline. It is an important indicator reflecting the interception and purification function of the riverbank.	Length of natural zone/(length of natural zone + length of artificial riverside zone) × 100%
	Cultural landscape function	Habitat representation of rare species	It mainly refers to whether the habitat reflects the characteristics of rare fish and important cultural landscape within the region, and whether it includes key species, rare and endangered species, and key protected species of natural ecosystem. It is an important indicator to reflect the function of cultural landscape.	Expert opinion method (Delphi method)
	Ecological protection investment	Ecological protection investment index	Proportion of environmental protection investment in regional GDP. According to the experience of developed countries, in the period of rapid economic growth, if a country wants to effectively control pollution, the investment in environmental protection should continuously and stably account for 1.5% of the gross national product within a certain period of time. Only when the investment in environmental protection reaches a certain proportion can it maintain good and stable environmental quality while rapid economic development.	Environmental protection investment/regional GDP × 100%
	Pollution control and environmental protection	Standard rate of industrial wastewater discharge	It refers to the proportion of the total amount of industrial wastewater discharged by key industrial enterprises within the scope of towns and townships through the sewage outlet and stably reaching the pollution discharge standard in the total amount of discharged industrial wastewater. It is an important indicator to reflect pollution control.	(Up to standard discharge of industrial wastewater/discharge of industrial wastewater) × 100%
		Centralized treatment rate of urban domestic sewage	The proportion of domestic sewage treated by the sewage treatment plant and meeting the discharge standard in the total discharge of urban domestic sewage. The reason for selection is same as above.	Treatment capacity of urban sewage treatment plant/total urban sewage generationer) × 100%
		Rural domestic sewage treatment rate	It refers to the proportion of rural domestic sewage treated by sewage treatment facilities and meeting the discharge standards in the total discharge of rural domestic sewage. The reason for selection is same as above.	Rural domestic sewage treatment capacity/total rural domestic sewage discharge × 100%
		Water and soil erosion control rate	Water and soil erosion refers to the migration and deposition process of water and soil caused by flowing water, gravity, or human action. The water and soil erosion control rate refers to the water and soil loss control area divided by the original water and soil loss area within a certain area and a certain period of time. It is an important indicator to reflect environmental protection.	Control area of water and soil erosion in a certain period/original water and soil erosion area × 100%

Table A1. *Cont.*

Factors	Involved Aspects	Indicators	Definition and Reasons for Selection	Method of Calculation (within the Study Area)
	Regulatory capacity	Regulatory capacity index	The ability to supervise and manage the ecological environment in the watershed. It is mainly composed of the degree of standardized construction of drinking water sources, environmental monitoring capacity, environmental monitoring standardization construction capacity, scientific and technological support capacity, etc. It is an important index to reflect human protection of the environment through regulation and management.	Expert opinion method (Delphi method)
	Long term mechanism	Construction of long-term management mechanism	An institutional system that can ensure the normal operation of the environmental protection system and play its expected functions for a long time. It is mainly composed of laws, regulations, policies, unified management institutions in the watershed, market-oriented long-term investment, and financing system, etc. The reason for selection is same as above.	Expert opinion method (Delphi method)

Table A2. Data acquisition process.

Indicators	Data Sources and Processing	Source of Original Data
Population density	Statistical yearbook 2016 of study area	Data available from government departments.
Population growth rate	Same as above	Same as above.
Per capita GDP	Same as above	Data available from government departments.
Human activity intensity index	Graphic translation of land use types	Purchased satellite images from the satellite remote sensing image Department and used Arc GIS 10.3 for remote sensing interpretation.
Area source COD load of per unit area	Statistical yearbook 2016 of study area, data calculation	We took the data published by government departments as the basis and reference and calculated it by the method in Table A1.
Area source TN load of per unit area	Same as above	Same as above.
Area source TP load of per unit area	Same as above	Same as above.
Point source COD load of per unit area	Same as above	Same as above.
Point source TN load of per unit area	Same as above	Same as above.
Point source TP load of per unit area	Same as above	Same as above.
Dissolved oxygen (water)	Sampling, detection, and analysis	In June 2017, 16 surface water samples were collected on site and tested in field (Hereinafter referred to as "sampling and test in field"). The test method was electrochemical probe method.
Transparency (water)	Same as above	Sampling and test in field, and the test method was Saybolt Disk Method.
Ammonia nitrogen (water)	Same as above	In June 2017, 16 surface water samples were collected on site, which were tested in the Analysis and Testing Center of Beijing Normal University (Hereinafter referred to as "sampling and test in laboratory"), and the test method was salicylic acid spectrophotometry.

Table A2. *Cont.*

Indicators	Data Sources and Processing	Source of Original Data
Total phosphorus (water)	Same as above	Sampling and test in laboratory. The test method was molybdic acid spectrophotometry.
Total nitrogen (water)	Same as above	Sampling and test in laboratory. The test method was alkaline potassium persulfate digestion UV spectrophotometry.
Permanganate index (water)	Same as above	Sampling and test in laboratory. The test method was acid method.
Mineralization (water)	Same as above	Sampling and test in laboratory. The test method was 180 °C dry gravimetric method.
Chlorophyll a (water)	Same as above	Sampling and test in laboratory. The test methods were acetone extraction and spectrophotometer determination.
Comprehensive nutritional index (water)	Calculated according to the test results	According to the test results of the samples, it was calculated by the method in attached Table A1.
Total nitrogen (sediment)	Sampling, detection, and analysis	Sampling and test in laboratory. The test method was the Kai's Nitrogen Determination Method.
Total phosphorus (sediment)	Same as above	Sampling and test in laboratory. The test methods were perchloric acid and sulfuric acid digestion.
Organic matter (sediment)	Same as above	Sampling and test in laboratory. The test method was potassium dichromate method.
Heavy metal risk index (sediment)	Calculated according to the test results	According to the test results of the samples, it was calculated by the method in attached Table A1.
Standard rate of centralized drinking water quality	Local water resources bulletin of study area 2015	Data available from government departments.
Forest and grass coverage	Interpretation of satellite remote sensing images	Purchased satellite images from the satellite remote sensing image Department and used Arc GIS 10.3 for remote sensing interpretation.
Natural shoreline rate of River (or reservoir)	Same as above	Same as above.
Habitat representation of rare species	Research data of the Chinese Academy of Sciences	Data from the Chinese Academy of Sciences.
Ecological protection investment index	Compilation of performance evaluation data of study area eco-environmental protection project in 2015–2016	Data from local environmental protection department and financial department.
Standard rate of industrial wastewater discharge	Calculated according to the environmental system data of study area	Data from local environmental protection department.
Centralized treatment rate of urban domestic sewage	Same as above	Same as above.
Rural domestic sewage treatment rate	Same as above	Same as above.
Water and soil erosion control rate	Local water and soil conservation Bulletin	Data available from government departments.
Regulatory capacity index	Compilation of performance evaluation data of study area eco-environmental protection project in 2015–2016	Data from local environmental protection department.
Construction of long-term management mechanism	Same as above	Same as above.

Table A3. Weight of each factor and indicator.

Factors	Weight	Indicators	Weight
Socioeconomic impact	0.21	Population density	0.020
		Population growth rate	0.020
		Per capita GDP	0.020
		Human activity intensity index	0.027
		Area source COD load of per unit area	0.020
		Area source TN load of per unit area	0.020
		Area source TP load of per unit area	0.020
		Point source COD load of per unit area	0.020
		Point source TN load of per unit area	0.020
		Point source TP load of per unit area	0.020
Water ecological health	0.36	Dissolved oxygen (water)	0.001
		Transparency (water)	0.022
		Ammonia nitrogen (water)	0.026
		Total phosphorus (water)	0.046
		Total nitrogen (water)	0.009
		Permanganate index (water)	0.009
		Mineralization (water)	0.025
		Chlorophyll a (water)	0.042
		Comprehensive nutritional index (water)	0.001
		Total nitrogen (sediment)	0.025
		Total phosphorus (sediment)	0.114
		Organic matter (sediment)	0.014
		Heavy metal risk index (sediment)	0.027
Ecological service function	0.23	Standard rate of centralized drinking water quality	0.058
		Forest and grass coverage	0.058
		Natural shoreline rate of River (or reservoir)	0.058
		Habitat representation of rare species	0.058
Regulation and management	0.2	Ecological protection investment index	0.029
		Standard rate of industrial wastewater discharge	0.029
		Centralized treatment rate of urban domestic sewage	0.029
		Rural domestic sewage treatment rate	0.029

Table A4. Reference standard and basis for determination of each indicator.

Indicators	Reference Value	Unit	Determination Basis
Population density	193.5	person/km^2	Statistical bulletin of local national economic and social development in 2016, regional level
Population growth rate	6.08	‰	Same as above
Per capita GDP	69565	¥	Same as above
Human activity intensity index	0.2	Dimensionless	Consulting experts, regional level
Area source COD load of per unit area	20	kg/(hm^2·a)	Lake ecological security strategy
Area source TN load of per unit area	5	kg/(hm^2·a)	Same as above
Area source TP load of per unit area	0.5	kg/(hm^2·a)	Same as above
Point source COD load of per unit area	40	kg/(hm^2·a)	Same as above
Point source TN load of per unit area	1.5	kg/(hm^2·a)	Same as above
Point source TP load of per unit area	0.1	kg/(hm^2·a)	Same as above
Dissolved oxygen (water)	7.5	mg/L	Environmental quality standard for surface water, Class I
Transparency (water)	1	m	Same as above

Table A4. *Cont.*

Indicators	Reference Value	Unit	Determination Basis
Ammonia nitrogen (water)	0.15	mg/L	Same as above
Total phosphorus (water)	0.02	mg/L	Same as above
Total nitrogen (water)	0.2	mg/L	Same as above
Permanganate index (water)	2	mg/L	Same as above
Mineralization (water)	1	μg/L	Lake ecological security strategy
Chlorophyll a (water)	300	mg/L	Comprehensive background value, groundwater standards, and drinking water standards
Comprehensive nutritional index (water)	30	Dimensionless	Comprehensive nutritional index classification, take the poor nutrition level
Total nitrogen (sediment)	700	mg/kg	Consulting experts, take a very safe level
Total phosphorus (sediment)	500	mg/kg	Same as above
Organic matter (sediment)	1.69	%	Same as above
Heavy metal risk index (sediment)	150	Dimensionless	Classification of heavy metal ecological risk index, take the level of slight ecological hazard
Standard rate of centralized drinking water quality	100	%	Consulting experts, take a very safe level
Forest and grass coverage	75	%	Same as above
Natural shoreline rate of River (or reservoir)	100	%	Same as above
Habitat representation of rare species	0.7	Dimensionless	Same as above
Ecological protection investment index	3	%	Environmental and economic benefit analysis data
Standard rate of industrial wastewater discharge	100	%	Consulting experts, take a very safe level
Centralized treatment rate of urban domestic sewage	90	%	The 13th five-year plan for environmental protection
Rural domestic sewage treatment rate	80	%	Same as above
Water and soil erosion control rate	2.45	%	2014 water and soil conservation Bulletin
Regulatory capacity index	5	Dimensionless	Consulting experts, regional level
Construction of long-term management mechanism	5	Dimensionless	Same as above

References

1. Zambrano-Monserrate, M.A.; Ruano, M.A.; Ormeno-Candelario, V.; Sanchez-Loor, D.A. Global ecological footprint and spatial dependence between countries. *J. Environ. Manag.* **2020**, *272*, 111069. [CrossRef]
2. Saxena, G.; Chandra, R.; Bharagava, R.N. Environmental Pollution, Toxicity Profile and Treatment Approaches for Tannery Wastewater and Its Chemical Pollutants. *Rev. Environ. Contam Toxicol.* **2017**, *240*, 31–69. [CrossRef]
3. Suk, W.A.; Ahanchian, H.; Asante, K.A.; Carpenter, D.O.; Diaz-Barriga, F.; Ha, E.H.; Huo, X.; King, M.; Ruchirawat, M.; da Silva, E.R.; et al. Environmental Pollution: An Under-recognized Threat to Children's Health, Especially in Low- and Middle-Income Countries. *Environ. Health Perspect.* **2016**, *124*, A41–A45. [CrossRef]
4. Qu, C.; Shi, W.; Guo, J.; Fang, B.; Wang, S.; Giesy, J.P.; Holm, P.E. China's Soil Pollution Control: Choices and Challenges. *Environ. Sci. Technol.* **2016**, *50*, 13181–13183. [CrossRef]
5. Hou, D.; Ok, Y.S. Soil pollution—Speed up global mapping. *Nature* **2019**, *566*, 455. [CrossRef]
6. Schwarzenbach, R.P.; Egli, T.; Hofstetter, T.B.; von Gunten, U.; Wehrli, B. Global Water Pollution and Human Health. *Annu. Rev. Environ. Resour.* **2010**, *35*, 109–136. [CrossRef]
7. Mekonnen, M.M.; Hoekstra, A.Y. Global Anthropogenic Phosphorus Loads to Freshwater and Associated Grey Water Footprints and Water Pollution Levels: A High-Resolution Global Study. *Water Resour. Res.* **2018**, *54*, 345–358. [CrossRef]
8. Jáuregui, E.; Luyando, E. Global radiation attenuation by air pollution and its effects on the thermal climate in Mexico City. *Int. J. Climatol.* **1999**, *19*, 683–694. [CrossRef]
9. Li, X.; Jin, L.; Kan, H. Air pollution: A global problem needs local fixes. *Nature* **2019**, *570*, 437–439. [CrossRef]
10. Rieuwerts, J. *The Elements of Environmental Pollution*; Nutritional Management of Digestive Disorders: London, UK, 2015. [CrossRef]
11. Cooke, C.A.; Bindler, R. *Lake Sediment Records of Preindustrial Metal Pollution*; Springer: Singapore, 2015; pp. 101–119. [CrossRef]
12. Wang, Z.; Deng, X.; Song, W.; Li, Z.; Chen, J. What is the main cause of grassland degradation? A case study of grassland ecosystem service in the middle-south Inner Mongolia. *Catena* **2017**, *150*, 100–107. [CrossRef]

13. Tudorache, D. *Sustainable Development in European Union as Expression of Social, Human, Economic, Technological and Environmental Progress*; ZBW-Leibniz Information Centre for Economics: Berlin, Germany, 2020.
14. Bai, Y.; Wong, C.; Jiang, B.; Hughes, A.; Wang, M.; Wang, Q. Developing China's Ecological Redline Policy using ecosystem services assessments for land use planning. *Nat. Commun.* **2018**, *9*, 3034. [CrossRef]
15. Nguyen, K.A.; Liou, Y.A. Global mapping of eco-environmental vulnerability from human and nature disturbances. *Sci. Total Environ.* **2019**, *664*, 995–1004.
16. Zhang, H.; Xu, E. An evaluation of the ecological and environmental security on China's terrestrial ecosystems. *Sci. Rep.* **2017**, *7*, 811. [CrossRef]
17. Luo, Y.W.; Wei, M. Integrated Reform Plan for Promoting Ecological Progress and its Enlightment to the Establishment of China's National Park System. *Landsc. Archit.* **2016**, *12*, 90–94. [CrossRef]
18. Zhang, Q.; Bai, D.; Peng, X. Research status and future trends of natural resources and sustainable development in China: Visual analysis based on citespace. *J. Resour. Ecol.* **2021**, *12*. [CrossRef]
19. Liu, J.; Raven, P.H. China's Environmental Challenges and Implications for the World. *Crit. Rev. Environ. Sci. Technol.* **2010**, *40*, 823–851.
20. Norton, S.B.; Rodier, D.J.; Schalie, W.; Wood, W.P.; Slimak, M.W.; Gentile, J.H. A framework for ecological risk assessment at the EPA. *Environ. Toxicol. Chem.* **2010**, *11*, 1663–1672.
21. Han, B.; Liu, H.; Wang, R. Urban ecological security assessment for cities in the Beijing-Tianjin-Hebei metropolitan region based on fuzzy and entropy methods. *Ecol. Model.* **2015**, *318*, 217–225.
22. Zhang, X.; Chen, M.; Guo, K.; Liu, Y.; Liu, Y.; Cai, W.; Wu, H.; Chen, Z.; Chen, Y.; Zhang, J. Regional Land Eco-Security Evaluation for the Mining City of Daye in China Using the GIS-Based Grey TOPSIS Method. *Land* **2021**, *10*, 118.
23. Wang, S.R.; Meng, W.; Jin, X.C.; Zheng, B.H.; Zhang, L.; Xi, H.Y. Ecological security problems of the major key lakes in China. *Environ. Earth Sci.* **2015**, *74*, 3825–3837.
24. Liu, C.; Wu, X.; Wang, L. Analysis on land ecological security change and affect factors using RS and GWR in the Danjiangkou Reservoir area, China. *Appl. Geogr.* **2019**, *105*, 1–14.
25. Nathwani, J.; Lu, X.; Wu, C.; Fu, G.; Qin, X. Quantifying security and resilience of Chinese coastal urban ecosystems. *Sci. Total Environ.* **2019**, *672*, 51–60.
26. Gao, Y.; Zhang, C.; He, Q.; Liu, Y. Urban Ecological Security Simulation and Prediction Using an Improved Cellular Automata (CA) Approach—A Case Study for the City of Wuhan in China. *Int. J. Environ. Res. Public Health* **2017**, *14*, 643.
27. Gari, S.R.; Newton, A.; Icely, J.D. A review of the application and evolution of the DPSIR framework with an emphasis on coastal social-ecological systems. *Ocean Coast. Manag.* **2015**, *103*, 63–77.
28. Levrel, H.; Kerbiriou, C.; Couvet, D.; Weber, J. OECD pressure-state-response indicators for managing biodiversity: A realistic perspective for a French biosphere reserve. *Biodivers. Conserv.* **2009**, *18*, 1719.
29. Borja, Á.; Galparsoro, I.; Solaun, O.; Muxika, I.; Tello, E.M.; Uriarte, A.; Valencia, V. The European Water Framework Directive and the DPSIR, a methodological approach to assess the risk of failing to achieve good ecological status. *Estuar. Coast. Shelf Sci.* **2006**, *66*, 84–96.
30. Kelble, C.R.; Loomis, D.K.; Lovelace, S.; Nuttle, W.K.; Ortner, P.B.; Fletcher, P.; Cook, G.S.; Lorenz, J.J.; Boyer, J.N. The EBM-DPSER Conceptual Model: Integrating Ecosystem Services into the DPSIR Framework. *PLoS ONE* **2013**, *8*, e70766. [CrossRef]
31. Cooper, P. Socio-ecological accounting: DPSWR, a modified DPSIR framework, and its application to marine ecosystems. *Ecol. Econ.* **2013**, *94*, 106–115.
32. Pinto, R.; Jonge, V.; Neto, J.M.; Domingos, T.; Marques, J.C.; Patricio, J. Towards a DPSIR driven integration of ecological value, water uses and ecosystem services for estuarine systems. *Ocean Coast. Manag.* **2013**, *72*, 64–79.
33. Pirrone, N.; Trombino, G.; Cinnirella, S.; Algieri, A.; Bendoricchio, G.; Palmeri, L. The Driver-Pressure-State-Impact-Response (DPSIR) approach for integrated catchment-coastal zone management: Preliminary application to the Po catchment-Adriatic Sea coastal zone system. *Reg. Environ. Change* **2005**, *5*, 111–137.
34. Lewison, R.L.; Rudd, M.A.; Al-Hayek, W.; Baldwin, C.; Beger, M.; Lieske, S.N.; Jones, C.; Satumanatpan, S.; Junchompoo, C.; Hines, E. How the DPSIR framework can be used for structuring problems and facilitating empirical research in coastal systems. *Environ. Sci. Policy* **2016**, *56*, 110–119.
35. Svarstad, H.; Petersen, L.K.; Rothman, D.; Siepel, H.; Wätzold, F. Discursive biases of the environmental research framework DPSIR. *Land Use Policy* **2008**, *25*, 116–125.
36. Kohsaka, R. Developing biodiversity indicators for cities: Applying the DPSIR model to Nagoya and integrating social and ecological aspects. *Ecol. Res.* **2010**, *25*, 925–936.
37. Binimelis, R.; Monterroso, I.; Rodríguez-Labajos, B. Catalan agriculture and genetically modified organisms (GMOs)—An application of DPSIR model. *Ecol. Econ.* **2009**, *69*, 55–62. [CrossRef]
38. Sun, S.; Wang, Y.; Jing, L.; Cai, H.; Xu, L. Sustainability Assessment of Regional Water Resources Under the DPSIR Framework. *J. Hydrol.* **2016**, *532*, 140–148.
39. Tscherning, K.; Helming, K.; Krippner, B.; Sieber, S.; y Paloma, S.G. Does research applying the DPSIR framework support decision making? *Land Use Policy* **2012**, *29*, 102–110.
40. Newton, A.; Weichselgartner, J. Hotspots of coastal vulnerability: A DPSIR analysis to find societal pathways and responses. *Estuar. Coast. Shelf Sci.* **2014**, *140*, 123–133.

41. Atkins, J.P.; Burdon, D.; Elliott, M.; Gregory, A.J. Management of the marine environment: Integrating ecosystem services and societal benefits with the DPSIR framework in a systems approach. *Mar. Pollut. Bull.* **2011**, *62*, 215–226. [CrossRef]
42. Rudy, V. Using DPSIR and Balances to Support Water Governance. *Water* **2018**, *10*, 118.
43. Akbari, M.; Memarian, H.; Neamatollahi, E.; Shalamzari, M.J.; Zakeri, D. Prioritizing policies and strategies for desertification risk management using MCDM–DPSIR approach in northeastern Iran. *Environ. Dev. Sustain.* **2021**, *23*, 2503–2523.
44. Bidone, E.; Lacerda, L.D. The use of DPSIR framework to evaluate sustainability in coastal areas. Case study: Guanabara Bay basin, Rio de Janeiro, Brazil. *Reg. Environ. Change* **2004**, *4*, 5–16.
45. Jun, K.S.; Chung, E.S.; Sung, J.Y.; Lee, K.S. Development of spatial water resources vulnerability index considering climate change impacts. *Sci. Total Environ.* **2011**, *409*, 5228–5242.
46. Kristensen, P. *The DPSIR Framework*; National Environmental Research Institute: Roskilde, Denmark, 2004.
47. Barton, D.N.; Kuikka, S.; Varis, O.; Uusitalo, L.; Henriksen, H.J.; Borsuk, M.; Hera, A.; Fa Rmani, R.; Johnson, S.; Linnell, J.D. Bayesian networks in environmental and resource management. *Integr. Environ. Assess. Manag.* **2012**, *8*, 418–429.
48. Hepelwa, A. Dynamics of Watershed Ecosystem Values and Sustainability: An Integrated Assessment Approach. *Int. J. Ecosyst.* **2014**, *4*, 43–52.
49. Skoulikidis, N.T. The environmental state of rivers in the Balkans—A review within the DPSIR framework. *Sci. Total Environ.* **2009**, *407*, 2501–2516.
50. Ness, B. Erratum to "Integrated research on subsurface environments in Asian urban areas". *Sci. Total Environ.* **2009**, *409*, 3076–3088.
51. Zhao, R.; Fang, C.; Liu, H.; Liu, X. Evaluating urban ecosystem resilience using the DPSIR framework and the ENA model: A case study of 35 cities in China. *Sustain. Cities Soc.* **2021**, *72*, 102997.
52. Jago-On, K.; Kaneko, S.; Fujikura, R.; Fujiwara, A.; Imai, T.; Matsumoto, T.; Zhang, J.; Tanikawa, H.; Tanaka, K.; Lee, B. Urbanization and subsurface environmental issues: An attempt at DPSIR model application in Asian cities. *Sci. Total Environ.* **2009**, *407*, 3089–3104.
53. Maxim, L.; Spangenberg, J.H.; O'Connor, M. An analysis of risks for biodiversity under the DPSIR framework. *Ecol. Econ.* **2009**, *69*, 12–23.
54. Alexakis, D.; Kagalou, I.; Tsakiris, G. Assessment of pressures and impacts on surface water bodies of the Mediterranean. Case study: Pamvotis Lake, Greece. *Environ. Earth Sci.* **2013**, *70*, 687–698.
55. Skulmoski, G.J.; Hartman, F.T.; Krahn, J. The Delphi Method for Graduate Research. *J. Inf. Technol. Educ.* **2007**, *6*, 1–21.
56. Zou, Z.H.; Yun, Y.; Sun, J.N. Entropy method for determination of weight of evaluating indicators in fuzzy synthetic evaluation for water quality assessment. *J. Environ. Sci.* **2006**, *18*, 1020–1023. [CrossRef]
57. Zhang, E.; Liu, E.; Jones, R.; Langdon, P.; Yang, X.; Shen, J. A 150-year record of recent changes in human activity and eutrophication of Lake Wushan from the middle reach of the Yangze River, China. *J. Limnol.* **2010**, *69*, 235–241.
58. Yang, G.; Ding, Y.; Zhou, L. Impact of Land-Use Changes on Intensity of Soil Erosion in the Mountainous Area in the Upper Reach of Shiyang River in Arid Northwest China. *Geosci. Remote Sens. Symposium Igarss IEEE Int.* **2008**, *4*, 639–642.
59. El Banna, M.M.; Frihy, O.E. Natural and anthropogenic influences in the northeastern coast of the Nile delta, Egypt. *Environ. Geol.* **2009**, *57*, 1593–1602.
60. Córdoba, E.; Martínez, A.; Ferrer, E.V. Water quality indicators: Comparison of a probabilistic index and a general quality index. The case of the Confederación Hidrográfica del Júcar (Spain). *Ecol. Indic.* **2010**, *10*, 1049–1054.
61. Hakanson, L. An ecological risk index for aquatic pollution control.a sedimentological approach. *Water Res.* **1980**, *14*, 975–1001.

water

MDPI

Article

Research on Water Resources Allocation System Based on Rational Utilization of Brackish Water

Dasheng Zhang [1,2,3], Xinmin Xie [2], Ting Wang [1,2], Boxin Wang [3] and Shasha Pei [3,*]

[1] State Key Laboratory of Simulation and Regulation of Water Cycle in River Basin, China Institute of Water Resources and Hydropower Research, Beijing 100038, China; 18501317443@163.com (D.Z.); wangt90@iwhr.com (T.W.)

[2] Department of Water Resources, China Institute of Water Resources and Hydropower Research, Beijing 100038, China; xiexm@iwhr.com

[3] Hebei Institute of Water Science, Shijiazhuang 050051, China; wangboxin293@163.com

* Correspondence: peishasha187@163.com; Tel.: +86-188-1176-7831

Abstract: The rational utilization of unconventional water sources is of great significance to areas where conventional water resources are scarce, and water resource allocation is an important way to realize the rational distribution of multiple water sources. This paper constructs a water resources allocation system integrating model data parameter database, water resources supply and demand prediction module, groundwater numerical simulation module and water resources allocation module. Taking brackish water as the main research object and final goal of achieving the best comprehensive optimization of social, economic and ecological benefits. The brackish water is incorporated as an independent water source into the water resource allocation model, and the stratum structure model and groundwater numerical model are constructed to simulate the brackish water level in the planning target year. The water resources allocation system is applied to Guantao County, China. The results show that increasing the development and utilization of brackish water under the recommended scheme can significantly reduce the water supply pressure of local fresh water resources in agriculture and industry. Compared with the current year, the overall water shortage in the region will be reduced by 4.493×10^6 m^3 in 2030, and meanwhile, the brackish water level will be decreased by 12.69 m in 2035, which plays a positive role in improving soil salinization.

Keywords: water allocation; brackish water; groundwater numerical simulation; general algebraic modeling system

Citation: Zhang, D.; Xie, X.; Wang, T.; Wang, B.; Pei, S. Research on Water Resources Allocation System Based on Rational Utilization of Brackish Water. *Water* **2022**, *14*, 948. https://doi.org/10.3390/w14060948

Academic Editors: Yuanzheng Zhai, Jin Wu, Huaqing Wang and Pankaj Kumar

Received: 30 December 2021
Accepted: 10 March 2022
Published: 17 March 2022

Publisher's Note: MDPI stays neutral with regard to jurisdictional claims in published maps and institutional affiliations.

1. Introduction

Water resources are one of the important natural resources that protect people's livelihood and promote social and economic development, especially for developing countries that have a large population and need to improve their water-saving level and practices. Since the 21st century, rapid population growth and rapid development of the economy has increased the urgent demand for the quantity and quality of water resources [1]. The increasing demand for water has gradually exceeded the supply capacity of local water resources, resulting in a large-scale water resources shortage [2,3]. Especially for agriculture, under the current situation of insufficient local surface water, it is often necessary to use large amounts of groundwater to maintain the normal growth of crops and meet the needs of agricultural production, which also causes problems such as groundwater level decline and land subsidence [4], threatening the safety of human water use and the benign operation of the natural water cycle. There is an urgent need to balance water supply and use, in order to eliminate the increasingly fierce competition of water use and deterioration of the water ecological environment [5,6].

Water resources allocation refers to the planning for macro-control of water resources according to historical water inflow data and user water demand. It is considered to be one of the most effective methods to coordinate the balance of social, economic, and ecological water use [7]. With researchers globally conducting in depth research, the water resources allocation model has been continuously developed [8,9]. Minsker, et al. [10] constructed a multi-objective allocation model of water resources based on genetic algorithm through the uncertainty of hydrological factors, and characterized the uncertain factors in the process of water resources allocation. Rosegrant, et al. [11] coupled the hydrological model with the economic development model to evaluate social and economic benefits brought by the improvement of water resources allocation and utilization efficiency. Chen, et al. [12] formulated the interval multi-level classified water resources allocation model to optimize water resources allocation to the municipal scale, and the allocation objects are industry, hydropower, and agriculture. In addition, with the development of computer technology, optimization algorithms, artificial intelligence, and space technology are increasingly used in the field of water resources allocation, such as artificial fish school algorithm [13,14], particle swarm optimization algorithm [15,16], and neuro-fuzzy reinforcement learning method [17], which greatly optimizes the scientific nature and accuracy of the water resources allocation model. In addition, it also drives the water resources allocation model to gradually transit to multi-disciplinary integration, so as to serve the solution of multi-objective problems. In the study of optimal allocation of unconventional water resources, Li, et al. [18] developed an interval number hierarchical planning model to integrate water into the water allocation system, taking into account the interests of decision-makers at different levels and the uncertainty of the water allocation system, and to guide the conventional and unconventional water supply sectors to supply water rationally and efficiently. Mooselu, et al. [19] proposed a method for optimal allocation of wastewater reuse water with the objective of minimizing the cost of water supply and regional water shortage. Gao, et al. [20] proposed an optimal allocation method for wastewater reuse water with the objective of minimizing the consumption of freshwater resources and the energy consumption of the water supply process by incorporating medium and desalinated water into the water allocation study, and used NSGA-II. The optimal water and energy-saving scheme is given on the Pareto boundary by solving this multi-objective problem using NSGA-II.

Brackish water refers to the water resources buried underground with a salinity of 2–5 g/L [21]. With the increasingly severe shortage of conventional water resources, it is urgent to bring brackish water into the optimal water resources allocation. China is rich in brackish water resources. According to statistics, the amount of brackish water resources in China is 27.7 billion m^3, of which 13 billion m^3 can be utilized. Underground shallow brackish water is mainly distributed in the Northern China Plain, northwest arid area and eastern coastal area, with a buried depth of about 10–100 m, which is easy to be exploited and utilized [22]. As one of the main grain production areas in China, the use of brackish water for agricultural irrigation in the Northern China Plain has very important practical significance [23]. The use of brackish water for farmland irrigation has a long history worldwide. In the southwest of the United States, the use of brackish water for irrigation of cotton, sugar beet and other crops basically does not affect the crop yields [24]. Israel, Australia, Spain, Tunisia, and other countries all have carried out practices related to brackish water irrigation and achieved good results [25]. In addition to the relevant research on the impact of brackish water irrigation on crop yield, other scholars have also carried out relevant research on the impact of brackish water salinity on soil infiltration characteristics [26], soil water and salt transport model [27], soil water and salt regulation measures [28–30], further expanding the depth and breadth of the research on the internal mechanism of the impact of brackish water irrigation on soil and crop growth.

In summary, this paper provides a comprehensive review of the research results on water resources allocation, non-conventional water resources allocation and brackish water utilization, etc. The limitations of the existing water resources allocation research are mainly reflected in the following two aspects: Firstly, the focus of the current research on optimal water resources allocation is to reasonably allocate the water supply from conventional water sources, medium water and desalinated water through various engineering and technical measures, so as to meet the demand of different users. Secondly, in terms of brackish water utilization, the current research mainly focuses on the impact of brackish water irrigation on crop growth, yield, and land quality by conducting irrigation experiments, while there are relatively few studies on the prediction and simulation of water level changes after the large-scale application of brackish water.

Based on the above discussion, the main objective of this study is to construct a water resources optimization allocation system, the core of which is to couple numerical groundwater simulation with water resources optimization allocation, to reasonably allocate brackish water as an independent water source, to realize the efficient utilization of brackish water resources in agricultural irrigation and industrial water use, and to input the allocation results into the numerical groundwater model to predict the changing trend of brackish water level. This study has certain reference value for reducing the water supply pressure of conventional water sources and inter-basin water transfer in agricultural irrigation and industrial water use, improving the development and utilization rate of brackish water resources, and realizing the synergistic allocation of conventional water resources and brackish water resources in water-scarce areas.

2. Material and Method

2.1. Study Area

Guantao County is located in the south of the Northern China Plain (longitude 115°06′–115°40′, latitude 36°27′–36°47′), with a total area of 456.3 km^2. It borders Shandong Province with Wei Canal in the east, Guangping, Quzhou and Qiu Counties in the west, Daming County in the south and Linxi County in the north. The main river in the territory is Wei Canal. Its annual average surface water resource is 4.645 million m^3, its groundwater resource is 57.654 million m^3, and the total amount of water resources is 62.299 million m^3. By the end of 2019, the total population was 350,900, including 74,100 urban residents and 276,800 rural residents, with an urbanization rate of 21.1%. The gross national product is CNY 7.095 billion, and the proportion of the three industries is 6.2:6.3:11.1. The geographical location of Guantao County is shown in Figure 1.

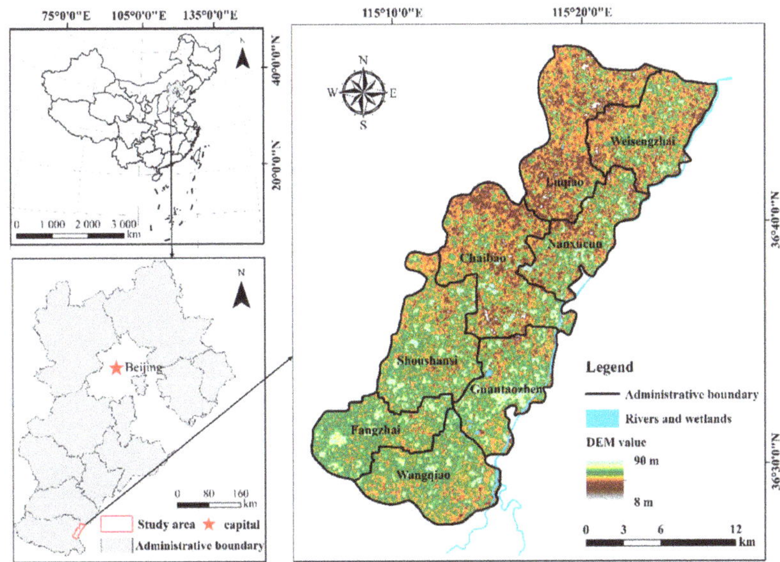

Figure 1. Geographical location of Guantao County.

Guantao County is rich in shallow brackish water reserves. According to the relevant results of Evaluation of Water Resources of Guantao County, Hebei Province, the salinity of Guantao County is 2 g/L < M $\leq$ 3 g/L, and the multi-year average groundwater resources of brackish water is 15.15 million m^3. The salinity is 3 g/L < M $\leq$ 5 g/L, and the multi-year average groundwater resources of brackish water is 4.05 million m^3. The distribution of brackish water in Guantao County is shown in Figure 2. It can be seen from the figure that groundwater with salinity M $\leq$ 2 g/L is mainly distributed in Shoushansi Township and Chaibao Town in the south-central part of Guantao County. The groundwater with salinity 3 g/L < M $\leq$ 5 g/L is mainly distributed in Luqiao Township and Weisengzhai Town in the north of Guantao County, Chaibao Town in the middle and Fangzhai Town and Wangqiao Township in the south. The salinity of other areas is 2 g/L < M $\leq$ 3 g/L.

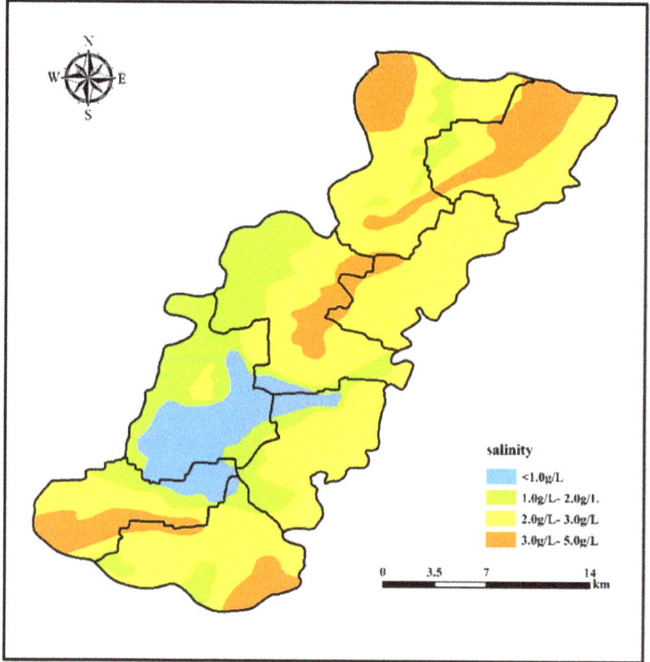

Figure 2. Distribution of brackish water in Guantao County.

2.2. Optimal Water Resources Allocation System

The optimal water resources allocation refers to the spatial and temporal allocation of limited water resources through various engineering or non-engineering measures in the basin or specific area, in accordance with the principle of natural sustainable development, so as to meet the water demand of each area to the maximum extent and coordinate the contradictions of water users without affecting the ecological environment. This aims to maximize the social and economic benefits of water resources, and to promote the sustainable and stable development of the river basin, regional economy, and the health and stability of ecological environment. In order to achieve the above objectives, this study constructs the optimal water resources allocation system on the premise of fully considering the water resources endowment conditions in the study area. The main work modules of the system include model data parameter database, water resources supply and demand prediction module, groundwater numerical simulation module, and water resources allocation module. The working relationship between modules is shown in Figure 3.

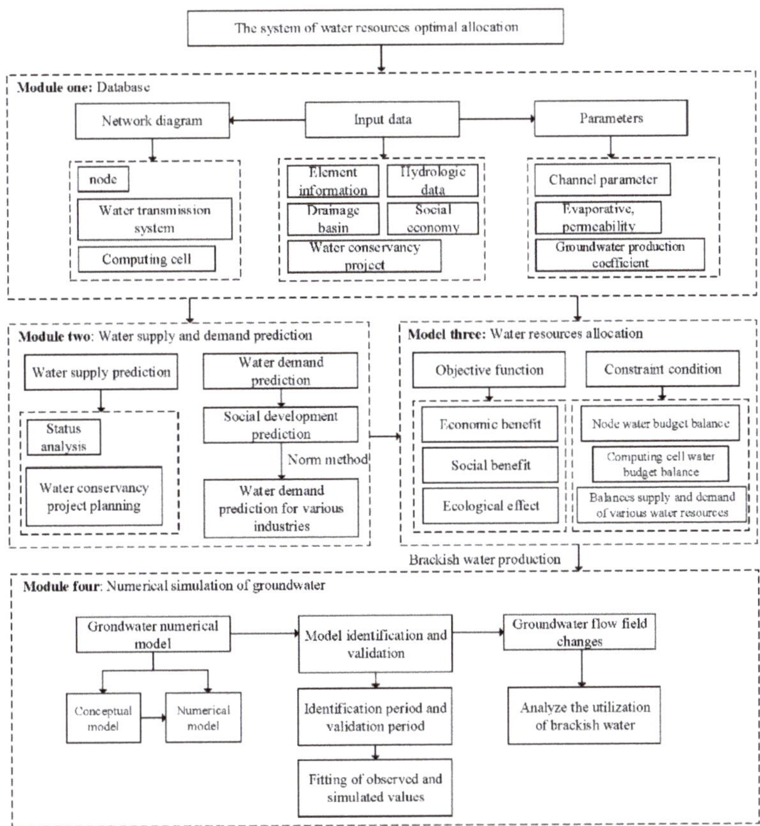

Figure 3. Working relationship of modules of optimal water resources allocation system.

2.2.1. Model Data Parameter Database

The model data parameter database is the data source of the optimal water resources allocation system as well as the basis for the normal operation of other modules. It mainly includes the following three contents:

1. Data input: it mainly includes hydrological data, social and economic data, river (canal) system network, river (canal) basic information, calculation unit information, basic information of hydraulic engineering, basin unit information, etc.
2. Model parameters: they mainly include the discharge capacity, evaporation and leakage coefficients, etc., of river, canal system and drainage channel, sewage discharge rate and reuse rate, upper and lower limit coefficients of annual and monthly exploitation of groundwater supply, and evaporation, leakage, and into-river proportion coefficient of irrigation canal system of calculation unit.
3. Water resources allocation system network diagram: its main function is to master the relationship among surface water, groundwater, external diverted water and unconventional water and urban life, rural life, agriculture, industry and tertiary industry, ecology and environment, and to clarify the supply, consumption and discharge of water resources among society, economy and ecological environment. It is an important basis for building the water resources allocation model [31]. The drawing principle of the water resources allocation network diagram is water balance principle, that is, any node on the diagram must conform to Equation (1):

$$Q_u - Q_s - Q_l - Q_d = \Delta Q \tag{1}$$

where Q_u is the upstream water inflow of the node, 10,000 m^3; Q_s is the water supply quantity of the node, 10-thousand m^3; Q_l is the water loss of the node, 10-thousand m^3; Q_d is the downstream water transmission of the node, 10,000 m^3; and ΔQ is the storage variable of the node, 10,000 m^3.

The water resources allocation system network diagram is the generalization and abstraction of water resources allocation. The river system is drawn by approximating the actual river distribution in the study area, river basin and administrative area sections, reservoir nodes, water diversion and lifting nodes, calculation units, lakes, etc., are marked in the river system. Reservoir nodes, water diversion and lifting nodes, calculation units, lakes and other nodes are used to connect local surface water sources, groundwater sources, unconventional water sources and external diverted water sources in the study area. The drainage line of the calculation unit is connected with the river to form a drainage system. The study area is partitioned to ensure that there are water transmission lines in each partition. On the basis of fully collecting the data of the study area and comprehensively considering the brackish water distribution and water use of various departments in Guantao County, Guantao County is generalized into 17 calculation units, five water lifting nodes, 12 control section nodes and one water plant node. The water resources allocation system network diagram in Guantao County is shown in Figure 4.

Figure 4. Water resources allocation system network diagram in Guantao County.

2.2.2. Water Resources Supply and Demand Prediction Module

The water resources supply and demand prediction module mainly includes a water demand prediction part and a water supply prediction part, wherein, the water demand prediction part adopts the combination of time series method and index quota method, and comprehensively considers the requirements of social and economic development and water conservation, so as to formulate the water demand schemes for different industries under different social development situations.

Urban and rural residents' domestic water demand forecast is calculated by the formula:

$$W_t = p_t q_t \tag{2}$$

where W_t is the projected water demand of urban and rural residents in level t, p_t is the projected population in level t, and q_t is the projected water quota in level t.

Industrial water demand forecast is calculated by the formula:

$$I_t = e_t r_t \tag{3}$$

where I_t is the predicted industrial water demand at level t, e_t is the predicted industrial value-added water consumption per million yuan at level t, and r_t is the predicted industrial value-added at level t.

Agricultural irrigation water demand forecast is calculated by the formula:

$$W_{mt} = \frac{\sum_{j=1}^{m} A_{tj}\omega_j}{\eta} \tag{4}$$

where W_{mt} is the gross irrigation water demand in level t, A_{tj} is the planted area of the jth crop in level t, ω_j is the irrigation quota of the jth crop, m is the crop type, and η is the integrated irrigation water effective utilization coefficient.

The water supply prediction part is based on the comprehensive investigation and analysis of the engineering layout, water supply capacity, operation status, development and utilization of water resources and existing problems of existing water supply facilities, and analyzes the prospect and potential of water resources development and utilization, so as to formulate water supply planning schemes for different target years, and predict the available water supply quantity of each planning scheme. The accuracy of water demand prediction and water supply prediction has a great impact on the results of water resources allocation. Therefore, it is necessary to closely combine the relevant requirements of social and economic development and water conservancy project planning in the study area so as to avoid the excessive deviation between the prediction results and development objectives.

2.2.3. Water Resources Allocation Module

The water resources allocation module is based on the binary water cycle theory, which takes the water resources system, the social economic system, and the ecological environment system as an organic whole, and efficiently promotes social and economic construction under the premise of ensuring the sustainable and sound development of the ecological environment. To promote the conservation and protection of water resources with the advancement of economy and technology, and the high-quality water resources saved can be distributed to various water users, realizing a virtuous circle of the organic whole. The water resources allocation module adopts the system analysis method based on linear programming, and establishes the balance equation and constraint equation of each control node, reservoir and calculation unit in the water resources allocation system network diagram according to the principle of water balance. It aims at the comprehensive optimization of social benefit goal, economic benefit goal and ecological environment benefit goal to construct the objective function of water resources allocation, and programs by using the general algebraic modeling system (GAMS). The optimal configuration scheme is solved by repeated iteration. The explanations of relevant variables are shown in Table 1, as the constructed objective Equations (2)–(6).

Table 1. Variable definition.

Number	Name	Definition	Number	Name	Definition
(5)	SC_j	Local surface water supplies the urban water supply	(3)	DA_j	External water supply for agricultural production
(5)	SI_j	Local surface water supplies industrial production	(3)	DR_j	External water supply rural water supply
(5)	SE_j	Local surface water supplies water to the ecological environment	(3)	BI_j	External water supply rural water supply
(5)	SA_j	Local surface water supplies agricultural production	(3)	BA_j	Brackish water supplies water for agriculture
(5)	SR_j	Local surface water supplies rural water supply	(5)	RI_j	Sewage treatment and reuse supply industrial production water
(5)	GC_j	Groundwater supplies water to cities and towns	(5)	RE_j	Sewage treatment and reuse supply ecological environment
(5)	GI_j	Groundwater supplies water for industrial production	(6)	AC_j	Urban life is short of water
(5)	GE_j	Groundwater supplies water to the ecological environment	(6)	AI_j	Industrial production is short of water
(5)	GA_j	Groundwater supplies water for agricultural production	(6)	AE_j	The ecological environment is short of water
(5)	GR_j	Groundwater supplies water for rural life	(6)	AA_j	Agricultural production is short of water
(5)	DC_j	External water supply urban living water	(6)	AR_j	Agricultural production is short of water
(5)	DI_j	External water supply for industrial production	(7)	WL_{wl}	Lakes and wetlands are short of water
(5)	DE_j	External water supply ecological environment water			

1. Social benefit goal

The social benefit goal is mainly reflected by the safety level of water supply. Each water allocation department shall carry out multi-source joint water supply to meet the requirements of each water user for water quantity and quality. Therefore, in the process of allocation, the water supply weight of each water source shall be reasonably set to determine the optimal safety level of water supply.

$$
\begin{aligned}
\text{Max } F_1 = {} & \sum_j \alpha_{sur} \times \left(SC_j + SI_j + SE_j + SA_j + SR_j \right) \\
& + \sum_j \alpha_{grd} \times \left(GC_j + GI_j + GE_j + GA_j + GR_j \right) \\
& + \sum_j \alpha_{div} \times \left(DC_j + DI_j + DE_j + DA_j + DR_j \right) \\
& + \sum_j \alpha_{bw} \times \left(BI_j + BA_j \right) \\
& + \sum_j \alpha_{rec} \times \left(RI_j + RE_j \right)
\end{aligned}
\tag{5}
$$

where α_{sur} is surface water supply coefficient, α_{grd} is groundwater supply coefficient, α_{div} is external diverted water supply coefficient, α_{bw} is brackish water supply coefficient, and α_{rec} is sewage disposal recycled water supply coefficient.

2. Economic benefit goal

The economic benefit goal is mainly reflected by water shortage rate. For industry and agriculture, the lower the water shortage rate is, the higher the economic benefit of water supply will be. The corresponding water shortage weight is allocated to each industry according to the industrial demand and the economic benefits generated by unilateral water, so as to minimize the regional water shortage and realize the coordinated development of the regional economy.

$$\text{Min } F_2 = \sum_j \left(\alpha_j^c AC_j + \alpha_j^I AI_j + \alpha_j^E AE_j + \alpha_j^A AA_j + \alpha_j^R AR_j \right) \tag{6}$$

where α_j^c is urban domestic water shortage coefficient; α_j^I is industrial water shortage coefficient; α_j^E is ecological environmental water shortage coefficient; α_j^A is agricultural water shortage coefficient; α_j^R is rural water shortage coefficient.

3. Ecological environment benefit goal

The ecological environment benefit goal is mainly reflected by lake water shortage. At present, Princess Lake in the study area has shrunk to a certain extent. Therefore, it is very necessary to protect the existing lake area and give full play to its due ecological function.

$$\text{Min } F_3 = \sum_{wl=1}^{n} \alpha_{wl} \times WL_{wl} \tag{7}$$

where α_{wl} is the water use coefficient of lake and wetland.

4. Final goal

The final goal is to achieve the optimal comprehensive benefits of water resources allocation. In the process of water resources allocation, it comprehensively considers the carrying capacity of water resources and water environment, and gives consideration to the water supply and consumption coordination between multiple water sources and multiple users, so as to meet the water requirements for the healthy development of social economy and ecological environment.

$$F = c_1 F_1 + c_2 F_2 + c_3 F_3 \tag{8}$$

where F is the final goal of water resources allocation, and c_1, c_2, and c_3 are the coefficients of social benefit goal, economic benefit goal, and ecological environment benefit goal, respectively.

5. Constraints

The constraints mainly include node water balance, supply and demand balance of calculation unit, water supply balance of various water sources (surface water, groundwater, external diverted water, brackish water, water for sewage treatment, etc.), river and canal overflow capacity, river network channel storage, return water balance of calculation unit, lake and wetland constraints, etc. Limited by the length of the article, see [32] for the specific constraint equation expression.

2.2.4. Groundwater Numerical Simulation Module

Numerical simulation is one of the most commonly used methods to solve the problem of groundwater flow. It mainly uses the approximation principle to divide the study area, and transforms the nonlinear partial differential equation into a difference equation through discretization, which can solve the complex groundwater evaluation problem on the premise of ensuring a certain accuracy. The differential equation of groundwater flow is as follows:

$$\begin{cases} \dfrac{\partial}{\partial x}\left[K(H-Z)\dfrac{\partial H}{\partial x}\right] + \dfrac{\partial}{\partial y}\left[K(H-Z)\dfrac{\partial H}{\partial y}\right] + \dfrac{\partial}{\partial z}\left[K(H-Z)\dfrac{\partial H}{\partial z}\right] + \varepsilon = \mu \dfrac{\partial H}{\partial t} \\[8pt] \dfrac{\partial}{\partial x}\left(KM\dfrac{\partial H}{\partial x}\right) + \dfrac{\partial}{\partial y}\left(KM\dfrac{\partial H}{\partial y}\right) + \dfrac{\partial}{\partial z}\left(KM\dfrac{\partial H}{\partial z}\right) + W + p = SM \dfrac{\partial H}{\partial t} \\[8pt] H(x,y,z)|_{t=0} = H_0(x,y,z) \\[4pt] H(x,y,z,t)|_{\Gamma_1} = H_1(x,y,z,t) & x,y,z \in \Gamma_1 \quad t>0 \\[4pt] KM\dfrac{\partial H}{\partial x}\Big|_{\Gamma_2} = q(x,y,t) & x,y,z \in \Gamma_2 \quad t>0 \end{cases} \tag{9}$$

where H is water level, m; Z is floor elevation of the first phreatic aquifer, m; K is hydraulic conductivity of aquifer, m/d; ε is rainfall infiltration and agricultural regression intensity, m/d; μ is water yield of the first phreatic aquifer; W is overflow strength, L/d; P is mining intensity per unit volume of aquifer, L/d; S is water storage rate of confined aquifer, L/m; H_0 is initial head, m; Γ_1 is class I head boundary; H_1 is class I boundary water level, m; Γ_2 is class II flow boundary; and q is boundary flow, m^2/d.

3. Results and Discussion

3.1. Establishment of Groundwater Numerical Model

According to the hydrogeological conditions of Guantao County, combining with the long-term observation data of groundwater and the actual exploitation of groundwater, the phreatic layer in Guantao County is generalized into a groundwater system with unified hydraulic connection by using the 3D groundwater flow numerical simulation software GMS, and the system is generalized into a heterogeneous and unstable groundwater system whose parameters change with space.

3.1.1. Stratigraphic Structure Generalization

The aquifer in the study area is a loose rock pore aquifer. Its lithology is mainly interbedded with fine sand, silty sand and silty clay, and the formation type is alluvial and lacustrine. The stratum thickness changes uniformly, and there is hydraulic connection between aquifers in the horizontal direction.

According to the spatial distribution of stratigraphic lithologic structure within the study area and combining with the horizon of existing monitoring wells in Guantao County, the strata in the study area are generalized into a three-layer structure composed of two aquifers and one aquitard which are interbedded. Referring to the deep mining depth of the groundwater mining layer in the study area, the bottom boundary depth of the model is set as 160 m.

3.1.2. Generalization of Aquifer Boundary Conditions

1. Lateral boundary conditions.

The southern part of the study area is mainly recharged by lateral runoff. In terms of boundary conditions, it is regarded as class II flow inflow boundary. There is almost no water exchange between the north of the study area and the surrounding counties, and the boundary is perpendicular to the water level line, which can be generalized as a water barrier boundary. Since the overall flow direction of shallow groundwater in Guantao County is from northeast to southwest, the west and the east of the study area can be generalized as constant head boundaries.

2. Vertical boundary generalization.

The upper part of the model is the phreatic surface boundary, which is recharged by atmospheric precipitation, river leakage and agricultural irrigation return, and discharged by phreatic surface evaporation and manual mining. Therefore, the upper boundary is generalized as the water exchange boundary. The bottom of the model has relatively impermeable clay and bedrock, so it is generalized as an impermeable boundary.

3.1.3. Determination of Hydrogeological Parameter Partitions and Initial Values

The hydrogeological parameters of Guantao County are determined according to the relevant hydrogeological maps and hydrogeological exploration data of Guantao County. The phreatic aquifer in Guantao County is a loose rock pore aquifer. Its lithology is mainly composed of fine sand, with NE trending silty sand bands scattered. The study area is divided into four areas. The initial values of hydraulic conductivity and water yields are shown in Table 2, and the hydrogeological parameter partitions are shown in Figure 5.

Table 2. Hydrogeological parameters.

Parameters/District	I	II	III	IV
K (m/d)	1.5	1.2	2.0	3.5
μ	0.1	0.08	0.07	0.12

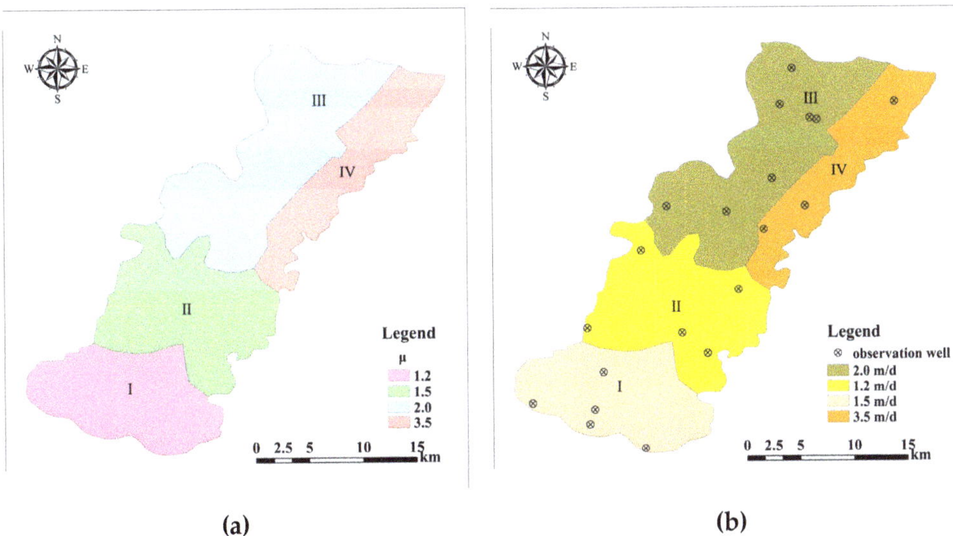

Figure 5. Permeability coefficient partition (**a**); hydraulic conductivity zoning and observation well locations in the study area (**b**).

3.1.4. Construction and Solution of Numerical Model

1. Spatial subdivision.

The study area is divided into rectangular grid elements by GMS software. After spatial subdivision, the study area is divided into rectangular grid elements, of which about 11,000 rectangular grid elements are in each layer. See Figure 6a for model element subdivision results.

2. Initial flow field.

In this study, the initial time of 1 January 2018 was taken. According to the groundwater level of groundwater monitoring wells, Kriging interpolation method is used to draw the initial water level contour of the study area, as shown in Figure 6b. Indicated by Figure 6b, due to the influence of topography, the groundwater level is generally high in the southwest and low in the northeast. Due to the long-term exploitation of groundwater, an obvious depression cone has been formed in Shoushansi area in the west and Fangzhai Town in the south of the study area.

3. Processing of source and sink items.

The main recharge items include: precipitation infiltration recharge, lateral recharge, irrigation recharge, etc. The main discharge items include: artificial mining, evaporation and lateral discharge. Darcy's Law is used to determine the lateral supply and discharge at the boundary. The precipitation is multiplied by the precipitation infiltration recharge coefficient to obtain the precipitation infiltration recharge. The determination method of irrigation infiltration recharge and evaporation discharge is the same as that of precipitation infiltration recharge. The mining yield is input into the model in a well form.

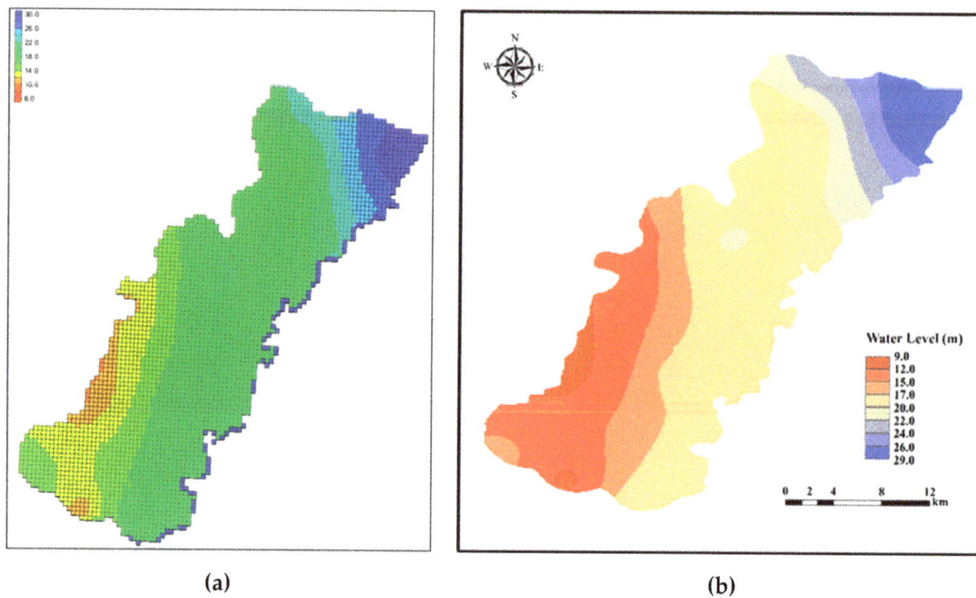

Figure 6. Schematic diagram of network segmentation (**a**); the initial flow field in the study area (**b**).

3.1.5. Model Identification and Verification

The identification period of the model is from May 2018 to May 2019. The identification period of the model is an inversion process. According to the formation lithology characteristics of each parameter partition, and combining with the obtained initial hydrogeological parameters, the calculated water level at the monitoring well is consistent with the measured water level through appropriate adjustment of hydrogeological parameters within a reasonable parameter value range, and the water level change trend is generally consistent with the measured water level change trend. The verification period of the model is from May 2019 to December 2019. The model verification period is a forward process, that is, the parameters are kept unchanged, and observe whether the calculated water level of the monitoring well and the trend are consistent. If the two are consistent and the trend is consistent, it indicates that the model can objectively reflect the actual situation of the study area.

After identification and verification, the comparison of historical changes between the calculated water level and the measured water level of some groundwater monitoring wells in the study area is shown in Figure 7. It can be seen from the Figure that the calculated water level of the monitoring well is well fitted with the measured water level, and the constructed groundwater seepage numerical model can objectively reflect the actual situation of the study area. The model simulation error statistics are shown in Table 3. From Table 3, it can be seen that the model MAE and RMSE are both lower than 0.5 m, with high accuracy, and can be used to predict the groundwater level situation in this region.

Table 3. Groundwater model simulation error statistics results.

Monitoring Wells	MAE/m	RMSE/m
Houshiyu	0.2473	0.3240
Houfudu	0.2974	0.3842
Xucunxlang	0.2836	0.3574
Fangzhaixiang	0.2127	0.2926

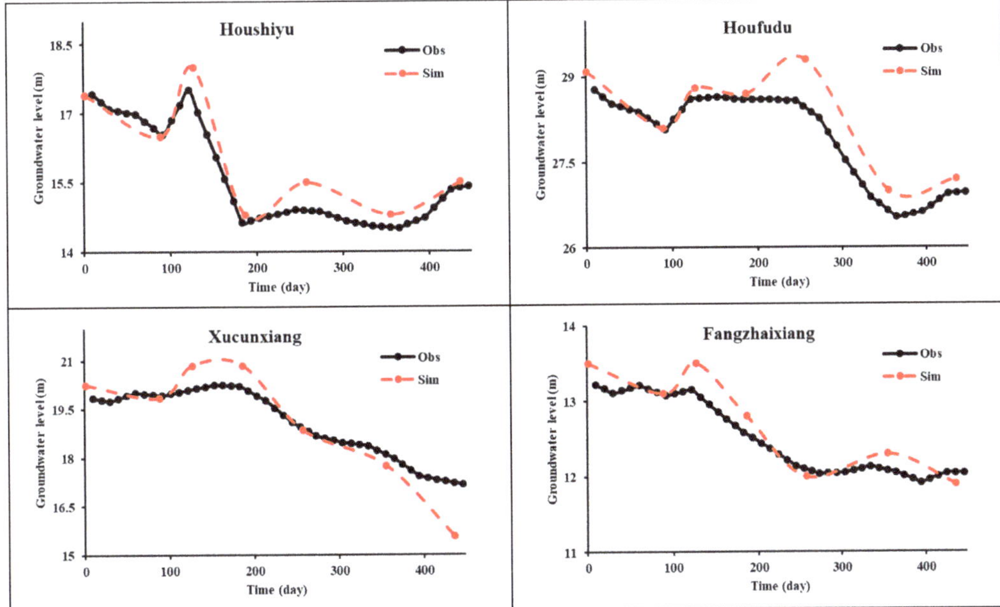

Figure 7. Comparison between measured and simulated groundwater level in partial typical monitoring wells.

3.2. Prediction Results of Social and Economic Development and Water Demands

This study analyzes the water supply source and water supply quantity of Guantao County in the base year, and predicts the water supply capacity of the study area in the planning target year according to the Surface Water Allocation and Utilization Plan of Hebei Province. Comprehensively considering the local social and economic development plan, total water consumption control index and water efficiency control index, this study adopts the scenario analysis method to formulate two schemes of high growth and moderate growth, respectively, according to the two scenarios of enhanced water saving and moderate water saving. A total of four water demand scenarios are combined to analyze and predict the social and economic development and water demands of Guantao County. See Table 4 for the setting of water demand prediction scheme.

Table 4. Setting of water demand forecasting scheme.

Development Plan	Moderate Water -Saving Scenario	Strengthen Water-Saving Scenarios
High growth	Scheme I	Scheme II
Moderate growth	Scheme III	Scheme IV

3.2.1. Analysis and Prediction of Water Supply

Water supply quantity refers to the sum of gross water supply quantity provided by various water source projects to users, including water transmission loss. The total water supply quantity of the existing water supply facilities in Guantao County in 2019 is 84.7 million m^3. Among them, the water supply quantities of surface water, groundwater, external diverted water, brackish water and reclaimed water are 38.25 million m^3, 32.5 million m^3, 4.95 million m^3, 7 million m^3, and 2 million m^3, respectively, accounting for 45.16%, 38.37%, 5.84%, 8.26%, and 2.36% of the total water supply quantity, respectively.

The surface water supply source mainly comes from the Wei Canal and the Yellow River Diversion Project in the east of the study area, the groundwater source comes from 9019 motor-pumped wells in the study area, the external diverted water source comes from the South-to-North Water Transfer Middle Route Project, and the reclaimed water source comes from the reclaimed water of Guantao County sewage treatment plant. The water supply capacity of the current water supply project in the study area is 103 million m^3. According to the Surface Water Allocation and Utilization Plan of Hebei Province, by the end of 2025, the Weixi New Canal Connection Project and Shennong Canal Project will be added in Guantao County, and the Matou, Xiaocun and Shenjie pump stations will be started. It is estimated that the water supply capacity of 5.6 million m^3 will be increased. Therefore, the estimated water supply capacity of Guantao County in the planning target year is 108.6 million m^3.

3.2.2. Prediction of Social and Economic Development

The main contents of prediction of social and economic development include population prediction, agricultural irrigation area prediction and industrial added value prediction. The short-term (2021–2025), medium-term (2026–2030), and long-term (2031–2035) social and economic development prediction results of Guantao County are shown in Figure 8 under the scenarios of high growth and moderate growth. The results show that in terms of population, Guantao Town, Shoushansi and other areas with relatively high urbanization rates have a large population and a relatively large growth rate; in terms of irrigation area, the agricultural irrigation area in Guantao County is widely distributed on both sides of the Weixi Main Canal, of which the length is long in Luqiao Township and Chaibao Town, so the irrigation area is significantly higher; and in terms of industrial added value, the main industrial enterprises in Guantao County are distributed in Guantao Town and Shoushansi, and the industrial enterprises in the other townships are relatively few.

3.2.3. Prediction of Water Demand

Based on the prediction results of social and economic development, this study adopts the quota method to predict the water demands of different industries in Guantao County. The results are shown in Figure 9, which indicates that the agricultural water demand under each scenario is significantly higher than that of other industries, followed by the urban domestic water demand. With the continuous improvement of China's attention to the ecological environment in recent years, the proportion of water demand of ecological environment is also relatively high. With the development of urbanization and industrialization, the water demands of urban life, industry, and ecological environment are gradually increasing. At the same time, the popularization of agricultural water-saving technology makes the water demand in agriculture and rural areas gradually decline. In addition, due to the relatively small scope of the study area, the water demand in the high growth scenario increases by 2–3% compared with that of the moderate growth scenario, and the water demand in the moderate water-saving scenario increases by 2–4% compared with that of the enhanced water-saving scenario.

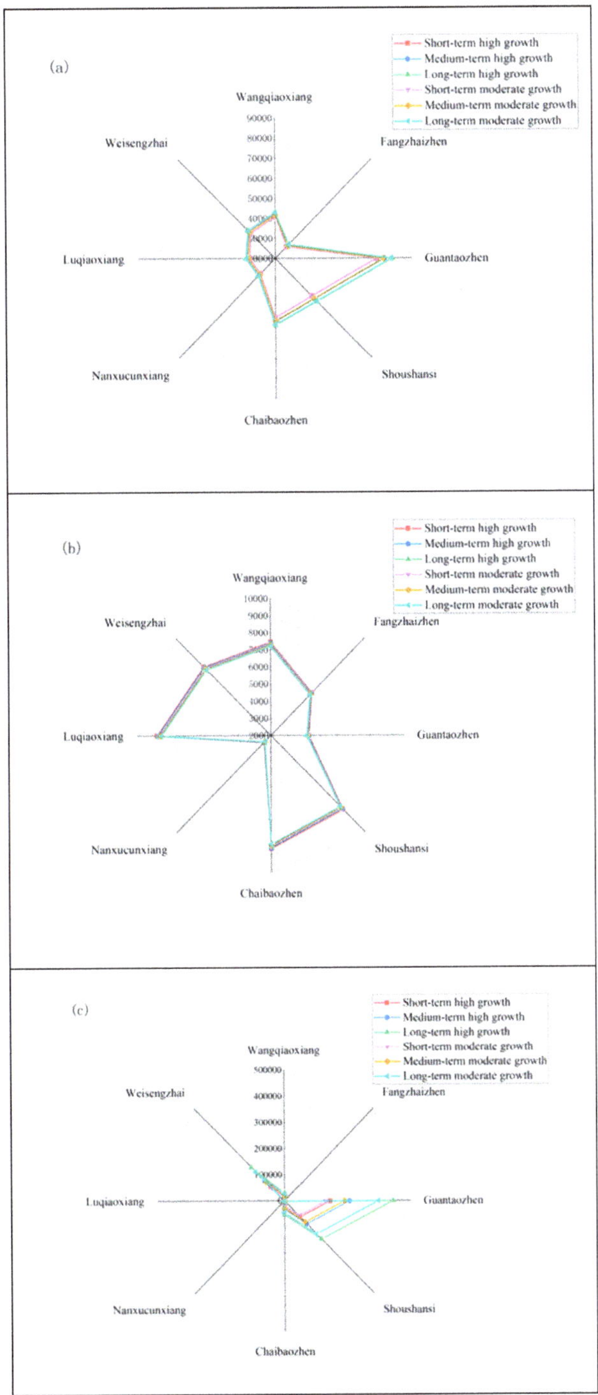

Figure 8. Social and economic development prediction: (**a**) Population prediction; (**b**) Agricultural irrigation area prediction; (**c**) Industrial added value prediction.

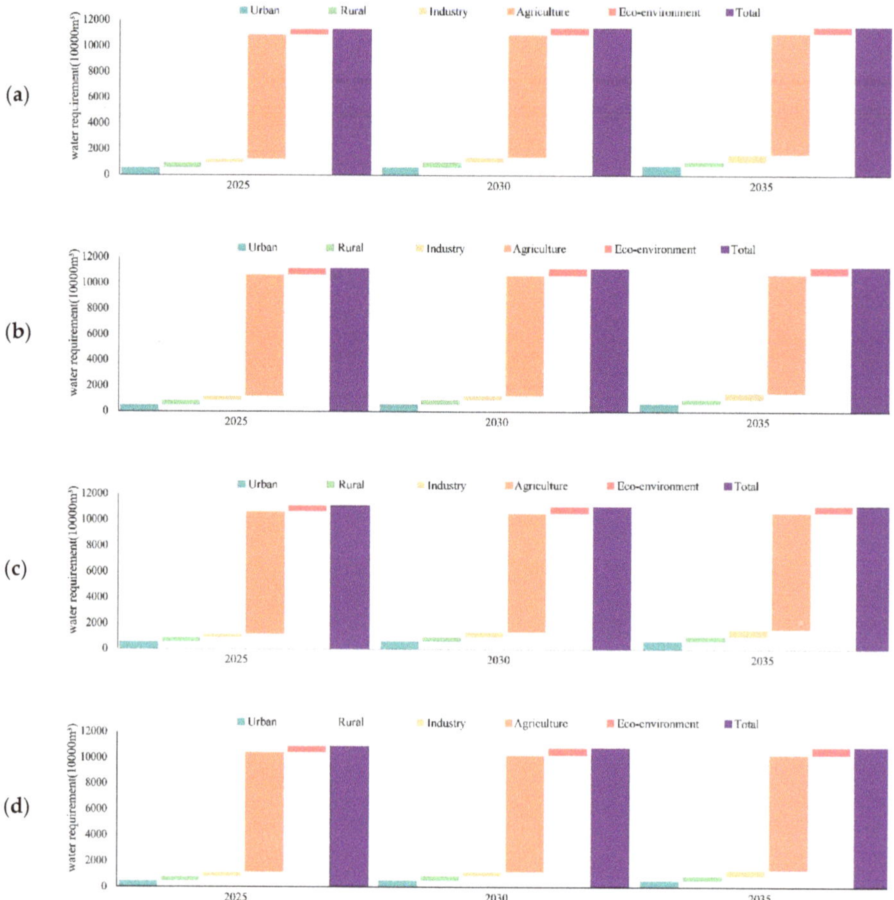

Figure 9. Prediction results of water demand in different industries under different scenarios: (**a**) High growth and moderate water-saving scenario; (**b**) Moderate growth and moderate water-saving scenario; (**c**) High growth intensifies water-saving scenario; (**d**) Moderate growth intensifies water-saving scenario.

3.3. Analysis of Optimal Water Resource Allocation Results

According to the proposed three water demand schemes in different planning target years, which are respectively the high scheme (high growth and moderate water saving), medium scheme (high growth and enhanced water saving) and low scheme (moderate growth and moderate water saving), the water resources allocation results of different schemes in the planning target year of the study area are obtained through the calculation with the water resources allocation module, as shown in Table 5. The relationship between water supply and demand between different water sources and different users is shown in Figure 10.

The overall supply–demand structure of the low scheme is basically reasonable through the reasonable allocation of water sources. The depth of water shortage is the least of the three schemes. It belongs to the way of "determining production by water and taking connotative development". In the planning year, it inhibits the agricultural development of the county so as to control water shortage, and it is a better scheme from the perspective of water resource supply–demand balance. However, the scheme limits the social and economic development of Guantao County to a certain extent and hinders the development strategy of "striving to build Guantao County into a sub-central city with great radiation in the border area of Hebei, Shandong and Henan and the most influential rural tourism destination in the central south plain of Hebei", thus, it is not the preferred allocation scheme.

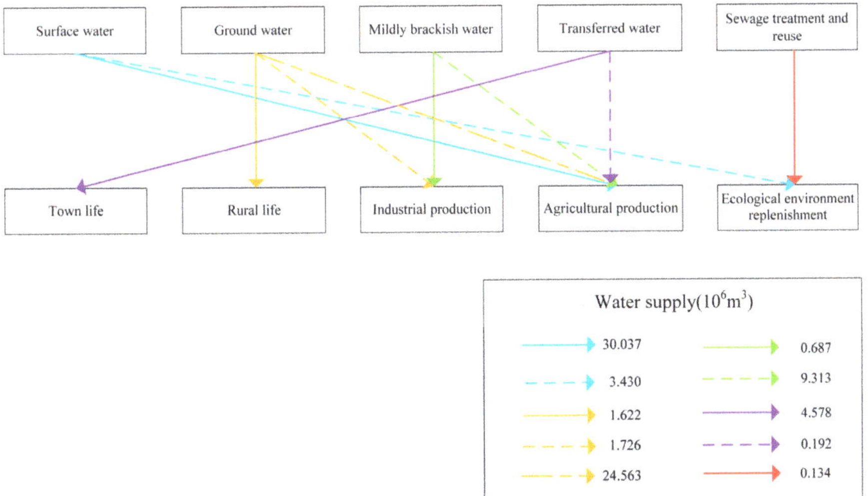

Figure 10. The relationship between water supply and demand between different water sources and different users in 2030 under the second scenario.

The high scheme adopts the mode of "determining water consumption by production and taking epitaxial development". Under the development goal of fully supporting social and economic development and building a sub-central city and ecotourism destination, it controls agricultural water consumption and increases industrial water consumption. Therefore, it is also the scheme with the highest water shortage. Although the quantity of water diverted from Wei Canal and the Yellow River and the water network coverage can be increased by building the water system connection project in the planning target year, considering the difficulty and complexity of the external water transfer project, there are still a lot of feasibility studies and preliminary demonstration work to be invested. Therefore, it is difficult to realize the large surface water supply quantity and scale by 2035. Thus, this scheme is not recommended as a priority scheme but a reservation scheme.

The medium scheme adopts the mode of "supply and demand coordination and steady and healthy development". In the planning target year, it improves the water use efficiency of various industries and comprehensively realizes the balance of regional groundwater exploitation and recharge by controlling groundwater overexploitation. In this way, under the water supply conditions of the established South-to-North water transfer project, it can not only meet the new social and economic, industrial and agricultural water demands, but also save costs and make the rational use of various resources. Therefore, it can be used as a recommended scheme. The results of water resources allocation of each calculation unit in 2030 under the medium scheme are shown in Table 6. It can be seen from Table 6 that by increasing the development and utilization of brackish water, the brackish water agricultural area of Nanxu Village, the brackish water agricultural Chaibao area 2 and the industrial area of Guantao Town will achieve the balance between supply and demand of water resources. For Shoushansi water-saving agricultural area which does not have brackish water resources, the local conventional water resources cannot support the regional agricultural economic development, so it is also the area with the highest water shortage.

Table 5. Water resources allocation results under different schemes (10^6 m^3).

Horizontal Year	Plan	Water Supply							Water Consumption					Water Shortage
		Water Demand	Surface Water	Groundwater	Brackish Water	Wastewater Reuse	Transfer Water	Total	Town Life	Rural Life	Industrial Production	Agricultural Production	Ecological Environment Replenishment	
2019	/	114.057	37.560	30.816	7.000	0.164	3.853	79.392	5.625	3.022	1.962	66.266	2.518	34.665
2025	low	106.099	34.917	28.638	9.000	0.192	4.727	77.473	5.122	2.384	2.512	63.950	3.505	28.621
	middle	106.633	33.142	28.651	9.000	0.135	4.655	75.584	4.051	1.660	2.065	64.578	3.230	31.049
	high	114.874	38.047	30.482	9.000	0.256	4.765	82.549	5.774	2.603	2.783	67.766	3.623	32.313
2030	low	104.886	34.613	28.330	10.000	0.224	4.761	77.927	5.787	2.242	3.136	62.841	3.921	26.954
	middle	106.454	33.467	27.911	10.000	0.134	4.770	76.282	4.578	1.622	2.413	64.105	3.564	30.172
	high	113.757	36.520	29.648	10.000	0.266	4.808	81.643	6.672	2.383	3.675	64.867	4.047	32.105
2035	low	105.761	35.807	27.055	12.000	0.337	4.761	79.959	6.373	2.166	4.546	62.697	4.177	25.792
	middle	106.643	34.526	26.170	12.000	0.223	4.703	77.622	5.017	1.516	3.373	63.904	3.812	29.014
	high	115.291	38.303	28.907	12.000	0.431	4.816	84.457	7.611	2.110	5.680	64.696	4.360	30.837

Table 6. The water resources allocation results of each computing unit in the plan in 2030 (10^6 m^3).

Unit	Water Supply							Water Consumption					Water Shortage
	Water Demand	Surface Water	Groundwater	Brackish Water	Wastewater Reuse	Transfer Water	Total	Town Life	Rural Life	Industrial Production	Agricultural Production	Ecological Environment Replenishment	
Luqiao Brackish Water Agriculture Zone 1	4.826	2.711	0.749	0.661	0.000	0.166	4.287	0.042	0.039	0.000	4.206	0.000	0.539
Luqiao Brackish Water Agriculture Second District	11.209	5.359	3.571	1.306	0.000	0.368	10.604	0.114	0.106	0.000	9.916	0.468	0.605
Weisengzhai Brackish Water Agriculture District No. 1	7.456	3.366	1.908	1.359	0.000	0.302	6.935	0.141	0.093	0.357	6.344	0.000	0.521
Weisengzhai Brackish Water Agriculture Second District	6.916	0.871	1.617	0.351	0.000	0.084	2.923	0.147	0.096	0.000	2.412	0.268	3.993
Brackish Water Agricultural Area of Nanxu Village	5.186	0.466	3.918	0.575	0.002	0.225	5.186	0.503	0.284	0.012	4.183	0.204	0.000
Chaibao Freshwater Agricultural Area	4.548	1.665	0.603	0.000	0.000	0.450	2.718	0.081	0.044	0.057	2.535	0.000	1.830
Chaibao Brackish Water Agriculture District I	6.150	2.365	0.468	0.824	0.000	0.181	3.838	0.071	0.039	0.000	3.729	0.000	2.312
Chaibao Brackish Water Agriculture Second District	5.296	3.140	0.852	1.094	0.000	0.210	5.296	0.083	0.045	0.000	5.096	0.072	0.000
Shoushan Temple Industrial Zone	3.297	0.711	0.304	0.000	0.000	0.262	1.277	0.161	0.006	0.155	0.955	0.000	2.020
Shoushan Temple Water-saving Agricultural Area	11.686	1.733	4.027	0.000	0.000	0.351	6.111	0.011	0.070	0.012	5.908	0.110	5.575
Guantao Town Ecological Zone	4.603	0.994	0.417	0.115	0.089	0.490	2.105	0.489	0.031	0.274	0.985	0.326	2.498
Guantao Town Agricultural District	4.834	3.053	0.698	0.353	0.015	0.334	4.453	0.791	0.044	0.587	2.438	0.593	0.381
Guantao Town Industrial Zone	4.401	2.677	0.846	0.472	0.015	0.391	4.401	0.896	0.045	0.823	1.841	0.796	0.000
Fangzhai Brackish Water Agriculture Zone 1	4.403	0.992	1.023	0.585	0.004	0.165	2.769	0.155	0.117	0.000	2.432	0.065	1.634
Fangzhai Brackish Water Agriculture Second District	6.620	1.090	2.260	0.643	0.006	0.228	4.227	0.238	0.186	0.000	3.708	0.095	2.393
Wangqiao Brackish Water Agriculture District 1	3.763	0.782	0.374	0.572	0.000	0.155	1.883	0.167	0.091	0.040	1.466	0.119	1.880
Wangqiao Brackish Water Agriculture Second District	11.260	1.492	4.276	1.090	0.003	0.408	7.269	0.488	0.286	0.096	5.951	0.448	3.991
Total	106.454	33.467	27.911	10.000	0.134	4.770	76.282	4.578	1.622	2.413	64.105	3.564	30.172

4. Discussion

4.1. Results of Brackish Water Allocated According to Industry under the Recommended Scheme

The water resources allocation model constructed in this paper can calculate the quantity of water supplied by different water sources to water users in various industries in different calculation units. Taking the brackish water allocation according to industry of the recommended scheme as an example, the brackish water allocation results in 2030 are shown in Figure 11 which indicate that only the brackish water in Guantao Town is supplied to industry, and the brackish water in other areas is only supplied to agricultural irrigation. The main reason is that the main industrial enterprises in Guantao County are located in Guantao Town and Shoushansi, and there are no brackish water resources in Shoushansi. Therefore, only brackish water in Guantao Town is used for industrial production. At the same time, brackish water consumption is positively correlated with crop yield in saline water areas. In areas with high crop yields (such as Luqiao Township, Weisengzhai Town, Chaibao Town, Wangqiao Township, etc.), the brackish water consumption is also high. The exploitation of brackish water not only relieves the water supply pressure of shallow groundwater and Wei Canal to a certain extent and alleviates the contradiction between supply and demand of water resources, but also reduces the water level of brackish water

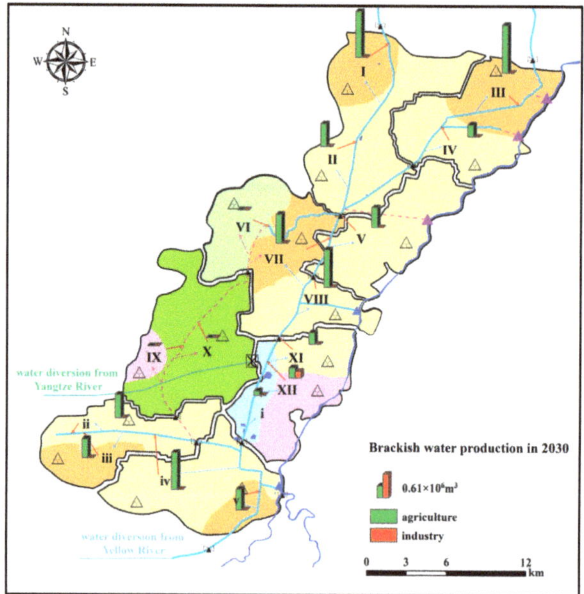

Figure 11. Brackish water configuration in 2030 under the second scenario, reduces soil salinization and improves the soil ecological environment.

4.2. Analysis on Variation Characteristics of Brackish Water Level under Recommended Scheme

The brackish water allocation results of the medium scheme are substituted into the groundwater model for water level simulation, and the variation results of brackish water level in each planning target year are shown in Figure 12. Figure 12 shows that the brackish water level in the study area decreases from −19.57 m in 2019 to −32.26 m in 2035. In 2035, the highest brackish water level in the study area is located in Weisengzhai Town, with a water level of −24.21 m, and the lowest water level is located in Nanxu Village, with a water level of −53.36 m, forming three brackish water cone areas: Chaibao Town, Guantao Town, and Wangqiao Township. The decline of brackish water level can enhance the regulation and storage capacity of soil aeration zone, promote the vertical alternating movement with shallow groundwater, and realize the desalination of brackish water, so as to increase the quantity of fresh water resources [33]. Therefore, the development and utilization of brackish water under the recommended scheme appropriately reduces the brackish water level, which is conducive to alleviating the demand for fresh water resources for agricultural irrigation and industrial production.

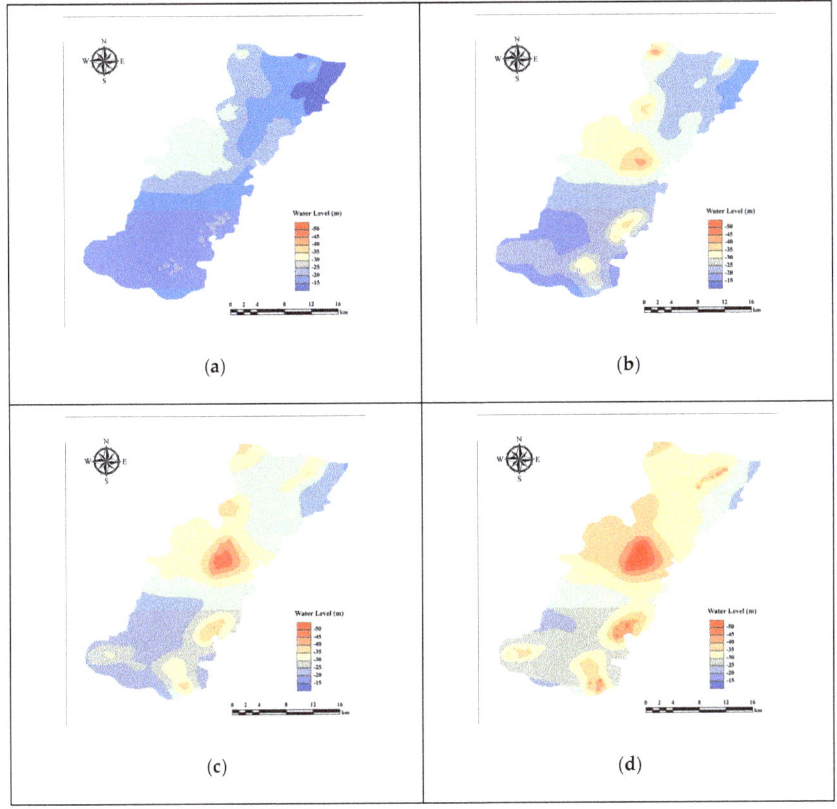

Figure 12. Groundwater flow field under high growth and enhanced water saving scheme: (**a**) at the end of 2019; (**b**) at the end of 2025; (**c**) at the end of 2030; (**d**) at the end of 2035.

5. Conclusions

(1) This paper proposes an optimal water resources allocation system integrating model data parameter database, water resources supply and demand prediction module, groundwater numerical simulation module and water resources allocation module, expounds the principle of coordinated work among all modules and the relationship between data input and output, and puts forward a method for predicting the change of brackish water level under the allocation scheme by groundwater numerical simulation.

(2) Taking the brackish water resources that are easy to develop and rich in reserves in unconventional water sources as the key allocation object, this paper takes the brackish water resource as the independent water source, aims to meet the comprehensive optimization of social, economic and ecological environmental benefits, and takes the water quantity balance and supply–demand balance as constraints, so as to construct the water resources allocation model based on the rational utilization of brackish water. Then it obtains the quantity of brackish water resources supplied

to agricultural irrigation and industrial production of each calculation unit in the planning target year, so as to realize accurate water resources allocation.

(3) This paper applies the water resources allocation system to Guantao County, Handan City. Through scheme comparison, the medium scheme is finally selected as the optimal allocation scheme. Under this scheme, in 2030, the brackish water agricultural area of Nanxu Village, the brackish water agricultural Chaibao area 2 and the industrial area of Guantao Town will achieve the balance between supply and demand of water resources. Compared with the current year, the overall water shortage in the study area is decreased by $4.493 \times 106 \text{ m}^3$. Meanwhile, under the recommended scheme, the brackish water level in the study area will drop by 12.69 m in 2035. Therefore, while alleviating the tension of water supply of local conventional water sources, it also reduces soil salinization and realizes the coordinated development of society, economy, and ecology in the study area.

Author Contributions: D.Z., methodology, data curation, software; X.X., conceptualization, formal analysis, supervision; T.W., supervision, software; B.W., data curation; S.P., methodology, data curation. All authors have read and agreed to the published version of the manuscript.

Funding: This study was funded by Key R&D Program of Hebei Province (Grant No: 21374201D), National Key Research and Development Program of China (2021YFC3001005) and Hebei Province Water Conservancy Research Plan (2018-73).

Institutional Review Board Statement: Not applicable.

Informed Consent Statement: Not applicable.

Data Availability Statement: Data available upon request due to privacy and ethical restrictions.

Acknowledgments: The authors would like to express their gratitude to the editors and anonymous experts for their constructive comments.

Conflicts of Interest: The authors declare no conflict of interest.

References

1. Foster, S.S.D.; Perry, C.J. Improving groundwater resource accounting in irrigated areas: A prerequisite for promoting sustainable use. *Hydrogeol. J.* **2010**, *18*, 291–294. [CrossRef]
2. Cheng, B.; Li, H.; Yue, S.; Huang, K. A conceptual decision-making for the ecological base flow of rivers considering the economic value of ecosystem services of rivers in water shortage area of Northwest China. *J. Hydrol.* **2019**, *578*, 124126. [CrossRef]
3. Wang, Y.; Liu, L.; Guo, P.; Zhang, C.; Zhang, F.; Guo, S. An inexact irrigation water allocation optimization model under future climate change. *Stoch. Env. Res. Risk A* **2019**, *33*, 271–285. [CrossRef]
4. Pacheco-Martínez, J.; Hernandez-Marín, M.; Burbey, T.J.; González-Cervantes, N.; Ortíz-Lozano, J.Á.; Zermeño-De-Leon, M.E.; Solís-Pinto, A. Land subsidence and ground failure associated to groundwater exploitation in the Aguascalientes Valley, México. *Eng. Geol.* **2013**, *164*, 172–186. [CrossRef]
5. Lu, Z.; Wei, Y.; Xiao, H.; Zou, S.; Ren, J.; Lyle, C. Trade-offs between midstream agricultural production and downstream ecological sustainability in the Heihe River basin in the past half century. *Agr. Water Manag.* **2015**, *152*, 233–242. [CrossRef]
6. Kapetas, L.; Kazakis, N.; Voudouris, K.; McNicholl, D. Water allocation and governance in multi-stakeholder environments: Insight from Axios Delta, Greece. *Sci. Total Environ.* **2019**, *695*, 133831. [CrossRef] [PubMed]
7. Wang, T.; Liu, Y.; Wang, Y.; Xie, X.; You, J. A Multi-Objective and Equilibrium Scheduling Model Based on Water Resources Macro Allocation Scheme. *Water Resour. Manag.* **2019**, *33*, 3355–3375. [CrossRef]
8. Liu, D.; Guo, S.; Liu, P.; Zou, H.; Hong, X. Rational Function Method for Allocating Water Resources in the Coupled Natural-Human Systems. *Water Resour. Manag.* **2019**, *33*, 57–73. [CrossRef]
9. Abdulbaki, D.; Al-Hindi, M.; Yassine, A.; Najm, M.A. An optimization model for the allocation of water resources. *J. Clean Prod.* **2017**, *164*, 994–1006. [CrossRef]
10. Minsker., B.S.; Padera, B.; Smalley, J.B. Efficient methods for including uncertainty and multiple objectives in water resources management models using genetic algorithms. In Proceedings of the XIII International Conference on Computational Methods in Water Resources, Calgary, AB, Canada, 25–29 June 2000; pp. 567–572.
11. Rosegrant, M.W.; Ringler, C.; McKinney, D.C.; Cai, X.; Keller, A.; Donoso, G. Integrated economic-hydrologic water modeling at the basin scale: The Maipo river basin. *Agr. Econ.* **2000**, *24*, 33–46.
12. Chen, S.; Shao, D.; Gu, W.; Xu, B.; Li, H.; Fang, L. An interval multistage water allocation model for crop different growth stages under inputs uncertainty. *Agr. Water Manag.* **2017**, *186*, 86–97. [CrossRef]

13. Zhang, Y.Q. Research on water resources allocation in irrigation area under artificial fish swarm algorithm. *Water Resour. Plan. Des.* **2016**, *7*, 42–44.
14. Hou, J.; Sun, Y.; Sun, J. Optimal Allocation of Water Resources Based onthe Multi-Objective Fish-Ant Colony Algorithm. *Resour. Sci.* **2011**, *33*, 2255–2261.
15. Li, Y.; Zhao, T.; Wang, P.; Gooi, H.B.; Wu, L.; Liu, Y.; Ye, J. Optimal Operation of Multimicrogrids via Cooperative Energy and Reserve Scheduling. *IEEE Trans. Ind. Inform.* **2018**, *14*, 3459–3468. [CrossRef]
16. Afshar, A.; Shojaei, N.; Sagharjooghifarahani, M. Multiobjective Calibration of Reservoir Water Quality Modeling Using Multiobjective Particle Swarm Optimization (MOPSO). *Water Resour. Manag.* **2013**, *27*, 1931–1947. [CrossRef]
17. Abolpour, B.; Javan, M.; Karamouz, M. Water allocation improvement in river basin using Adaptive Neural Fuzzy Reinforcement Learning approach. *Appl. Soft Comput.* **2007**, *7*, 265–285. [CrossRef]
18. Li, J.; Song, S.; Ayantobo, O.O.; Wang, H.; Liang, J.; Zhang, B. Coordinated allocation of conventional and unconventional water resources considering uncertainty and different stakeholders. *J. Hydrol.* **2022**, *605*, 127293. [CrossRef]
19. Mooselu, M.G.; Nikoo, M.R.; Latifi, M.; Sadegh, M.; Al-Wardy, M.; Al-Rawas, G.A. A multi-objective optimal allocation of treated wastewater in urban areas using leader-follower game. *J. Clean. Prod.* **2020**, *267*, 122189. [CrossRef]
20. Gao, J.; Li, C.; Zhao, P.; Zhang, H.; Mao, G.; Wang, Y. Insights into water-energy cobenefits and trade-offs in water resource management. *J. Clean. Prod.* **2019**, *213*, 1188–1203. [CrossRef]
21. Yang, S.Q. Prediction and Research of Water-Soil Environment Effect Under Light-Saline Water Irrigation Based on Visual MODFLOW and SWAP Coupling Model in Arid Area Inner Mongolia Agricultural University. Ph.D. Thesis, Inner Mongolia Agricultural University, Hohhot, China, 2005.
22. Liu, Y.; Fu, G. Utilization of Gentle Salty Water Resource in China. *Geogr. Geo-Inf. Sci.* **2004**, *20*, 57–60.
23. Shao, J. *Optimization of Unconventional Water Resources in Yinchuan City Based on Comprehensive Utillization*; Ningxia University: Ningxia, China, 2017.
24. Dutt, R.G.; Pennington, D.A.; Turner, F. Irrigation as a Solution to Salinity Problems of River Systems. In *Salinity in Watercourses and Reservoirs*; Ann Arbor Science: Ann Arbor, MI, USA, 1984; pp. 465–472.
25. Wang, J.; Shan, Y. Review of Research Development on Water and Soil Regulation with Brackish Water Irrigation. *Trans. Chin. Soc. Agric. Mach.* **2015**, *46*, 117–126.
26. Wu, Z.; Wang, J. Effect on Both SoilI nfiltration Characteristics and Ion Mobility Features by Mineralization Degree of Infiltration Water. *Trans. Chin. Soc. Agric. Mach.* **2010**, *41*, 64–69.
27. Phogat, V.; Yadav, A.K.; Malik, R.S.; Kumar, S.; Cox, J. Simulation of salt and water movement and estimation of water productivity of rice crop irrigated with saline water. *Paddy Water Environ.* **2010**, *8*, 333–346. [CrossRef]
28. Chen, X.; Yang, J.; Yang, C. Irrigation-drainage management and hydro-salinity balance in Weigan River Irrigation District. *Trans. Chin. Soc. Agric. Eng.* **2008**, *4*, 59–65.
29. Pang, H.C.; Li, Y.Y.; Yang, J.S.; Liang, Y.S. Effect of brackish water irrigation and straw mulching on soil salinity and crop yields under monsoonal climatic conditions. *Agr. Water Manag.* **2010**, *97*, 1971–1977. [CrossRef]
30. Liu, Y.; Feng, Y.; Huang, J. Effects of modifiers on saline soil salt distribution under brackish water irrigation conditions. *Agric. Res. Arid. Areas* **2015**, *33*, 146–152.
31. Wei, J.; Wang, H. Network diagram of regional water resource allocation system. *J. Hydraul. Eng.* **2007**, *9*, 1103–1108.
32. Wei, C.; Han, J.; Han, S. *Key Technologies and Demonstrations for Optimal Allocation of All Elements of River Basin/Regional Water Resources*; China Water Power Press: Beijing, China, 2012.
33. Zhang, Y.; Shen, Y.; Wang, Y. Distribution characteristics and exploitation of brackish water in hebei Plain. *J. Agric. Resour. Environ.* **2009**, *26*, 29–33.

water

MDPI

Article

Groundwater Recharge Modeling under Water Diversion Engineering: A Case Study in Beijing

Mingyan Zhao, Xiangbo Meng *, Boxin Wang, Dasheng Zhang, Yafeng Zhao and Ruyi Li

Hebei Institute of Water Science, Shijiazhuang 050051, China; mingyanzhao99@126.com (M.Z.); wangboxin293@163.com (B.W.); 18501317443@163.com (D.Z.); sjz791979@126.com (Y.Z.); liruyi0117@126.com (R.L.)
* Correspondence: mengxb1994@163.com; Tel.: +86-130-3103-8799

Abstract: The influence of surface water resource exploitation and utilization projects on groundwater has been widely studied. Surface water diversion projects lead to a reduction in river discharge, which changes the recharge of groundwater systems. In this study, the numerical simulation method is used to predict the variation in groundwater level under different diversion scale scenarios. The Zhangfang water diversion project in Beijing, China, was chosen for the case study. The downstream plain area of the Zhangfang water diversion project is modeled by MODFLOW to predict the influence of reducing water diversion on the dynamic change in the groundwater level in the downstream plain area. The model results show that the difference in groundwater recharge and discharge on the downstream plain of Zhangfang is 9,991,900 m^3/a, which is in a negative water balance state, and the groundwater level continues to decrease. Reducing the amount of water diverted by the Zhangfang water diversion project to replenish groundwater is beneficial to the rise of the groundwater level in the downstream plain area. The results indicate that the groundwater flow model in the downstream plain area of Zhangfang performed well in the influence assessment of surface water resource exploitation and utilization projects on groundwater. This study also provides a good example of how to coordinate the relationship between surface water resources and groundwater resources.

Keywords: water diversion project; groundwater; numerical simulation; MODFLOW

Citation: Zhao, M.; Meng, X.; Wang, B.; Zhang, D.; Zhao, Y.; Li, R. Groundwater Recharge Modeling under Water Diversion Engineering: A Case Study in Beijing. *Water* **2022**, *14*, 985. https://doi.org/10.3390/w14060985

Academic Editor: Juan José Durán

Received: 29 January 2022
Accepted: 17 March 2022
Published: 21 March 2022

Publisher's Note: MDPI stays neutral with regard to jurisdictional claims in published maps and institutional affiliations.

1. Introduction

Groundwater plays an important role in social development and in maintaining the ecological environment [1]. River leakage recharge is an important part of groundwater recharge, so the interaction between the river and aquifer has become a research hotspot [2–4]. The interaction between surface water and groundwater is constantly changing due to the different spatial and temporal distribution of the exchange between surface water and groundwater [5,6]. A water diversion project will change the interaction between the surface water and groundwater, and directly reduce the discharge of the river channel, which is one of the man-made factors causing the change of regional groundwater level and quality [7,8]. Among them, there are many studies on the impact of external water transfer in the water receiving area, including the impact of the South-to-North water transfer project on the groundwater systems of the North China plain [9] and Beijing [10], the impact of watershed ecological water transfer on the hydrological situation in the lower reaches of the Heihe River basin [11], and the impact of the ecological water supplement on the section of Yongding River, Beijing plain [12]. Research shows that water transfer is conducive to the recovery of the groundwater level in the water receiving area. However, there have been few studies on the impact in the water transfer area. The water transfer project directly leads to the reduction of groundwater discharge in the water transfer area [13], resulting in a decline of the downstream groundwater level [14].

The Zhangfang water diversion project is located in the Fangshan district of Beijing, supplying water from the Juma River to Beijing through the Shengtian Canal. In September

1981, Beijing decided to divert water from the Juma River to relieve the pressure of a water shortage. In 2004, the Zhangfang water diversion project was completed, and supplied water to Beijing. Since the construction of the project, the Zhangfang water diversion project has diverted approximately 100 million m^3 of water from the Juma River every year. The large amount of water diversion has reduced the surface runoff of the Nanjuma River and the Beijuma River downstream of the Zhangfang water diversion project, which has caused a serious impact on downstream water resources and the water environment. Therefore, research on the influence of the Zhangfang water diversion project on downstream water resources is conducive to rational water diversion projects and can enrich the research on the influence of water diversion on the water diversion area.

The water diversion project inevitably leads to changes in hydrogeological conditions along the river and changes the supplement and discharge relationship between downstream groundwater and the surface. A variety of methods can be used to quantitatively describe the process: the water balance method [15], Darcy law [16], the groundwater level dynamic method [17], the isotope method [18] and the numerical simulation method [19]. Numerical simulation achieves the purpose of simulating the actual groundwater system by solving the mathematical model. It is one of the most popular methods for solving groundwater-related problems and has been widely used in groundwater resource evaluation [20], the evaluation of the impact of groundwater development [21], the impact of water conservancy projects on groundwater [22], the mutual conversion of groundwater and surface water, etc. Popular groundwater flow numerical simulation codes include the finite element groundwater flow system FEFLOW [23] and modular 3D groundwater flow model MODFLOW based on the finite difference method [24]. MODFLOW-2005 [25] is a recent popular core version of MODFLOW, which uses a block-centered finite difference method to simulate confined and unconfined groundwater flow, as well as external stresses that include regional recharge, pumping wells, evapotranspiration, rivers, etc.

In this study, we use MODFLOW-2005 as a solver to construct a transient three-dimensional groundwater flow numerical model in the downstream plain of Zhangfang and study the current groundwater flow of this area. Based on this model, different water diversion amounts from the Zhangfang water diversion project are designed to evaluate the impact of the Zhangfang water diversion project on the groundwater level in the downstream plain area. Additionally, this study reveals the importance of river infiltration for groundwater recharge and proposes suggestions for the rational development and utilization of downstream water resources.

2. Basic information of the Study Area

2.1. Meteorology of the Study Area

The study area is located between 39°35′57″ and 39°4′50″ east longitude and 115°30′40″ and 116°15′26″ north latitude. The total area is approximately 2603 km^2. The research scope is shown in Figure 1. The climate in this area is a temperate continental monsoon climate. The perennial mean temperature is 13.4 °C, the winter is severely cold from December to January, and the summer is hot from July to August. The mean annual precipitation is 489.9 mm.

2.2. Hydrogeological Conditions in the Study Area

The study area is located on multistage geological faults in front of the alluvial-proluvial fan of the Juma River. The main fault structures are the Huairou-Laishui deep fault and Donglutou-Qiaoliufan fault. West of the Huairou-Laishui deep fault, quaternary deposits directly cover Proterozoic carbonate rock, and the buried depth of the bedrock is generally 10–40 m. East of the fault to the line of the Donglutou-Qiaoliufan fault, underlying the quaternary strata are tertiary cemented conglomerate and glutenite, and the buried depth of the bedrock is generally 20–60 m. To the east of the Donglutou-Qiaoliufan fault line, quaternary deposits suddenly thicken to 170–550 m. The hydrogeological profile in the study area is shown in Figure 2b.

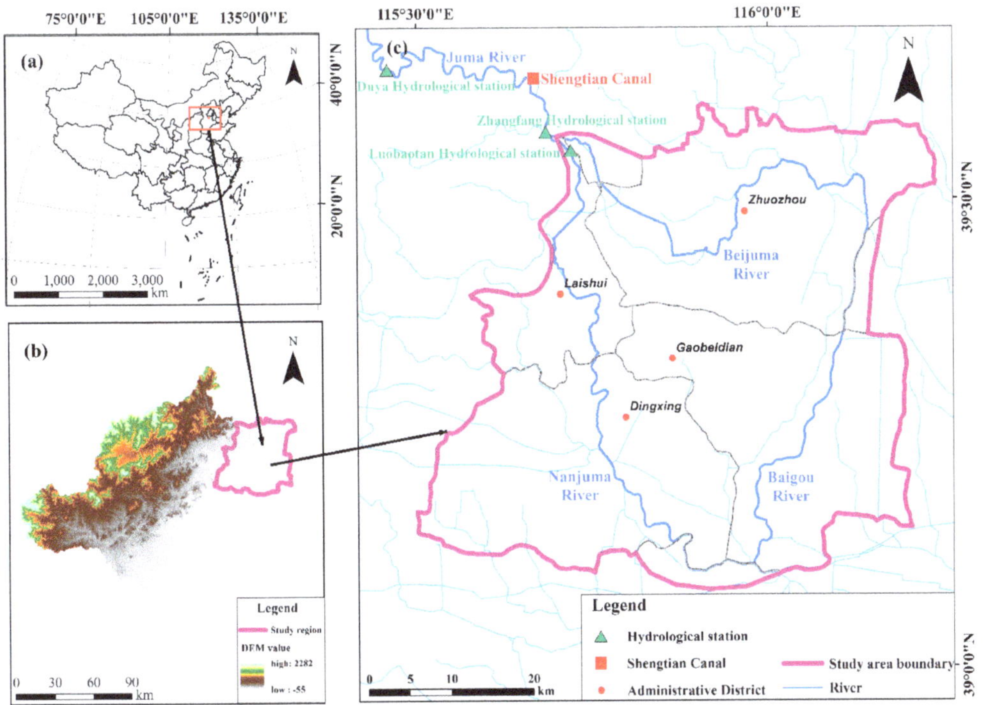

Figure 1. (**a**,**b**) Location and terrain distribution of the study area. (**c**) The location of the river system and hydrological stations in the study area.

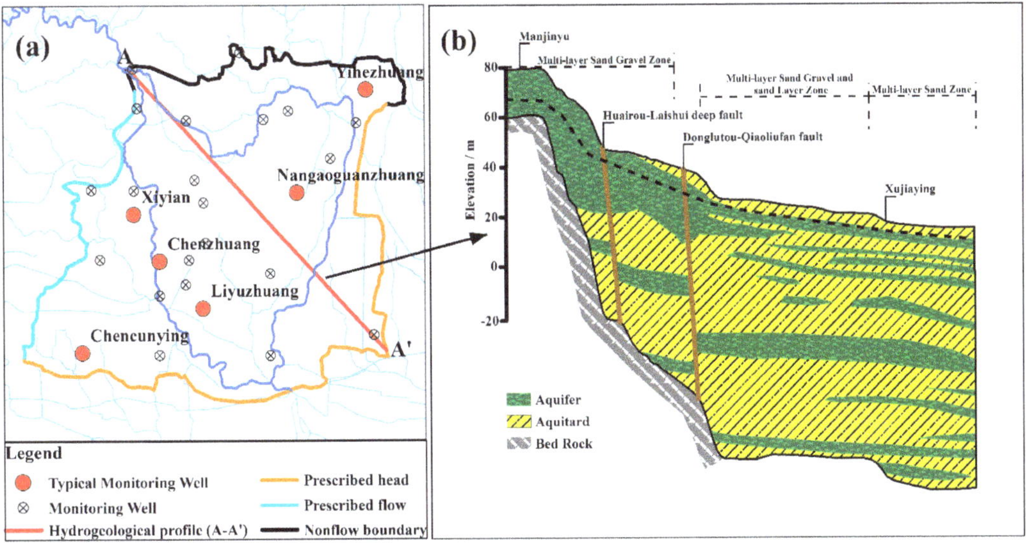

Figure 2. (**a**) The distribution of boundary conditions and monitoring well locations in the study area. (**b**) Hydrogeological profile in the study area.

The aquifer is mainly recharged by means of lateral runoff in the piedmont zone, and precipitation infiltration and water infiltration of the Nanjuma River and Beijuma River; it is discharged by means of artificial mining and lateral outflow of the aquifer. Affected by topography, groundwater flow in the study area mainly flows from northwest to southeast. In recent years, as a result of the large scale of water through the Zhangfang water diversion project and human exploitation of groundwater, the aquifer has almost drained.

2.3. The Water Diversion Information of the Zhangfang Water Diversion Project

The Zhangfang water diversion project supplies water to Beijing through the Shengtian Canal, with an annual flow of approximately 100 million m^3 from the Juma River. The water diversion amounts can be estimated through the observation data of hydrologic stations. The location of three hydrological stations on the Juma River is shown in Figure 1. Zhangfang hydrological station is located downstream of the Duya hydrological station, with Shengtian Canal between them. Figure 3 shows the monthly average flow data of each hydrological station. According to Figure 3, the discharge data of the Duya Hydrological station is significantly higher than that of the Zhangfang hydrological station, which is caused by the interception of a large amount of water from Juma River by the Shengtian Canal. The Luobaotan hydrological station is located on the Juma River and downstream of the Zhangfang hydrological station. Except for flood season, the flow of Luobaotan hydrological station is almost 0, indicating that the Nanjuma River is in a state of perennial flow failure. The observation data of the hydrological stations show that the Zhangfang diversion project has caused a serious impact on water resources and aquatic ecology in the downstream.

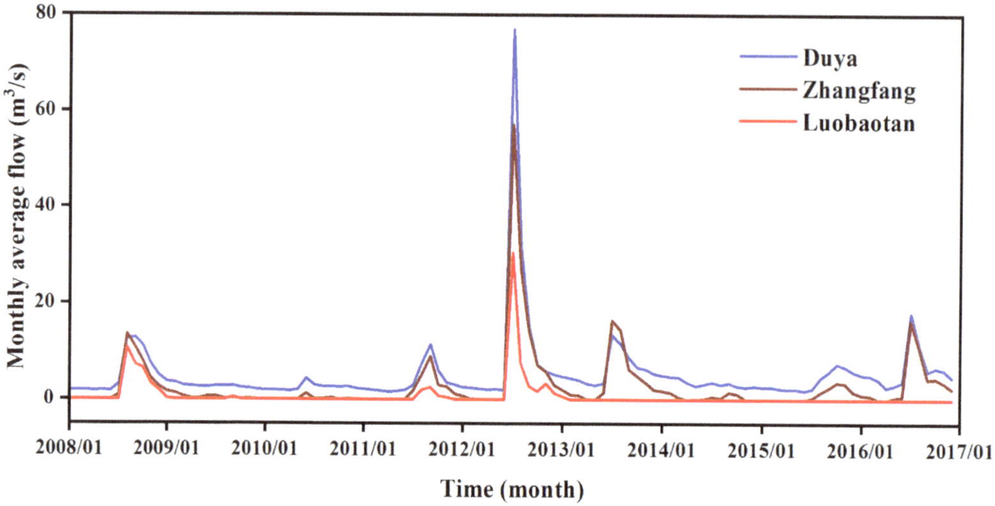

Figure 3. Monthly average flow of hydrological station from 2008 to 2016.

3. Methods

3.1. Groundwater Flow Numerical Model

The western boundaries of the study area are the junction of a mountainous area and plain area, which mainly receives lateral recharge in front of the mountain and is generalized as prescribed flow boundaries. The northern boundaries of the study area are perpendicular to the multiyear average groundwater level contour, which can be generalized as no-flow boundaries. The eastern and southern boundaries are mainly the administrative boundaries of cities and counties. Observation wells are distributed near the boundaries, and the variation range of the groundwater level is small; therefore, the boundary can be generalized as prescribed head boundaries. The generalization of boundary conditions is shown in Figure 2a. Based on the study objectives and the distribution characteristics of the aquifer, we generalized the model as a single-layer model. The upper boundary is the phreatic surface, which accepts external replenishment. Due to the continuous increase in groundwater exploitation, the depth of the groundwater table is greater than 10 m; therefore, the evaporative water loss of the diving surface can be ignored. The bottom boundary of the phreatic aquifer is selected as the model boundary and regarded as a nonflow boundary.

In this study, the flow field on 1 January 2017 was taken as the initial condition of the model. Due to topography, groundwater mainly flows from northwest to southeast. Figure 4a shows the initial flow field in the study area. The study area is located on the alluvial fan of the Juma River, and the groundwater type is quaternary loose rock pore water. According to the survey data, the northwest of the study area is mainly gravel, medium sand, and coarse sand, while the southeast is interbedded with medium sand, fine sand, and silty clay. According to formation lithology and the geological structure, the study area is divided into five regions of hydrogeological parameters. Figure 4b shows the five regions of the hydrogeological parameters. Table 1 shows the value range of hydrogeological parameters in different partitions.

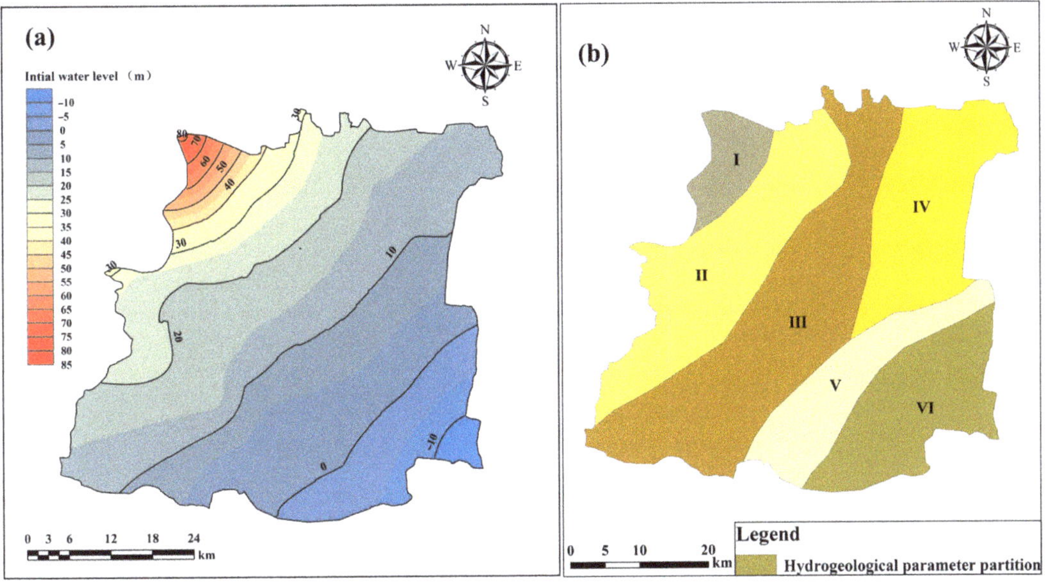

Figure 4. (**a**) The initial flow field in the study area. (**b**) The hydrogeological parameter partition of the study area.

Table 1. Value range of hydrogeological parameters of the study area.

Partition Number	Lithology	K (m/d)	μ
I	gravel	45–55	0.15–0.2
II	Coarse sand	25–35	0.11–0.18
III	Medium-coarse sand	15–25	0.10–0.15
IV	Medium sand	5–15	0.02–0.08
V	Medium sand	5–15	0.09–0.15
VI	Fine sand	0–5	0.05–0.1

Groundwater recharge in the study area includes precipitation infiltration, agricultural irrigation regression, lateral inflow, and river leakage. The precipitation infiltration coefficient in the study area is about 0.11–0.22 and the agricultural irrigation coefficient is about 0.1–0.15. According to the data of the Water Resources Bulletin, the average annual agricultural irrigation water consumption in the study area is about 3.74 million m^3/a, and the agricultural irrigation regression water amount is about 56 million m^3/a. The lateral inflow is calculated by Darcy's law according to the contour map of the groundwater level. The lateral inflow of piedmont and other inflow boundaries is calculated to be 85 million m^3/a. According to the investigation data, the river leakage coefficient in the study area is 0.45. Groundwater discharge in the study area includes exploitation and lateral outflow. According to the Water Resources Bulletin, the average groundwater exploitation in the study area is 461.56 million m^3/a. The lateral outflow is calculated by Darcy's law. The distribution of precipitation infiltration, agricultural irrigation and groundwater exploitation is divided according to the administrative boundaries of each county. This is mainly determined by the data of the Water Resources Bulletin that we collected. Figure 1 shows the administrative divisions of Zhuozhou, Yuyuan, Gaobeidian and Dingxing.

According to the variation in the groundwater flow field, the model is extended to a three-dimensional transient groundwater flow model with heterogeneous anisotropy. To simulate the change in groundwater level in the study area, MODFLOW-2005 is used to solve the three-dimensional numerical groundwater flow model in the study area. MODFLOW-2005 uses the finite difference method to calculate the hydraulic conduction between cells. First, the asymmetric conduction matrix is generated, and then the conjugate gradient method is used to solve the problem. MODFLOW-2005 is realized through the GMS platform developed by the Environmental Modeling Research Laboratory of Brigham Young University in the United States. In this study, we set the basic grid size of the model to 500 m×500 m. In the vertical direction, the region is discretized into one layer by using a deformed mesh. A total of 2592 grids exists in the model.

The simulation period is from January 2017 to December 2023, which is divided into 29 stress periods. Each stress period from 2017 to 2018 is one month, with 24 stress periods in total and a time step of 10 days. This time period is used as the identification and verification period of the model. From 2019 to 2023, each stress period is one year, and the time step is 30 days.

3.2. Model Calibration and Validation

Model calibration is the process of adjusting and improving the model parameter structure and parameter values. The model in this study is corrected by the trial and error method to continuously adjust the hydraulic conductivity and recharge input parameters [26,27] to minimize the error between the simulated value and the observed value [28,29]. The measured groundwater flow field on 26 May 2017 is used as the identification and verification flow field. Figure 5a shows that the simulated flow field and the measured flow field have the same trend and flow patterns. Except on the southwestern side of the study area, which has a poor simulation effect, the other parts all reflect the actual groundwater flow trend.

A total of 27 monitoring wells are used in this calibration process, and their distribution locations are shown in Figure 2a. In this study, Yihezhuang, Xiyi'an, Nangaoguanzhuang, Chenzhuang, Liyuzhuang and Chencunying are selected as six typical monitoring holes for display (the rest are not shown). Figure 5b shows the process fitting curve between the calculated water levels and the measured water levels of the selected monitoring wells. The simulated groundwater level process line of the typical monitoring wells is consistent with the measured groundwater level, which accurately reflects the process of groundwater before and after water replenishment.

The parameter partition after identification and correction is shown in Table 2. The hydraulic conductivity changes show a clear trend of gradually decreasing from the upper part of the alluvial fan to the downstream plain region, and the corrected parameter is within its value range. Therefore, the established numerical model can reflect the variation characteristics of the groundwater in the plain area downstream of Zhangfang. The model can be used to simulate and predict the influence of the Zhangfang water diversion project on the groundwater level in the downstream plain.

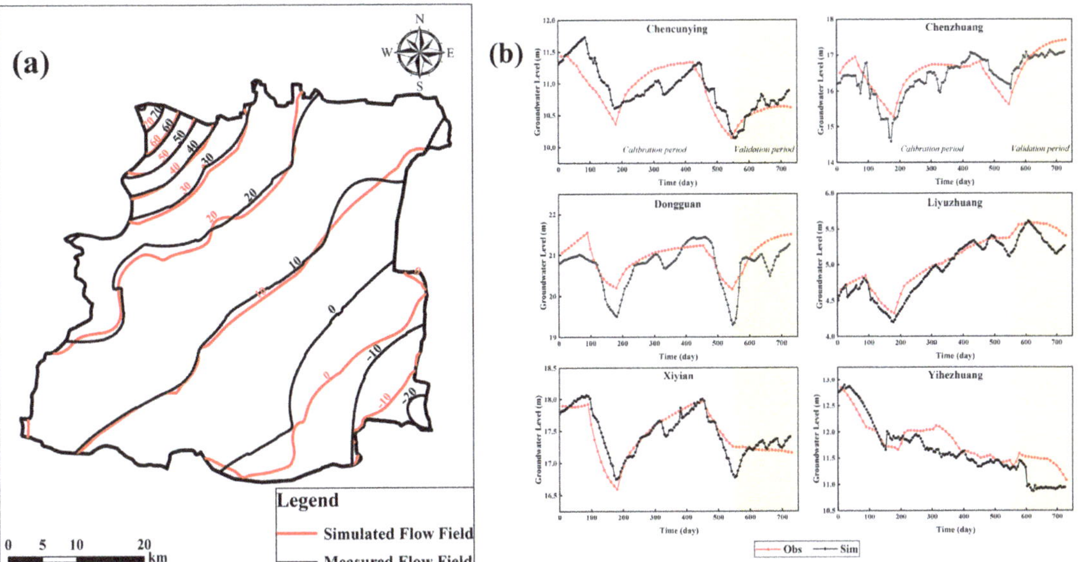

Figure 5. (a) Flow-field fitting diagram on 26 May 2017. (b) The fitting curve of typical monitoring wells in the study area.

Table 2. Calibrated hydrogeology parameters with the groundwater flow model.

Partition Number	Lithology	K (m/d)	μ
I	gravel	50	0.2
II	Coarse sand	30	0.17
III	Medium-coarse sand	17	0.15
IV	Medium sand	10	0.12
V	Medium sand	8	0.05
VI	Fine sand	5	0.1

4. Results and Discussion

4.1. Groundwater Budget Calculation

The groundwater budget calculation in the study area from January 2017 to December 2018 is obtained according to the operation results of the groundwater flow numerical model, as shown in Table 3. During the simulation period, the total recharge of groundwater in the study area is 466.8 million m^3, the total discharge is 476.79 million m^3, and the supplementary discharge difference is −9.99 million m^3. The groundwater system is in a negative water balance state. Among the recharge items, the precipitation infiltration is 257.35 million m^3, followed by river leakage of 68.77 million m^3. They account for 69.86% of the total groundwater recharge in the study area. The discharge is mainly assembled by artificial mining, which is 461.56 million m^3, which accounts for 96.81% of the total discharge.

Table 3. Mean groundwater budget of the study area during the simulation period (2017–2018).

	Budget Components	Volume ($\times 10^6$ m^3)	Percent (%)
	precipitation infiltration	257.35	55.13
	river leakage	68.77	14.73
IN	Lateral inflow	84.68	18.14
	agricultural irrigation	56.00	12.00
	Total IN	466.80	100
	Exploitation	461.56	96.81
OUT	Lateral outflow	15.23	3.19
	Evaporation	Negligible	0
	Total OUT	476.79	100
IN-OUT		−9.99	

4.2. Schemes of Reducing Water Diversion in the Zhangfang water diversion project

4.2.1. Consideration of Different Diversion Schemes of Zhangfang Water Diversion Project

Through model correction and validation, the groundwater flow model can be used to predict the dynamic changes of groundwater level in the downstream plain of Zhangfang in the future. According to the monitoring data of the hydrology station, the Zhangfang water diversion project diverted a large amount of water, which caused a serious impact to the downstream water resources and river ecology. Water diversion projects directly lead to a significant reduction in river discharge, coupled with a large degree of groundwater exploitation; surface water and groundwater interaction gradually developed into a single form of river recharge groundwater. Therefore, the Zhangfang water diversion project should consider the shortage of water resources in the downstream as a whole and make a reasonable allocation of water resources.

In this study, we considered three schemes to predict groundwater level changes in the study area from 2019 to 2023, we set the stress period as one year and the time step as one month in the prediction period.

Scheme 1 maintains the current water diversion volume of the Zhangfang water diversion project at 100 million m^3. This scheme can be used to evaluate the change of groundwater resources of the downstream plain of Zhangfang.

Scheme 2 considers reducing the amount of water diverted from the Zhangfang water diversion project by 50%, that is, to discharge 50 million m^3/a of water into the South Juma River and North Juma River. The discharge is divided between the South Juma River and North Juma River in a natural water ratio of 7:3. The direct utilization coefficient of surface water is 0.5, and the infiltration coefficient of the river is 0.45. Therefore, the river infiltration replenishment in the downstream plain area increases by 11.25 million m^3/a.

Scheme 3 assumes that the water diversion volume of the Zhangfang water diversion project is 0 and analyzes the dynamic influence of groundwater recharge on the groundwater level of the downstream plain after the water volume of the Beijuma River and Nanjuma River increases. In Scheme 3, groundwater infiltration recharge is increased to 22.5 million m^3/a.

4.2.2. Prediction of Groundwater Level of the Downstream Plain of Zhangfang in 2023 in Scheme 1

A groundwater flow field in the study area at the end of 2018 is shown in Figure 6a. The prediction of the groundwater level in 2023 is shown in Figure 6b. Figure 6c shows the simulation of groundwater level variation from 2019 to 2023 in Scheme 1. According to Figure 6c, if the amount of diversion of the Zhangfang water diversion project remains unchanged, the groundwater level in the downstream plain of Zhangfang will continue to decline. The water diversion of the Zhangfang water diversion project is about 100 million m^3/a, accounting for 1/5 of the total replenishment of the downstream plain area. We think that the amount of water diverted is catastrophic. The region with the largest drop in groundwater level is the northwest of the study area, where the aquifer is thin and the hydraulic conductivity is large. In the case of unfavorable upstream water inflow conditions, the groundwater in this region is very easy to lose. The maximum drop of water level in this area is more than 10 m.

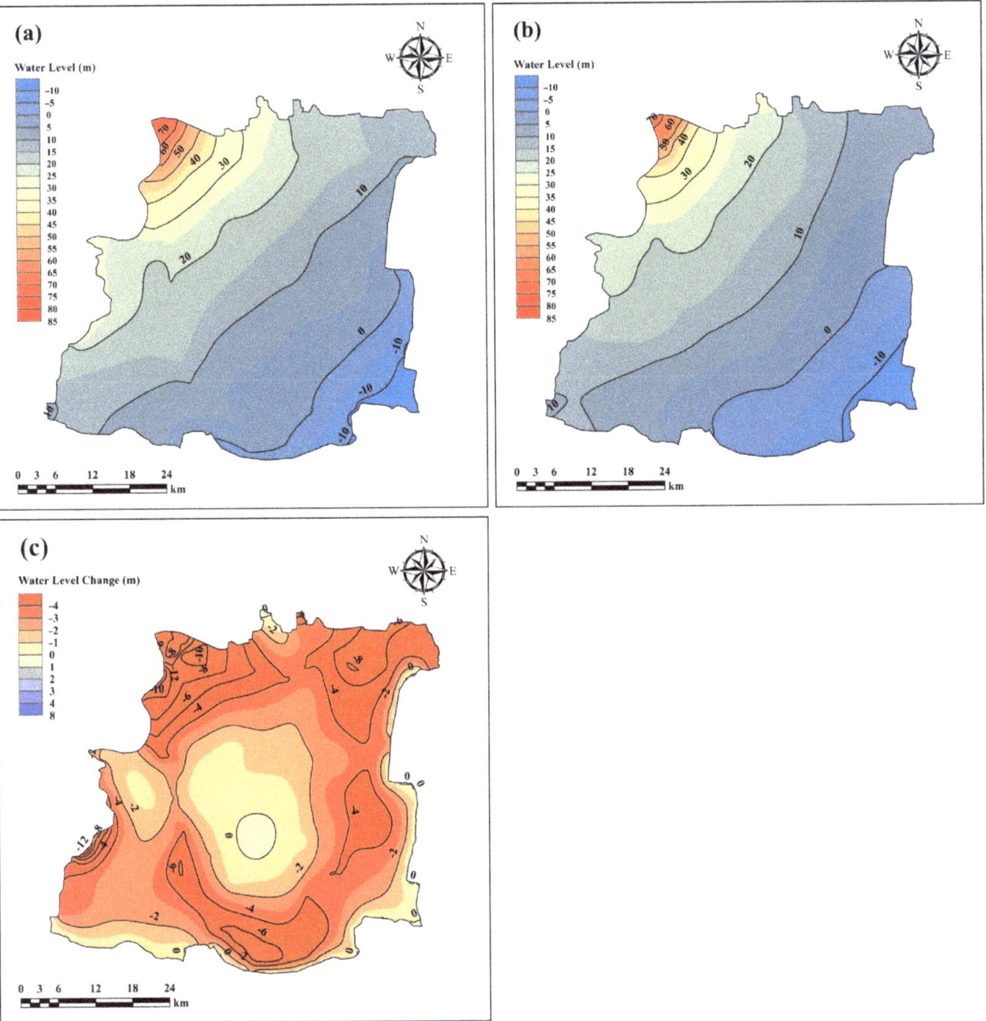

Figure 6. (**a**) contour map of groundwater level at the end of 2018. (**b**) contour map of groundwater level of Scheme 1 at the end of 2023. (**c**) Simulation of groundwater level variation from 2018 to 2023 of Scheme 1.

The current development and utilization pattern of water resources in the downstream plain of Zhangfang is extremely unreasonable. The diversion of the Zhangfang water diversion project reduces the recharge source of groundwater and intensifies the drop of groundwater level. At the same time, the Zhangfang water diversion project leads to the interruption of downstream river flow, which causes serious harm to water ecological security. Therefore, we should consider reducing the amount of water diverted by the Zhangfang water diversion project and rationally allocate the water resources.

4.2.3. Prediction of Groundwater Level of the Downstream Plain of Zhangfang in 2023 in Schemes 2 and 3

Figure 7a,b shows the contour maps of the groundwater levels at the end of 2023 of the Schemes 2 and 3. In order to quantitatively describe the water level changes from 2018 to 2023., Figure 8a,b shows the groundwater level variation from 2018 to 2023 of Schemes 2 and 3.

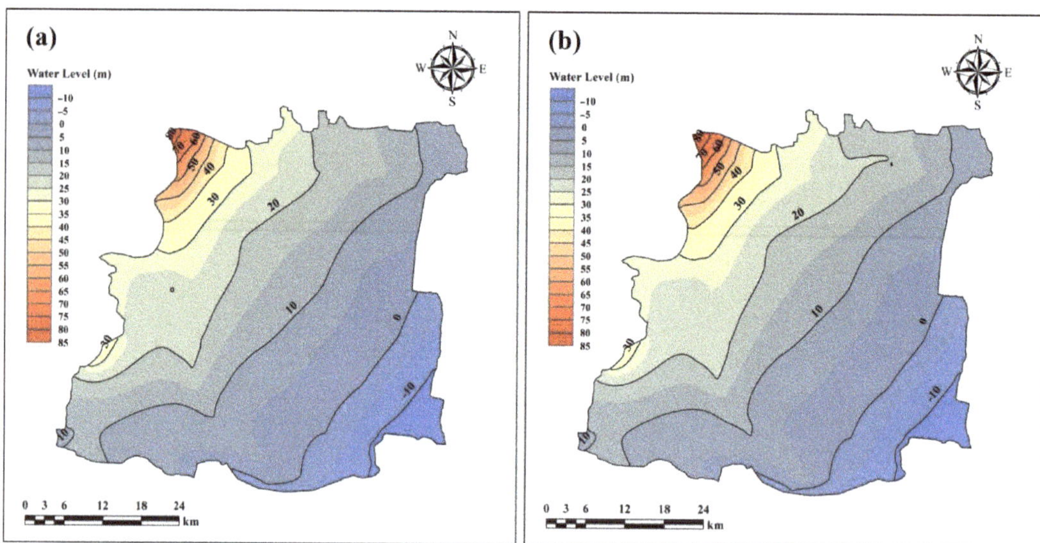

Figure 7. (**a**) contour map of groundwater level of Scheme 2 at the end of 2023. (**b**) contour map of groundwater level of Scheme 3 at the end of 2023.

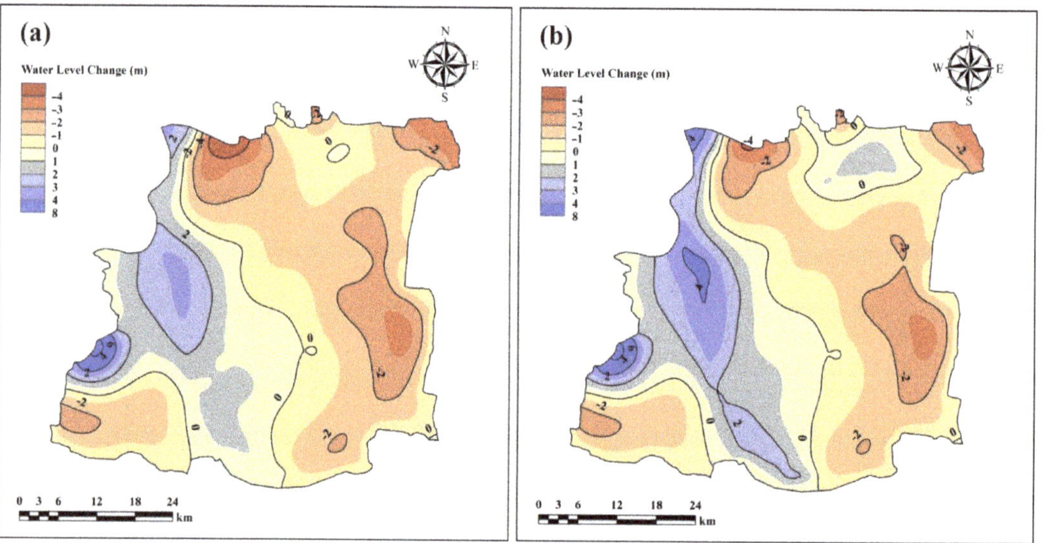

Figure 8. The simulation of groundwater level variation from 2018 to 2023: (**a**) Scheme 2; (**b**) Scheme 3.

According to Figure 8, compared with Scheme 1, reducing the water diversion amount of the Zhangfang water diversion project to recharge groundwater can alleviate the decline rate of the groundwater level in the whole study area. Although the water level in the eastern part of the research area still continues to decline, the maximum decline range is reduced to 2–4 m. The water level in the northwest of the study area increased greatly, and the groundwater level in the water receiving areas of the South Juma River and North Juma River also increased to a certain extent. Figure 8 shows that the unreasonable allocation of water resources in the Zhangfang water diversion project has caused serious impacts on the downstream water resources. The pressures resulting from the water shortage in the downstream can be alleviated by reasonably reducing the water diversion.

According to the operation results of Schemes 2 and 3, the calculation results of the groundwater budget in the study area are obtained. According to Table 4, the river leakage in Scheme 2 increases by 11.25 m^3/a, and the supplementary discharge difference is 0.13 million m^3/a. The groundwater system is in a positive water balance state. In Scheme 3, the river leakage increases by 22.5 million m^3/a, and the supplementary discharge difference is 8.31 million m^3/a. The groundwater system is in a positive water balance state. Therefore, in order to ensure the sustainable utilization of groundwater in the study area, we propose to reduce at the diversion water amount of the Zhangfang water diversion project by at least 50%to recharge the groundwater.

Table 4. Groundwater budget of the study area of Schemes 2 and 3.

	Budget Components	Scheme 2		Scheme 3	
		Volume ($\times 10^6$ m^3)	Percent (%)	Volume ($\times 10^6$ m^3)	Percent (%)
IN	precipitation infiltration	257.35	53.84	257.35	52.83
	river leakage	80.02	16.74	91.27	18.74
	Lateral inflow	84.68	17.71	82.48	16.93
	agricultural irrigation	56.00	11.71	56.00	11.50
	Total IN	478.05	100.00	487.10	100.00
OUT	Exploitation	461.56	96.58	461.56	96.40
	Lateral outflow	16.36	3.42	17.23	3.60
	Evaporation	Negligible	0.00	Negligible	0.00
	Total OUT	477.92	100.00	478.79	100.00
IN-OUT		0.13		8.31	

According to Figure 8b, when the water diversion amount of Zhangfang water diversion project is 0, the groundwater level of the South Juma River and North Juma River receiving area rises significantly, but the water level near Baigou River still drops. The reason for this is that we only recharge groundwater though natural rivers. Therefore, when a large number of surface water sources can be used to recharge the groundwater, the replenishment amount of the surface water increased by connecting channels.

5. Conclusions

By means of groundwater monitoring and numerical simulation, this paper studied the impact of the Zhangfang water diversion project on the downstream groundwater level. Through research, the following conclusions are obtained.

The groundwater flow model in the downstream plain area of Zhangfang is established. With the flow model, it is concluded that the Zhangfang water diversion project reduces the river discharge and the river infiltration recharge in the study area, which places the supplement and discharge difference in a negative water balance state.

If the Zhangfang water diversion project maintains the current water diversion amount of 100 million m^3/a, the water level of the downstream plain of Zhangfang will continue to decline. According to the prediction results, the groundwater level of the South Juma River and the North Juma River will rise to a certain extent by reducing the amount of the Zhangfang water diversion project to recharge the groundwater. The water level of the Baigou River will continue to decline, but the decreasing rate will be effectively alleviated.

For the sustainable development of the region, it is recommended to reduce the amount of water diverted by the Zhangfang water diversion project by at least 50%.

Through numerical simulation, the dynamic influence of the Zhangfang water diversion project on the change of groundwater levels in the downstream is discussed from the perspective of water resources allocation, which has practical significance. The prediction scheme of the model can provide suggestions for decision- making.

Author Contributions: M.Z., conceptualization, formal analysis, investigation, methodology, software, and writing–original draft; X.M., conceptualization, formal analysis, and supervision; B.W., data curation and methodology; D.Z., data curation and software; Y.Z., data curation and Investigation; R.L., software. All authors have read and agreed to the published version of the manuscript.

Funding: This study is funded by the Hebei Province Water Conservancy Science and Technology Project (Grant No.: JSKY2021-KY010). Research on key technology of water quality safety in underground hydraulic Mining and restoration in Miyun District, Beijing (2018000020124G037).

Institutional Review Board Statement: Not applicable.

Informed Consent Statement: Not applicable.

Data Availability Statement: Data available upon request due to privacy and ethical restrictions.

Acknowledgments: The authors would like to express their gratitude to the editors and anonymous experts for their constructive comments.

Conflicts of Interest: The authors declare that they have no conflict of interest.

References

1. Luo, Z.; Zhao, S.; Wu, J.; Zhang, Y.; Liu, P.; Jia, R. The influence of ecological restoration projects on groundwater in Yongding River Basin in Beijing, China. *Water Supply* **2019**, *19*, 2391–2399. [CrossRef]
2. Woessner, W.W. Stream and fluvial plain ground water interactions: Rescaling hydrogeologic thought. *Groundwater* **2000**, *38*, 423–429. [CrossRef]
3. Rodríguez, L.B.; Cello, P.A.; Vionnet, C.A. Modeling stream-aquifer interactions in a shallow aquifer, Choele Choel Island, Patagonia, Argentina. *Hydrogeol. J.* **2006**, *14*, 591–602. [CrossRef]
4. Kazezyelmaz-Alhan, C.M.; Medina, M.A. The effect of surface/ground water interactions on wetland sites with different characteristics. *Desalination* **2008**, *226*, 298–305. [CrossRef]
5. Kazezyelmaz-Alhan, C.M.; Medina, M.A.; Richardson, C.J. A wetland hydrology and water quality model incorporating surface water/groundwater interactions. *Water Resour. Res.* **2007**, *43*, W04434. [CrossRef]
6. Chen, X.; Shu, L. Stream-Aquifer Interactions: Evaluation of Depletion Volume and Residual Effects from Ground Water Pumping. *Groundwater* **2002**, *40*, 284–290. [CrossRef]
7. József, K.; László, M.; József, S.; Ilona, S.K. Detection and evaluation of changes induced by the diversion of River Danube in the territorial appearance of latent effects governing shallow-groundwater fluctuations. *J. Hydrol.* **2015**, *520*, 314–325.
8. Trásy, B.; Kovács, J.; Hatvani, I.G.; Havril, T.; Németh, T.; Scharek, P.; Szabó, C. Assessment of the interaction between surface- and groundwater after the diversion of the inner delta of the River Danube (Hungary) using multivariate statistics. *Anthropocene* **2018**, *22*, 51–65. [CrossRef]
9. Zhang, C.; Duan, Q.; Pat, J.F.Y.; Pan, Y.; Gong, H.; Di, Z.; Lei, X.; Liao, W.; Huang, Z.; Zheng, L.; et al. The Effectiveness of the South-to-North Water Diversion Middle Route Project on Water Delivery and Groundwater Recovery in North China Plain. *Water Resour. Res.* **2020**, *56*, e2019WR026759. [CrossRef]
10. Zhu, L.; Gong, H.; Chen, Y.; Wang, S.; Ke, Y.; Guo, G.; Li, X.; Chen, B.; Wang, H.; Teatini, P. Effects of Water Diversion Project on groundwater system and land subsidence in Beijing, China. *Eng. Geol.* **2020**, *276*, 105763. [CrossRef]
11. Lu, Z.; Feng, Q.; Xiao, S.; Xie, J.; Zou, S.; Yang, Q.; Si, J. The impacts of the ecological water diversion project on the ecology-hydrology-economy nexus in the lower reaches in an inland river basin. *Resour. Conserv. Recycl.* **2021**, *164*, 105154. [CrossRef]
12. Ji, Z.; Cui, Y.; Zhang, S.; Wang, C.; Shao, J. Evaluation of the Impact of EcologicalWater Supplement onGroundwater Restoration Based on Numerical Simulation: A Case Study in the Section of Yongding River, Beijing Plain. *Water* **2021**, *13*, 3059. [CrossRef]
13. Neville, C.M. *Impounded Rivers, Perspectives for Ecological Management: By Geoffrey E. Petts*; John Wiley and Sons: Chichester, UK; New York, NY, USA, 1984; Volume 34, p. 1.
14. Peng, Y.; Chu, J.; Peng, A.; Zhou, H. Optimization Operation Model Coupled with Improving Water-Transfer Rules and Hedging Rules for Inter-Basin Water Transfer-Supply Systems. *Water Resour. Manag.* **2015**, *29*, 3787–3806. [CrossRef]
15. Lerner, D.N. Groundwater recharge. In *Geochemical Process, Weathering and Groundwater Recharge in Catchments*; Saether, O.M., de Caritat, P., Eds.; AA Balkema: Rotterdam, The Netherlands, 1997; pp. 109–150.

16. Tenant, S.; Johannes, C.N.; Stefan, U. Comparison of groundwater recharge estimation methods for the semi-arid Nyamandhlovu area, Zimbabwe. *Hydrogeol. J.* **2009**, *17*, 1427–1441.
17. Richard, W.H.; Peter, G.C. Using groundwater levels to estimate recharge. *Hydrogeol. J.* **2002**, *10*, 91–109.
18. Constantz, J.; Thomas, C.L.; Zellweger, G. Influence of diurnal variations in stream temperature on streamflow loss and groundwater recharge. *Water Resour. Res.* **1994**, *30*, 3253–3264. [CrossRef]
19. Singh, V.P. *Computer Models of Watershed Hydrology*; Water Resources Publications: Highlands Ranch, CO, USA, 1995.
20. Zhang, F.; Wei, Y.; Qian, H.; Zhang, X. Numerical Simulation of the Groundwater in Bulang River- Red Stone Bridge Water Source. *Procedia Environ. Sci.* **2011**, *8*, 140–145.
21. Chandio, A.S.; Lee, T.S.; Mirjat, M.S. The extent of waterlogging in the lower Indus Basin (Pakistan)–A modeling study of groundwater levels. *J. Hydrol.* **2012**, *426–427*, 103–111. [CrossRef]
22. Palma, H.C.; Bentley, L.R. A regional-scale groundwater flow model for the Leon-Chinandega aquifer, Nicaragua. *Hydrogeol. J.* **2007**, *15*, 1457–1472. [CrossRef]
23. Diersch, H.J. *FEFLOW Finite Element Subsurface Flow and Transport Simulation System, Reference Manual*; WASY GmbH Institute for Water Resources Planning and System Research: Berlin, Germany, 2009.
24. McDonald, M.G.; Harbaugh, A.W. A modular three-dimensional finite difference groundwater flow model. *U. S. Geol. Surv. Open File Rep.* **1988**, *83*, 528–875.
25. Harbaugh, A.W. *MODFLOW-2005, the US Geological Survey Modular Ground-Water Model: The Ground-Water Flow Process*; US Department of the Interior, US Geological Survey: Reston, VA, USA, 2005.
26. Wu, J.; Li, J.; Teng, Y.; Chen, H.; Wang, Y. A partition computing-based positive matrix factorization (PC-PMF) approach for the source apportionment of agricultural soil heavy metal contents and associated health risks. *J. Hazard. Mater.* **2019**, *388*, 121766. [CrossRef] [PubMed]
27. Zhang, D.; Zhang, Y.; Liu, L.; Li, B.; Yao, X. Numerical Simulation of Multi-Water-Source Artificial Recharge of Aquifer: A Case Study of the Mi-Huai-Shun Groundwater Reservoir. *Water Resour.* **2020**, *47*, 399–408. [CrossRef]
28. Husam, B. Stochastic water balance model for rainfall recharge quantification in Ruataniwha Basin, New Zealand. *Environ. Geol.* **2009**, *58*, 85–93.
29. Husam, M.B. Modelling surface-groundwater interaction in the Ruataniwha basin, Hawke's Bay, New Zealand. *Environ. Earth Sci.* **2012**, *66*, 285–294.

MDPI
St. Alban-Anlage 66
4052 Basel
Switzerland
Tel. +41 61 683 77 34
Fax +41 61 302 89 18
www.mdpi.com

Water Editorial Office
E-mail: water@mdpi.com
www.mdpi.com/journal/water